Infrared Characteristic Group Frequencies

Infrared Characteristic Group Frequencies

G. SOCRATES
Brunel University, Uxbridge, Middlesex

A Wiley–Interscience Publication

JOHN WILEY & SONS
Chichester · New York · Brisbane · Toronto

Library of Congress Cataloging in Publication Data:

Socrates, George.
Infrared characteristic group frequencies.

'A Wiley–Interscience publication.'
Includes index.
1. Infrared spectrometry. I. Title.
QC457.S69 543'.085 79-1406

ISBN 0 471 27592 1

Typeset in Great Britain by John Wright & Sons Ltd.,
at the Stonebridge Press, Bristol and printed at The Pitman Press, Bath

Contents

List of Charts and Figures

Symbols used

R	alkyl	asym	asymmetric
Ar	aromatic	sym	symmetric
G	aliphatic or aromatic	str	stretching
vw	very weak	def	deformation
w	weak	vib	vibration
m	medium	sat	saturated
s	strong	unsat	unsaturated
vs	very strong	Me	methyl
v	variable	Et	ethyl
br	broad	Ph	phenyl
sh	sharp		

Preface

The purpose of this book is to provide a simple introduction to characteristic group frequencies so as to assist all who may need to interpret or examine infrared spectra. The characteristic absorptions of functional groups over the entire infrared region (including the far infrared) are given in tables as well as being discussed and amplified in the text.

In order to assist the beginner, three basic correlation charts are provided. Chart 1 may be used to deduce the absence of one or more classes of chemical compound by the absence of an absorption band in a given region. Chart 2 may be used to determine which groups may possibly be responsible for a band at a given position. Chart 3 may be used if the class of chemical is known (and hence the functional groups it contains) in order to determine at a glance the important absorption regions. Having identified a functional group as possibly being responsible for an absorption band, the information in the relevant chapter (or section) and table should both be used to confirm or reject this assumption. If the class of chemical is known then the relevant chapter may be turned to immediately. It may well be that information contained in more than one chapter is required, as, for example, in the case of aromatic amines, for which the chapters on aromatics and on amines should both be referred to. In order to assist the reader, absorptions of related groups, e.g. inorganic ions, may also be dealt with in a given chapter.

Throughout the text, tables, and charts, an indication of the absorption intensities is given. Strictly speaking, extinction coefficients should be quoted. However, there are insufficient data in the literature on the subject and, in any case, the intensity of an absorption of a given functional group may be affected by neighbouring atoms or groups as well as by the chemical environment (e.g. solvent, etc.). It should also be borne in mind that the term 'frequency' is often used in discussions on infrared spectra when the actual measurement being given is in wavenumbers. Since frequency is proportional to wavenumber, this is a widely accepted practice. The values of the characteristic group frequencies are given to the nearest 5 cm^{-1}.

Normally, the figures quoted for the absorption range of a functional group refer to the region over which the maximum of the particular absorption band may be found. In the main, the absorption ranges of functional groups are quoted for the spectra of dilute solutions using an inert solvent. Therefore, if the sample is not in this state, e.g. is examined as a solid, then depending on its nature some allowance in the band position(s) may need to be made.

The near infrared region is discussed briefly in a separate chapter as are the absorptions of inorganic compounds.

The references given at the end of each chapter and in the appendix provide a source of additional information.

I should like to acknowledge the encouragement given to me by Dr K. P. Kyriakou. There are no words which can adequately express my thanks to my wife, Jeanne, for the vast amount of assistance I have received not only in the form of helpful discussions and encouragement but also in the laborious task of typing and checking the manuscript.

CHAPTER 1
Introduction

Infrared spectroscopy is an extremely powerful analytical technique for both qualitative and quantitative analysis. However, it should not be used in isolation since other analytical methods may yield important complementary and/or confirmatory information regarding the sample. For example, simple chemical tests and elemental analysis should not be overlooked. Techniques such as chromatography, nuclear magnetic resonance, atomic absorptions, mass spectroscopy, ultraviolet and visible spectroscopy, thermal analysis, etc., may result in additional information being obtained.

If the sample is a mixture, as is often the case, separation of the individual components, or partial separation, will result in simpler infrared spectra being obtained. This separation may be accomplished by solvent extraction or by chromatographic techniques.

From the earliest days of infrared spectroscopy, it was noticed that certain aggregates of atoms (functional groups) could be associated with definite characteristic absorption bands, i.e. the absorption of infrared radiation over certain frequency intervals. The infrared data given in the correlation tables and charts have been derived empirically over many years by the careful and painstaking work of very many scientists.

The infrared spectrum of any given substance is interpreted by the use of these known group frequencies and thus it is possible to characterize the substance as one containing a given type of group or groups. Although group frequencies occur within 'narrow' limits, interference or perturbation may cause a shift of the characteristic bands due to (a) the electronegativity of neighbouring groups or atoms, (b) the spatial geometry of the molecule, or (c) the mechanical mixing of vibrational modes.

Functional groups sometimes have more than one characteristic absorption band associated with them. On the other hand, two or more functional groups may absorb in the same region and therefore, in general, can only be distinguished from each other by means of other characteristic infrared bands which occur in non-overlapping regions.

Absorption bands may be considered in the main as having two origins, these being the fundamental vibrations of (a) functional groups, e.g. C=O, C=C, C≡N, $-CH_2-$, $-CH_3$, and (b) skeletal groups, i.e. the molecular backbone or skeleton of the molecule, e.g. C—C—C—C. Absorption bands may also be considered as arising from *stretching* vibrations, i.e. vibrations involving bond-length changes, or *deformation* vibrations, i.e. vibrations involving bond-angle changes of the group. Each of these may, in some cases, be considered as arising from symmetric or asymmetric vibrations.

For a given functional group, the vibration bands due to stretching occur at higher frequencies than those due to deformation. This is because more energy is required to stretch the group than to deform it due to the bonding force directly opposing the change.

Two other types of absorption band may also be observed: overtone and combination bands. Overtone bands are observed at approximately twice the frequency of strong fundamental absorption bands (overtones of higher order have too low an intensity to be observed). Combination bands result from the combination (addition or subtraction) of two fundamental frequencies.

The intensity of an absorption band is dependent on the magnitude of the dipole change during the vibration, the larger the change, the stronger the absorption band.

The vibrational frequencies, relative intensities, and shapes of the infrared absorption bands may all be used in the qualitative characterization of a sample. The presence of a band at a particular frequency should not on its own be used as an indication of the presence of a particular functional group. Confirmation should always, where possible, be sought from other bands or other analytical techniques.

For example, if a sharp absorption is observed in the region 3100–3000 cm^{-1} (3.23–3.33 μm), the sample may contain an aromatic or an olefinic component and the absorption observed may be due to the carbon–hydrogen (=C—H) stretching vibration. If bands are not observed in regions where other aromatic absorptions are expected, then aromatic components are absent from the sample. The suspected alkene is tackled in the same manner. By examining the absorptions observed, it is possible to determine the type of aromatic or alkene component in the sample. It may, of course, be that both groups are present, or indeed absent, the band observed being due to another functional group that absorbs in the same region, e.g. an alkane group with a strong adjacent electronegative atom or group.

It should be noted that the observation of a band at a position predicted by what may be believed to be valid prior knowledge of the sample should not on its own be taken as conclusive evidence for the presence of a particular functional group.

Certain functional groups may not always give rise to absorption bands, even though they are present in the sample, since the particular energy transitions involved may be infrared inactive (due to symmetry). For example, symmetrical alkene groups do not have a C=C stretching vibration band. Therefore, the absence of certain absorption bands from a spectrum leads one to conclude that

(a) the functional group is not present in the sample, (b) the functional group is present but in too low a concentration to give a signal of detectable intensity, or (c) the functional group is present in the sample but is infrared inactive. In a similar way, the presence of an absorption band in the spectrum of a sample may be interpreted as indicating (a) that a given functional group is present (confirmed by other information), or (b) that although more than one type of the given functional group is present in the sample their absorption bands all coincide, or (c) that, although more than one type of the given functional group is present, all but one have an infrared inactive transition.

The shape of an absorption band can give useful information. For example, it may indicate the presence of hydrogen bonding.

The relative intensity of one band compared with another may, in some cases, give an indication of the relative amounts of two functional groups. The intensity of a band may also indicate the presence of certain atoms or groups adjacent to the functional group responsible for the absorption band.

Before running an infrared spectrum, it is advisable to check the wavelength calibration of the spectrometer. This may be done by examining a suitable reference substance (such as polystyrene film, ammonia gas, carbon dioxide gas, water vapour, or indene) which has sharp bands, the positions of which are accurately known in the region of interest.

Purity is, of course, very important. In general, the more components a sample has, the more complicated the spectrum and hence the more difficult the analysis. Care should always be taken not to contaminate the sample or the cells used. The limits of detectability of substances vary greatly and, in general, depend on the nature of the functional groups they contain. Obviously, the parameters used for scanning the wavelength range, e.g. scanning speed, slit width, etc., are also important. As a rough guide, below about 3 per cent by weight, detection becomes relatively difficult, although this may be too low a value for some substances. Silicones are common contaminants due to their presence in stop-cock and high-vacuum greases and their use in gas chromatography columns. Silicones have a sharp band at about 1265 cm^{-1} (7.9 μm) and a broad band in the region 1100–1000 cm^{-1} (9–10 μm). Obviously, incorrect conclusions may be drawn if the sample is contaminated. If a solvent has been used in the extraction or separation of a sample, this solvent must be thoroughly removed. The presence of a contaminating solvent may be detected by observing the spectrum for strong solvent bands which are then used to verify the progress of subsequent solvent removal.

Certain samples may react chemically in the cell compartment even while the spectrum is being run, and this may account for changes in spectra run at different times. Care should always also be taken that the sample does not react with the cell plates (or with the dispersion medium or solvent, if used). A common error is to examine wet samples on salt cell plates (e.g. NaCl or KBr) which are, of course, soluble in water. Chemical and physical changes may also occur as a result of the sample preparation technique, e.g. due to the grinding of the sample for the preparation of discs or mulls.

The infrared spectrum may be used as a fingerprint of the sample. A bank of the infrared spectra of the constituents of the type of samples encountered in a given laboratory should be made or purchased. Such reference spectra are of great assistance in the interpretation of the spectrum of an unknown sample. It may be, as is often the case, that all that is required is a simple confirmation of a sample. This may easily be achieved by comparing the spectrum of the sample and that of the known reference material. If the absorption bands are the same (i.e. in wavelength, relative intensities, and shapes) or nearly so, then it is reasonable to assume that the sample and reference are either identical or very similar in molecular structure.

Positive and Negative Spectral Interpretation

In the interpretation of infrared spectra there is no substitute for experience and, if possible, guidance from an expert in the field should be sought by the inexperienced.

The spectrum should be interpreted by (a) seeing which absorption bands are absent—negative spectral interpretation—and (b) examining those bands present—positive spectral interpretation.

Negative spectral interpretation

By examining a spectrum for the absence of bands in given regions, it is possible to eliminate particular functional groups and, hence, compounds containing these groups. In general, this type of interpretation is made by a search in a particular region where a given functional group always absorbs strongly. If no bands are observed in this region then this functional group may be excluded. For this purpose, Table 1.1 and the more detailed Chart 1 should be used. With a little experience, negative interpretation may be carried out at a glance.

Positive spectral interpretation

The technique of negative interpretation should, of course, be used in conjunction with the positive approach. It is important to be aware that correlation tables give the positions and intensities of bands characteristic of a large number of classes of compounds and groups. However, it may well be that bands appear in the spectrum of a particular sample which are not given in the tables. Assuming that these bands do belong to the sample and are not due to (a) solvent(s), (b) dispersive media, (c) air, (d) instrumental fault, or (e) operator error, then correlations involving these bands may not as yet have been made, or the bands are not characteristic of the class of compound or group considered. It may well be, for example, that the band or bands have arisen due to solid-state effects, e.g. due to different crystalline modifications of the compound. In general, it is not necessary to identify every single (weak) band that appears in a spectrum in order to characterize a sample and be in a position to propose a molecular structure.

Regions for Preliminary Investigation

There are no rigid rules for the interpretation of infrared spectra. However, a few general hints may be given.

Preliminary regions to examine

It is usually advisable to tackle the bands at the higher-frequency end of the spectrum, the most intense bands being looked at first and associated bands, occurring in other regions, thus also being identified. In the light of the information gained, the region between 900 cm^{-1} and 650 cm^{-1} (11.1 μm and 15.4 μm) can then be looked at. The origin of bands found in the so-called 'fingerprint' region 1350–900 cm^{-1} (7.4–11.1 μm) is usually difficult to decide on as the bands may arise in various ways, and similarly, below 650 cm^{-1} (above 15.4 μm), skeletal vibrations occur which are also often difficult to interpret. Hence these two regions are best avoided initially. Table 1.1 and Chart 1 may be used in reverse, i.e. to indicate the possible presence of a group which must then be confirmed.

Confirmation

It must be stressed again that the presence of a particular band should not, on its own, be used as an indication of the presence of a particular group. Confirmation should always be sought from the presence of other associated bands or from other independent techniques. The correlation Charts 2 and 3 should be used first and then the tables and text of relevant chapters employed for the detailed confirmation and identification. Having positively identified the first band looked at, the next band is approached in a similar fashion.

Chemical Modification

Quite often it is helpful for identification purposes to modify the sample chemically and compare the spectra of the original and modified samples. Isotope exchanges may be helpful in the assignment of bands. Deuterium exchange is very useful and the most common. Labile hydrogen atoms are replaced by deuterium atoms. On comparing the spectra of the original and the deuterated sample, bands shifted in frequency by a factor of approximately $1/\sqrt{2}$ compared with the original may be associated with vibrations due to the substituted labile hydrogen.

Chemical reactions may also be helpful for assignment purposes, e.g.

(a) conversion of an acid to its salt or ester;
(b) conversion of an amine or amino acid to its hydrochloride;
(c) hydrogenation of unsaturated bonds;
(d) saponification of esters, this being particularly useful in the identification of the monomers of a polyester resin.

Table 1.1. Negative spectral interpretation table

Absorption band absent in region		Type of vibration responsible for bands in this region	Type of group or compound absent
cm^{-1}	μm		
4000–3200	2.50–3.13	O—H and N—H stretch	Primary and secondary amines and alcohols, amides, organic acids and phenols
3310–3300	3.02–3.03	C—H stretch (unsaturated)	Alkynes
3100–3000	3.23–3.33	C—H stretch (unsaturated)	Aromatic and olefinic compounds
3000–2800	3.33–3.57	C—H stretch (aliphatic)	Methyl, methylene, methyne groups
2500–2000	4.00–5.00	X≡Y, X=Y=Z stretch†	Alkynes‡, allenes, cyanate, isocyanate, nitrile, isocyanides, azides, diazonium salts, ketenes, thiocyanates, isothiocyanates
1870–1550	5.35–6.45	C=O stretch	Esters, ketones, amides, carboxylic acids and their salts, acid anhydrides
1690–1620	5.92–6.17	C=C stretch	Olefinic compounds‡
1680–1610	5.92–6.21	N=O stretch	Organic nitrite compounds
1655–1610	6.04–6.21	—O—NO_2 asymmetric stretch	Organic nitrate compounds (the symmetric —O—NO_2 stretch occurs at 1300–1255 cm^{-1} (7.69–7.97 μm))
1600–1510	6.25–6.62	—NO_2 asymmetric stretch	Organic nitro- compounds (the symmetric —NO_2 stretch occurs at 1385–1325 cm^{-1} (7.22–7.55 μm))
1600–1450	6.25–6.90	C=C stretch (normally four bands)	Aromatic ring system
1490–1150	6.71–8.70	H—C—H bend	Methyl, methylene
1420–990	7.04–10.10	S=O stretch	Sulphoxides, sulphates, sulphites, sulphinic acids or esters, sulphones, sulphonic acids, sulphonates, sulphonamides, sulphonyl halides
1310–1020	7.63–9.80	C—O—C stretch	Ethers (aromatic, olefinic, or aliphatic)
1225–1045	8.16–9.67	C=S stretch	Thioesters, thioureas, thioamides, pyrothiones
1000–780	10.00–12.82	C=C—H deformation	Aliphatic unsaturation
900–670	11.11–14.93	C—H deformation	Substituted aromatics
850–500	11.76–20.00	C—X stretch§	Organohalogens
730–720	13.70–13.90	$(CH_2)_{n>3}$	Four or more consecutive methylene groups

† X, Y, and Z may represent any of the atoms C, N, O, and S.
‡ Band may be absent due to symmetry of functional group.
§ X may be Cl, Br, or I.

Chart 1 Negative correlation chart. The absence of a band in the position(s) marked indicates the absence of the group (or chemical class) specified. (Note the change of scale at 2000 cm^{-1} (5.0 μm)).

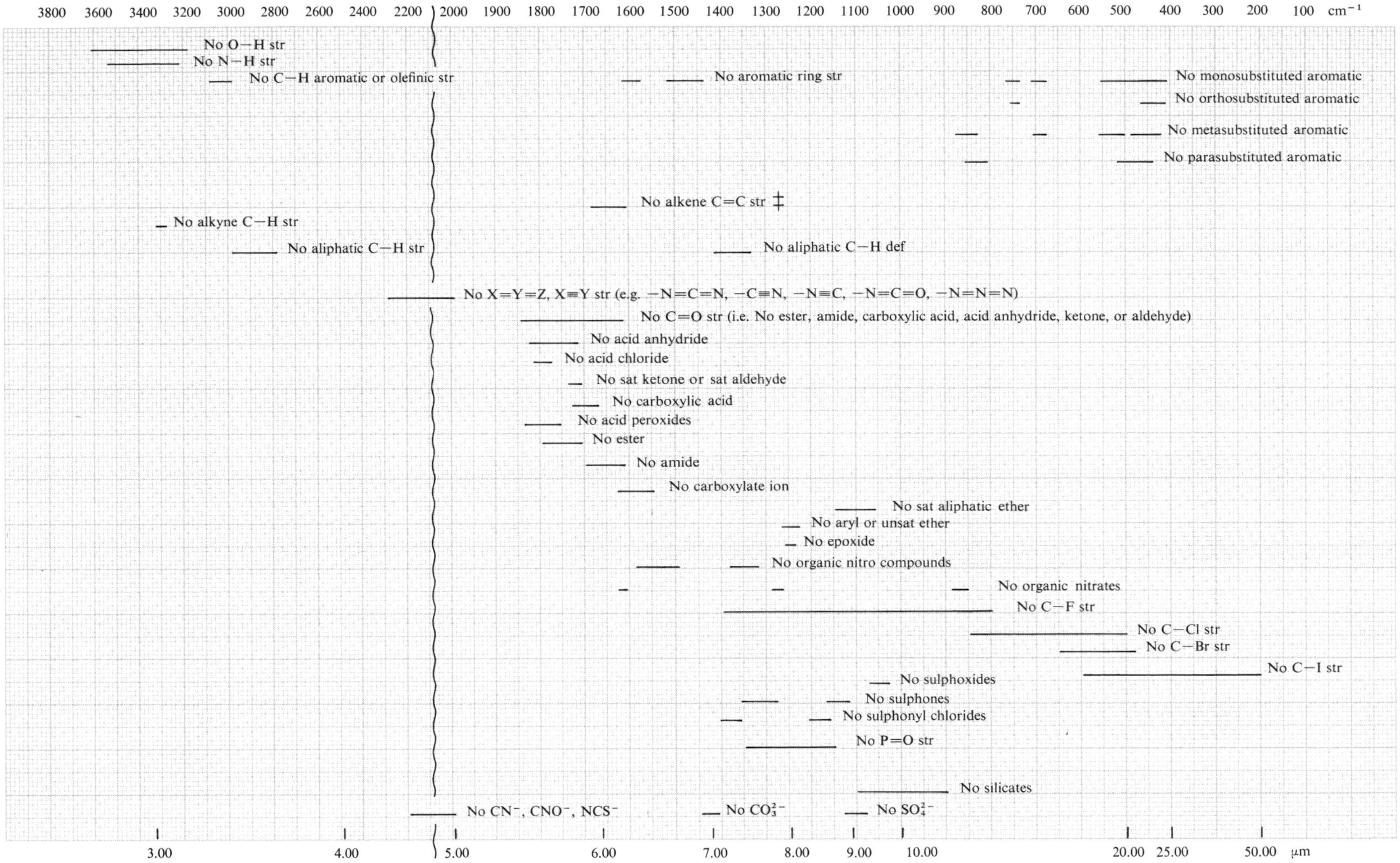

3800 3600 3400 3200 3000 2800 2600 2400 2200 2000 1900 1800 1700 1600 1500 1400 1300 1200 1100 1000 900 800 700 600 500 400 300 200 100 cm^{-1}
No O—H str
No N—H str
No C—H aromatic or olefinic str
No aromatic ring str
No monosubstituted aromatic
No orthosubstituted aromatic
No metasubstituted aromatic
No parasubstituted aromatic
No alkene C=C str ‡
No alkyne C—H str
No aliphatic C—H str
No aliphatic C—H def
No X=Y=Z, X≡Y str (e.g. —N=C=N, —C≡N, —N≡C, —N=C=O, —N=N=N)
No C=O str (i.e. No ester, amide, carboxylic acid, acid anhydride, ketone, or aldehyde)
No acid anhydride
No acid chloride
No sat ketone or sat aldehyde
No carboxylic acid
No acid peroxides
No ester
No amide
No carboxylate ion
No sat aliphatic ether
No aryl or unsat ether
No epoxide
No organic nitro compounds
No organic nitrates
No C—F str
No C—Cl str
No C—Br str
No C—I str
No sulphoxides
No sulphones
No sulphonyl chlorides
No P=O str
No silicates
No CN^-, CNO^-, NCS^-
No CO_3^{2-}
No SO_4^{2-}
3.00 4.00 5.00 6.00 7.00 8.00 9.00 10.00 20.00 25.00 50.00 μm

Chart 2 This chart may be used to identify the possible type of vibration responsible for a band at a given position. The range of the position of the maximum absorption of a functional group is given in order of decreasing wavenumber. The information given in both the text and tables of relevant chapters may then be used to confirm or eliminate a particular group. The relative intensities of bands are given.

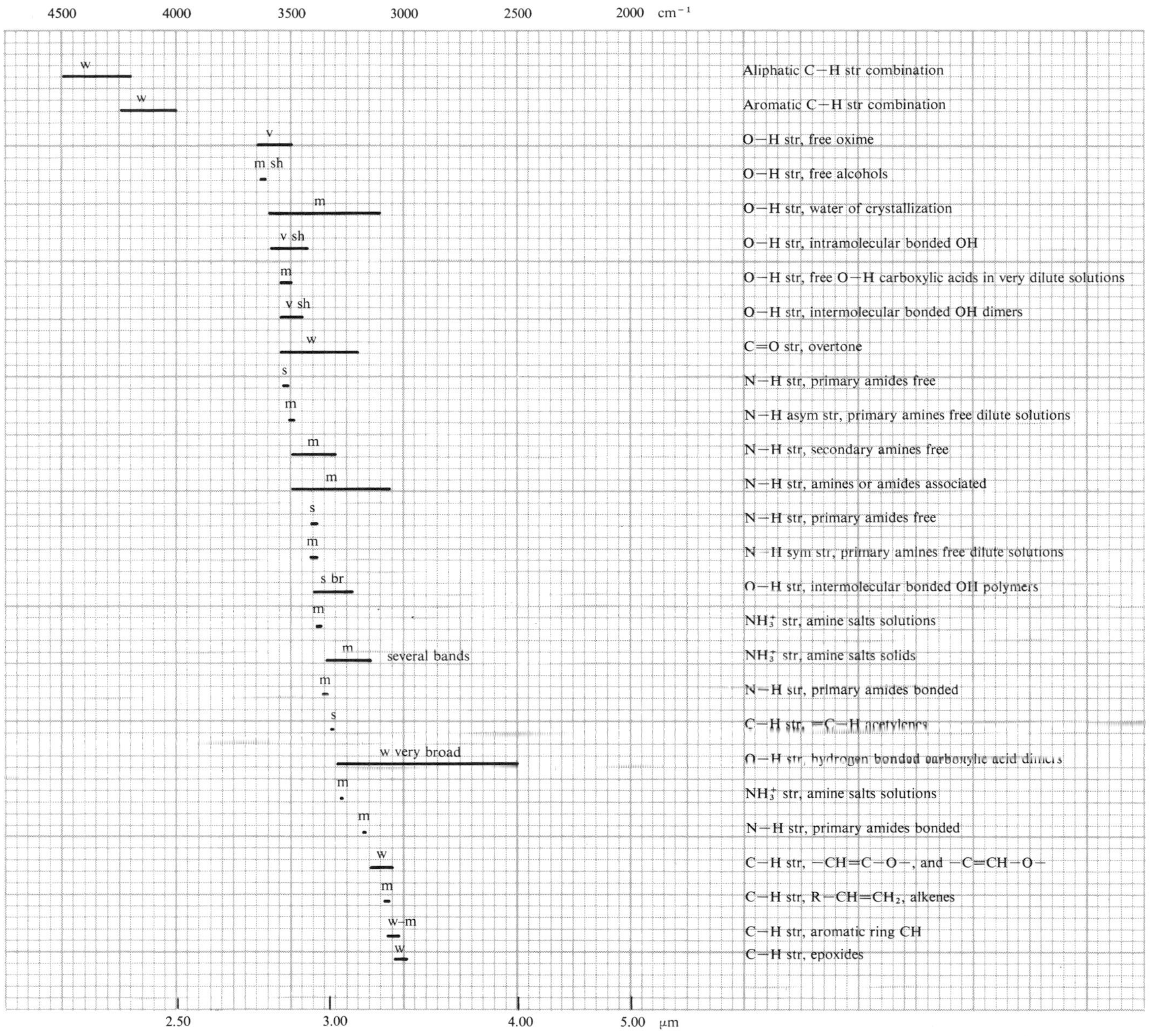
4500
4000
3500
3000
2500
2000 cm⁻¹
w
Aliphatic C—H str combination
w
Aromatic C—H str combination
v
O—H str, free oxime
m sh
O—H str, free alcohols
m
O—H str, water of crystallization
v sh
O—H str, intramolecular bonded OH
m
O—H str, free O—H carboxylic acids in very dilute solutions
v sh
O—H str, intermolecular bonded OH dimers
w
C=O str, overtone
s
N—H str, primary amides free
m
N—H asym str, primary amines free dilute solutions
m
N—H str, secondary amines free
m
N—H str, amines or amides associated
s
N—H str, primary amides free
m
N—H sym str, primary amines free dilute solutions
s br
O—H str, intermolecular bonded OH polymers
m
NH3+ str, amine salts solutions
m
several bands
NH3+ str, amine salts solids
m
N—H str, primary amides bonded
s
C—H str, ≡C—H acetylenes
w very broad
O—H str, hydrogen bonded carboxylic acid dimers
m
NH3+ str, amine salts solutions
m
N—H str, primary amides bonded
w
C—H str, —CH=C—O—, and —C=CH—O—
m
C—H str, R—CH=CH2, alkenes
w–m
C—H str, aromatic ring CH
w
C—H str, epoxides
2.50
3.00
4.00
5.00 μm

Chart 2 (*continued*)

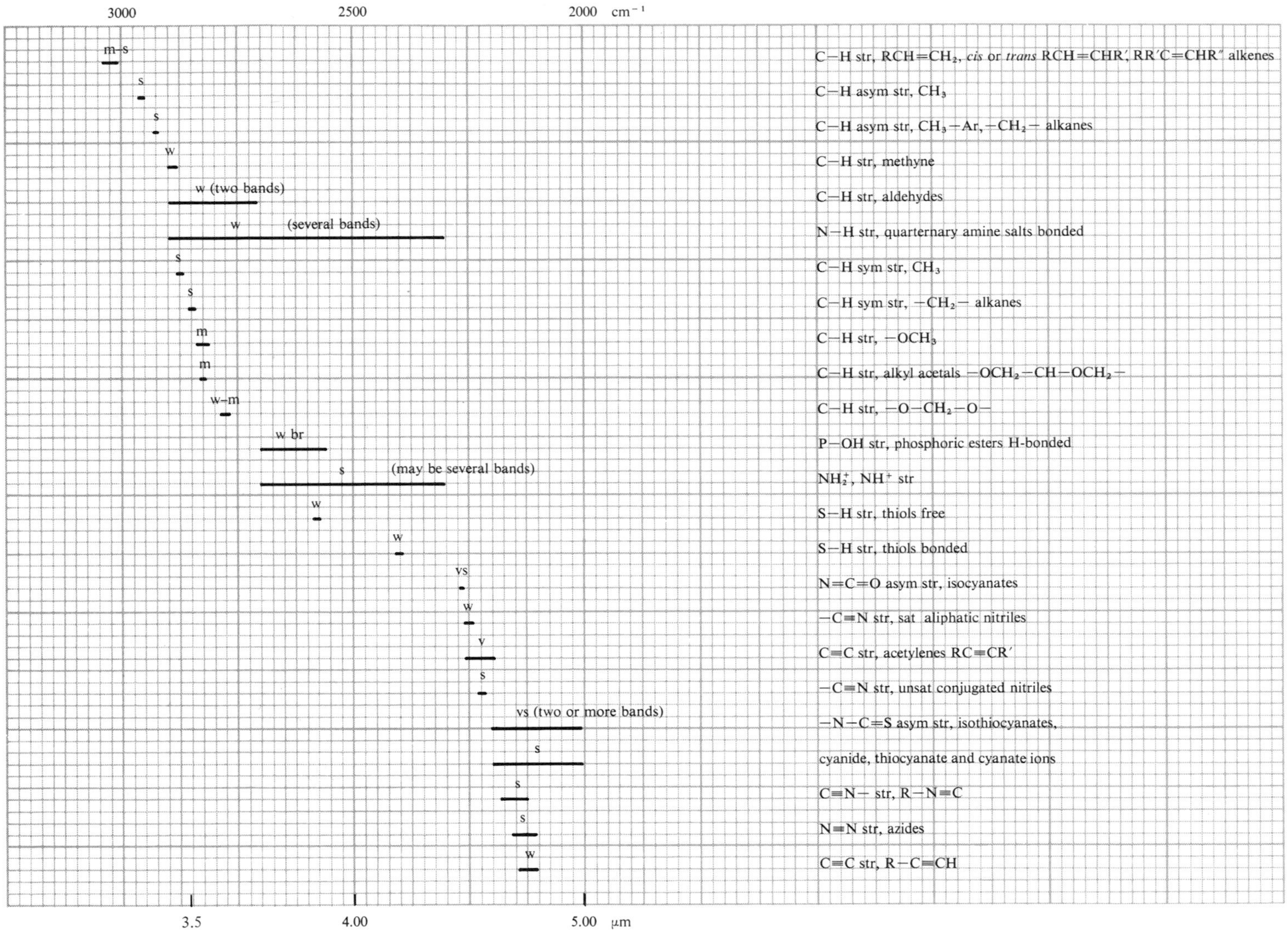
3000
2500
2000 cm^{-1}
m–s
s
s
w
w (two bands)
w (several bands)
s
s
m
m
w–m
w br
s (may be several bands)
w
w
vs
w
v
s
vs (two or more bands)
s
s
s
w
C—H str, $RCH{=}CH_2$, *cis* or *trans* $RCH{=}CHR'$, $RR'C{=}CHR''$ alkenes
C—H asym str, CH_3
C—H asym str, $CH_3{-}Ar$, $-CH_2-$ alkanes
C—H str, methyne
C—H str, aldehydes
N—H str, quarternary amine salts bonded
C—H sym str, CH_3
C—H sym str, $-CH_2-$ alkanes
C—H str, $-OCH_3$
C—H str, alkyl acetals $-OCH_2{-}CH{-}OCH_2-$
C—H str, $-O{-}CH_2{-}O-$
P—OH str, phosphoric esters H-bonded
NH_2^+, NH^+ str
S—H str, thiols free
S—H str, thiols bonded
$N{=}C{=}O$ asym str, isocyanates
$-C{\equiv}N$ str, sat aliphatic nitriles
$C{\equiv}C$ str, acetylenes $RC{\equiv}CR'$
$-C{\equiv}N$ str, unsat conjugated nitriles
$-N{-}C{=}S$ asym str, isothiocyanates,
cyanide, thiocyanate and cyanate ions
$C{\equiv}N-$ str, $R{-}N{\equiv}C$
$N{\equiv}N$ str, azides
$C{\equiv}C$ str, $R{-}C{\equiv}CH$
3.5
4.00
5.00 μm

Chart 2 (*continued*)

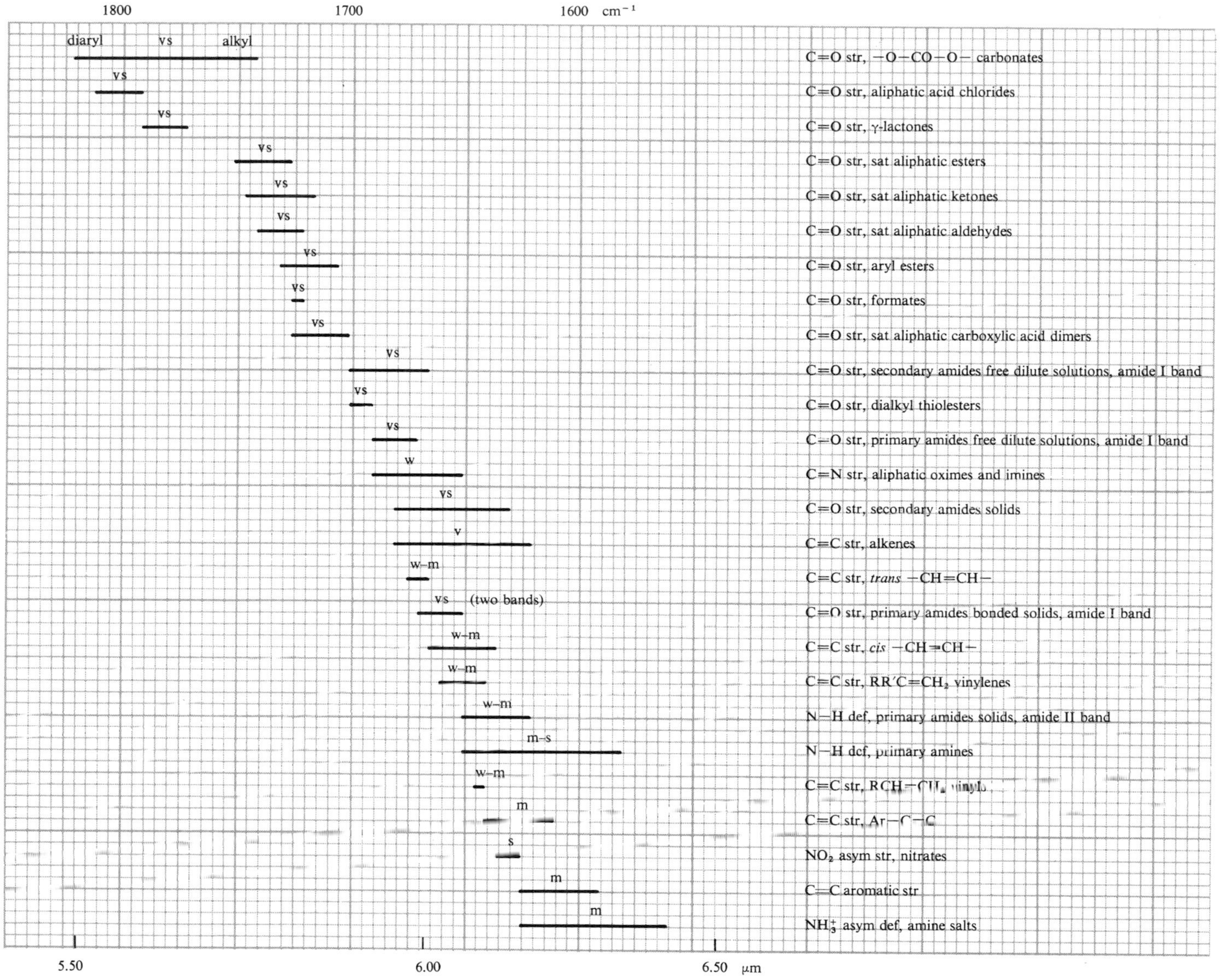
1800
1700
1600 cm^{-1}
diaryl
vs
alkyl
vs
vs
vs
vs
vs
vs
vs
vs
vs
vs
vs
w
vs
v
w–m
vs (two bands)
w–m
w–m
w–m
m–s
w–m
m
s
m
m
C=O str, —O—CO—O— carbonates
C=O str, aliphatic acid chlorides
C=O str, γ-lactones
C=O str, sat aliphatic esters
C=O str, sat aliphatic ketones
C=O str, sat aliphatic aldehydes
C=O str, aryl esters
C=O str, formates
C=O str, sat aliphatic carboxylic acid dimers
C=O str, secondary amides free dilute solutions, amide I band
C=O str, dialkyl thiolesters
C=O str, primary amides free dilute solutions, amide I band
C=N str, aliphatic oximes and imines
C=O str, secondary amides solids
C=C str, alkenes
C=C str, *trans* —CH=CH—
C=O str, primary amides bonded solids, amide I band
C=C str, *cis* —CH=CH—
C=C str, RR′C=CH_2 vinylenes
N—H def, primary amides solids, amide II band
N—H def, primary amines
C=C str, RCH=
C=C str, Ar—
NO_2 asym str, nitrates
C=C aromatic str
NH_3^+ asym def, amine salts
5.50
6.00
6.50 μm

Chart 2 (*continued*)

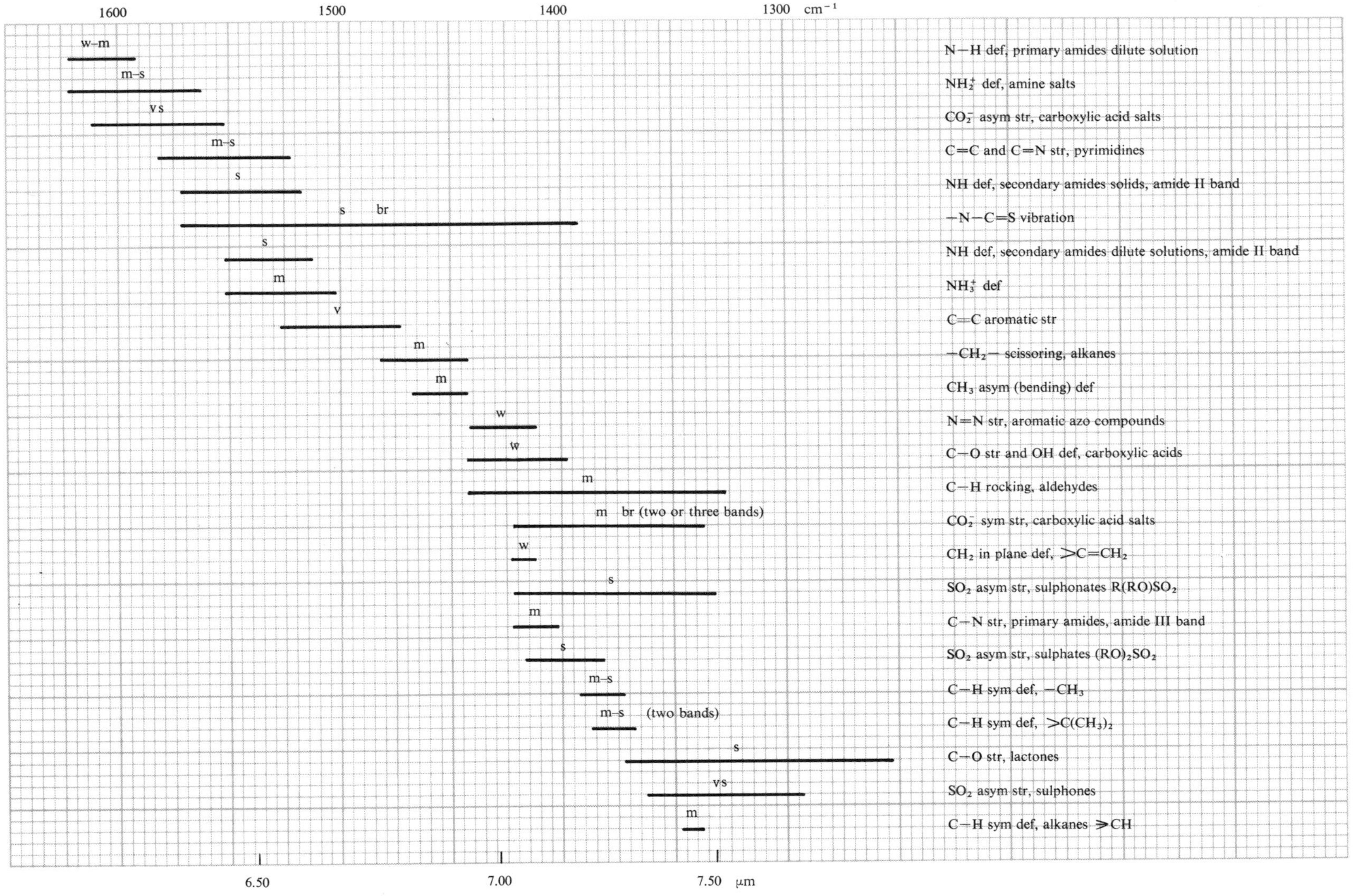
1600
1500
1400
1300 cm⁻¹
w–m
N—H def, primary amides dilute solution
m–s
NH₂⁺ def, amine salts
vs
CO₂⁻ asym str, carboxylic acid salts
m–s
C=C and C=N str, pyrimidines
s
NH def, secondary amides solids, amide II band
s br
—N—C=S vibration
s
NH def, secondary amides dilute solutions, amide II band
m
NH₃⁺ def
v
C=C aromatic str
m
—CH₂— scissoring, alkanes
m
CH₃ asym (bending) def
w
N=N str, aromatic azo compounds
w
C—O str and OH def, carboxylic acids
m
C—H rocking, aldehydes
m br (two or three bands)
CO₂⁻ sym str, carboxylic acid salts
w
CH₂ in plane def, >C=CH₂
s
SO₂ asym str, sulphonates R(RO)SO₂
m
C—N str, primary amides, amide III band
s
SO₂ asym str, sulphates (RO)₂SO₂
m–s
C—H sym def, —CH₃
m–s (two bands)
C—H sym def, >C(CH₃)₂
s
C—O str, lactones
vs
SO₂ asym str, sulphones
m
C—H sym def, alkanes ⋗CH
6.50
7.00
7.50 μm

Chart 2 (*continued*)

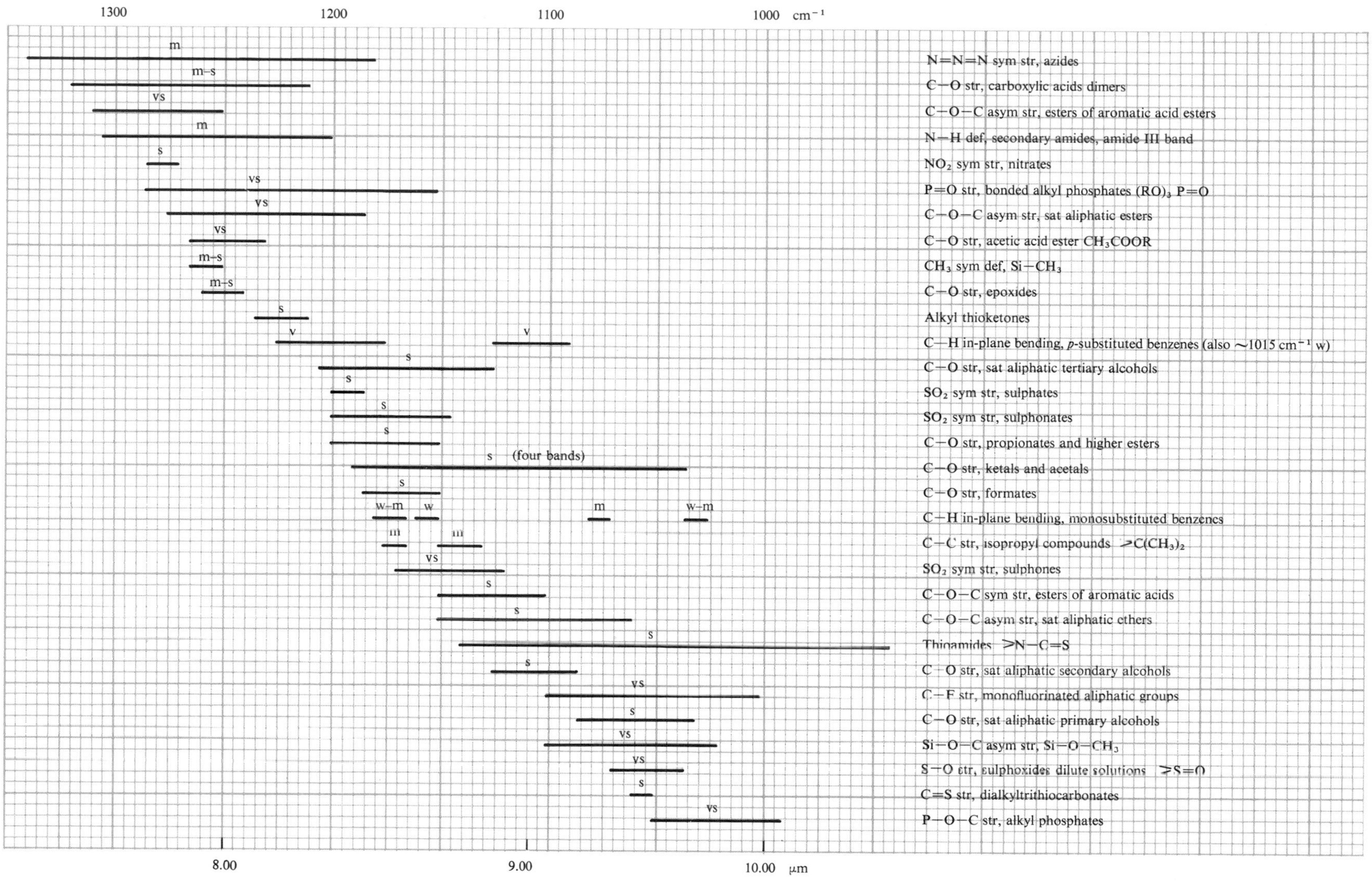
1300
1200
1100
1000 cm⁻¹
N=N=N sym str, azides
C—O str, carboxylic acids dimers
C—O—C asym str, esters of aromatic acid esters
N—H def, secondary amides, amide III band
NO_2 sym str, nitrates
P=O str, bonded alkyl phosphates $(RO)_3$ P=O
C—O—C asym str, sat aliphatic esters
C—O str, acetic acid ester CH_3COOR
CH_3 sym def, Si—CH_3
C—O str, epoxides
Alkyl thioketones
C—H in-plane bending, *p*-substituted benzenes (also ~1015 cm⁻¹ w)
C—O str, sat aliphatic tertiary alcohols
SO_2 sym str, sulphates
SO_2 sym str, sulphonates
C—O str, propionates and higher esters
C—O str, ketals and acetals
C—O str, formates
C—H in-plane bending, monosubstituted benzenes
C—C str, isopropyl compounds ⩾C$(CH_3)_2$
SO_2 sym str, sulphones
C—O—C sym str, esters of aromatic acids
C—O—C asym str, sat aliphatic ethers
Thioamides ⩾N—C=S
C—O str, sat aliphatic secondary alcohols
C—F str, monofluorinated aliphatic groups
C—O str, sat aliphatic primary alcohols
Si—O—C asym str, Si—O—CH_3
S—O str, sulphoxides dilute solutions ⩾S=O
C=S str, dialkyltrithiocarbonates
P—O—C str, alkyl phosphates
s (four bands)
8.00
9.00
10.00 μm

Chart 2 (*continued*)

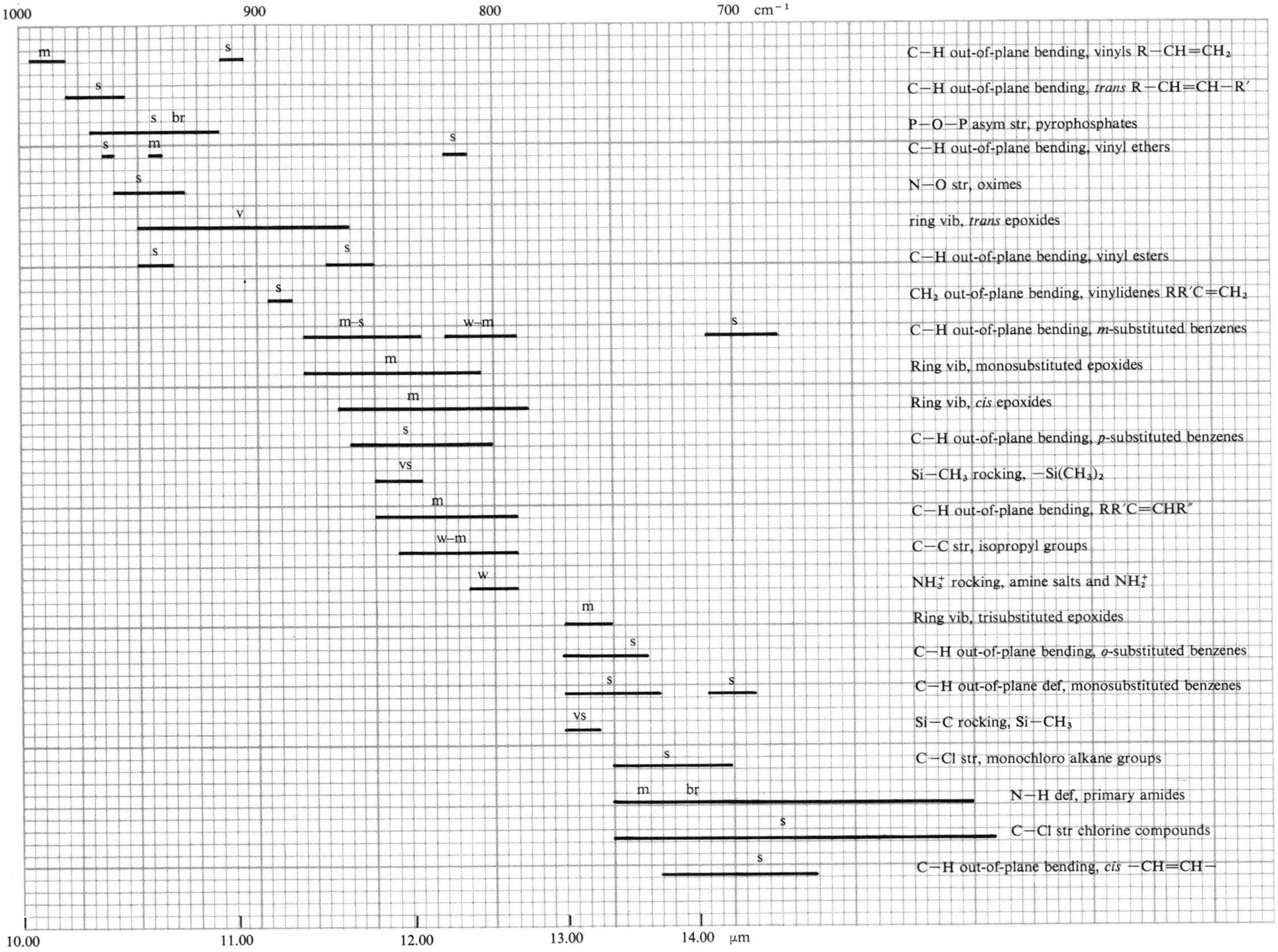
1000
900
800
700 cm⁻¹
C—H out-of-plane bending, vinyls R—CH=CH₂
C—H out-of-plane bending, *trans* R—CH=CH—R′
P—O—P asym str, pyrophosphates
C—H out-of-plane bending, vinyl ethers
N—O str, oximes
ring vib, *trans* epoxides
C—H out-of-plane bending, vinyl esters
CH₂ out-of-plane bending, vinylidenes RR′C=CH₂
C—H out-of-plane bending, *m*-substituted benzenes
Ring vib, monosubstituted epoxides
Ring vib, *cis* epoxides
C—H out-of-plane bending, *p*-substituted benzenes
Si—CH₃ rocking, —Si(CH₃)₂
C—H out-of-plane bending, RR′C=CHR″
C—C str, isopropyl groups
NH₃⁺ rocking, amine salts and NH₂⁺
Ring vib, trisubstituted epoxides
C—H out-of-plane bending, *o*-substituted benzenes
C—H out-of-plane def, monosubstituted benzenes
Si—C rocking, Si—CH₃
C—Cl str, monochloro alkane groups
N—H def, primary amides
C—Cl str chlorine compounds
C—H out-of-plane bending, *cis* —CH=CH—
10.00
11.00
12.00
13.00
14.00 μm

Chart 2 (*continued*)

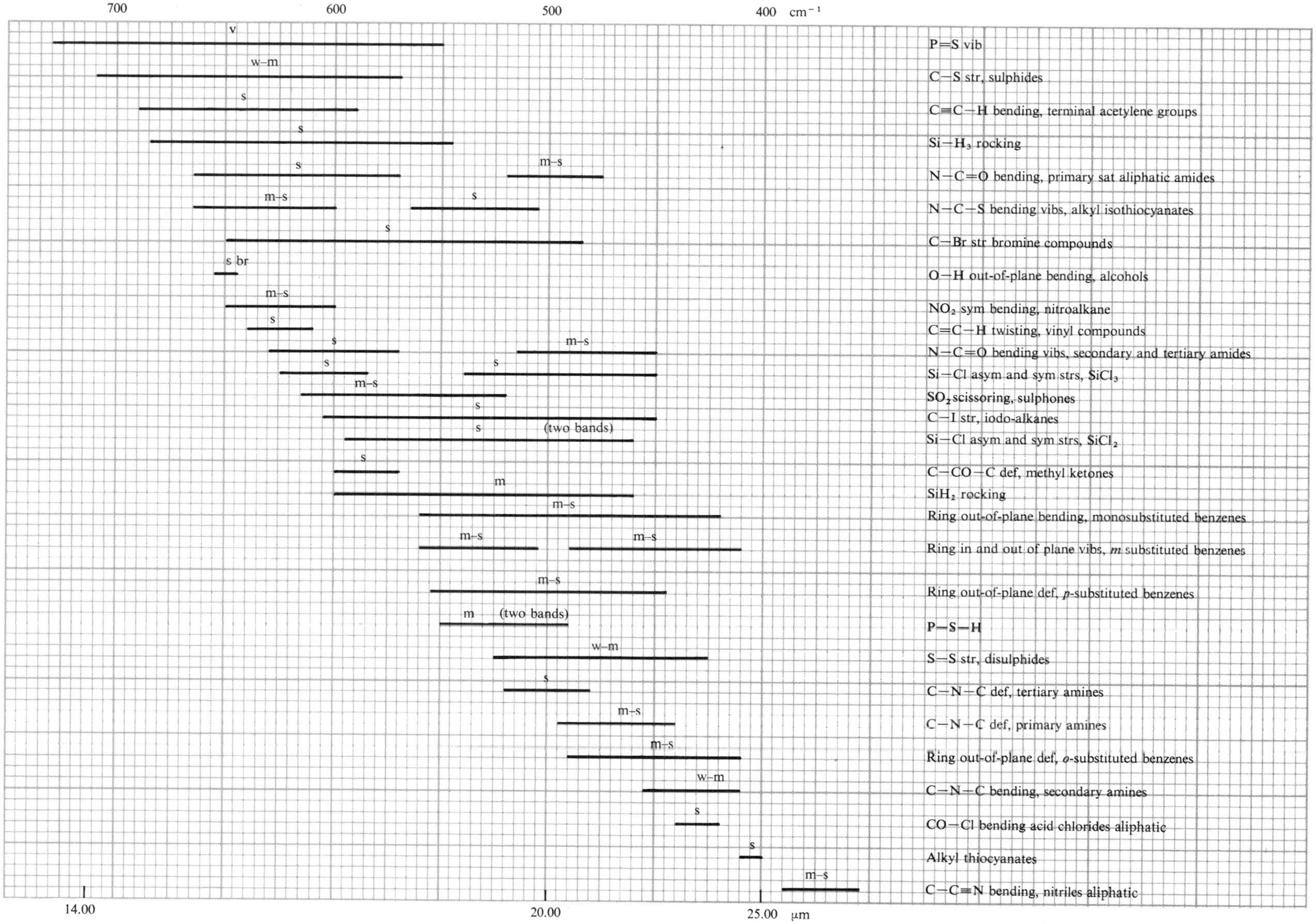
700
600
500
400 cm⁻¹
v
w–m
s
s
s
m–s
m–s
s
s
s br
m–s
s
s
m–s
s
s
m–s
s
s
(two bands)
s
m
m–s
m–s
m–s
m–s
m
(two bands)
w–m
s
m–s
m–s
w–m
s
s
m–s
P=S vib
C–S str, sulphides
C≡C–H bending, terminal acetylene groups
Si–H₃ rocking
N–C=O bending, primary sat aliphatic amides
N–C–S bending vibs, alkyl isothiocyanates
C–Br str bromine compounds
O–H out-of-plane bending, alcohols
NO₂ sym bending, nitroalkane
C=C–H twisting, vinyl compounds
N–C=O bending vibs, secondary and tertiary amides
Si–Cl asym and sym strs, SiCl₃
SO₂ scissoring, sulphones
C–I str, iodo-alkanes
Si–Cl asym and sym strs, SiCl₂
C–CO–C def, methyl ketones
SiH₂ rocking
Ring out-of-plane bending, monosubstituted benzenes
Ring in and out of plane vibs, m substituted benzenes
Ring out-of-plane def, p-substituted benzenes
P–S–H
S–S str, disulphides
C–N–C def, tertiary amines
C–N–C def, primary amines
Ring out-of-plane def, o-substituted benzenes
C–N–C bending, secondary amines
CO–Cl bending acid chlorides aliphatic
Alkyl thiocyanates
C–C≡N bending, nitriles aliphatic
14.00
20.00
25.00 μm

Chart 3 The ranges of the main characteristic bands of groups or classes of chemical compound are indicated by either thick or fine lines. The thick lines indicate important band ranges which either are completely specific for that group or can be used in those ranges to distinguish the group from similar groups. The thin lines indicate other important band regions which should be borne in mind.

The intensities of the band(s) in each region are given where these are known and reliable. The intensities of bands occurring in the regions represented by thin lines are as given previously in the chart for similar groups unless specifically shown. (Note the change of scale at 2000 cm⁻¹ (5.0 μm)).

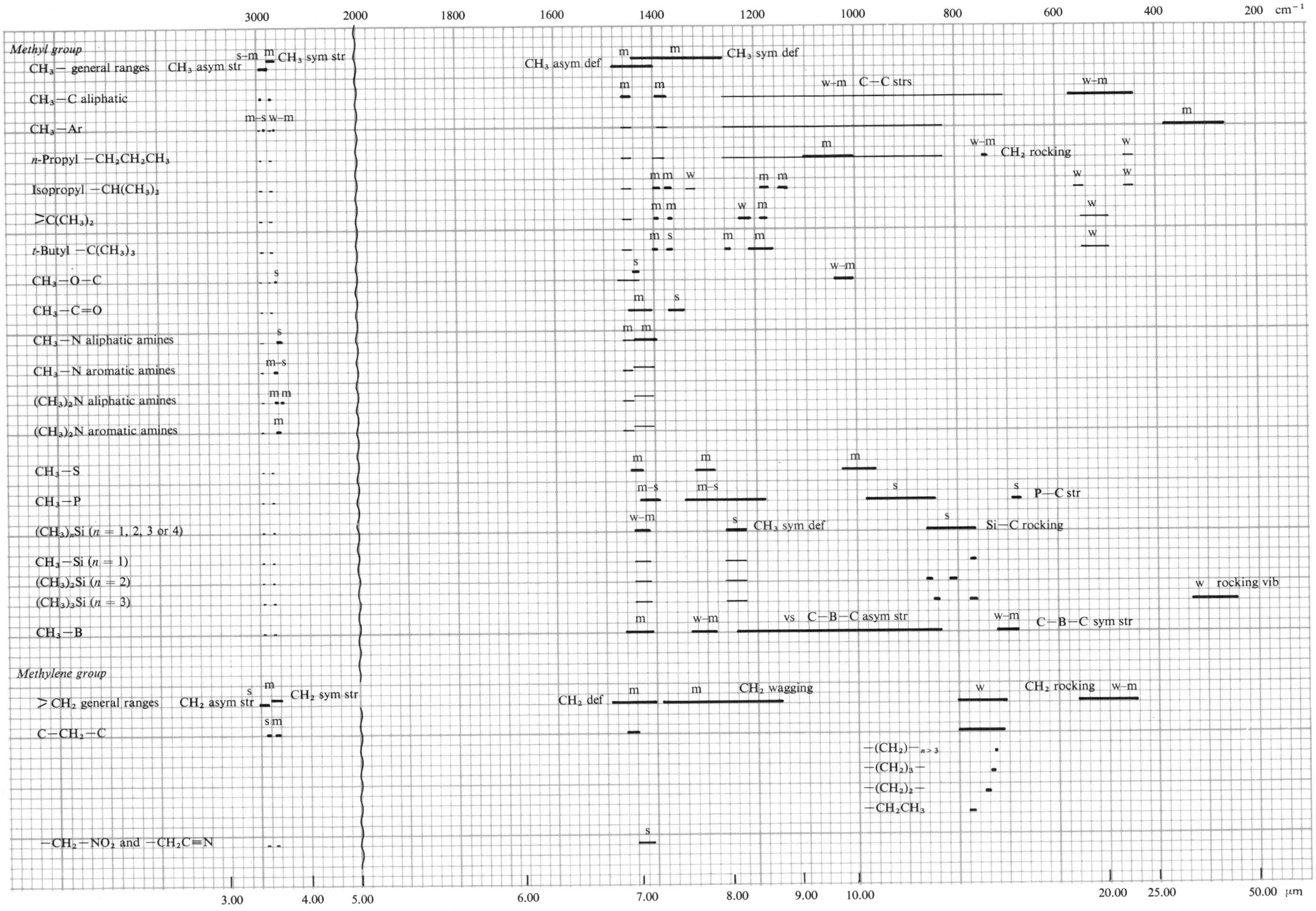

3000
2000
1800
1600
1400
1200
1000
800
600
400
200 cm⁻¹
Methyl group
CH₃— general ranges
CH₃ asym str
s–m
m
CH₃ sym str
CH₃ asym def
CH₃ sym def
CH₃—C aliphatic
w–m C—C strs
w–m
CH₃—Ar
m–s w–m
n-Propyl —CH₂CH₂CH₃
w–m
CH₂ rocking
Isopropyl —CH(CH₃)₂
>C(CH₃)₂
t-Butyl —C(CH₃)₃
CH₃—O—C
CH₃—C=O
CH₃—N aliphatic amines
CH₃—N aromatic amines
(CH₃)₂N aliphatic amines
(CH₃)₂N aromatic amines
CH₃—S
CH₃—P
P—C str
(CH₃)ₙSi (n = 1, 2, 3 or 4)
CH₃ sym def
Si—C rocking
CH₃—Si (n = 1)
(CH₃)₂Si (n = 2)
(CH₃)₃Si (n = 3)
w rocking vib
CH₃—B
vs C—B—C asym str
w–m
C—B—C sym str
Methylene group
>CH₂ general ranges
CH₂ asym str
CH₂ sym str
CH₂ def
CH₂ wagging
CH₂ rocking
w–m
C—CH₂—C
—(CH₂)—ₙ₊₃
—(CH₂)₃—
—(CH₂)₂—
—CH₂CH₃
—CH₂—NO₂ and —CH₂C≡N
3.00
4.00
5.00
6.00
7.00
8.00
9.00
10.00
20.00
25.00
50.00 μm

Chart 3 (*continued*)

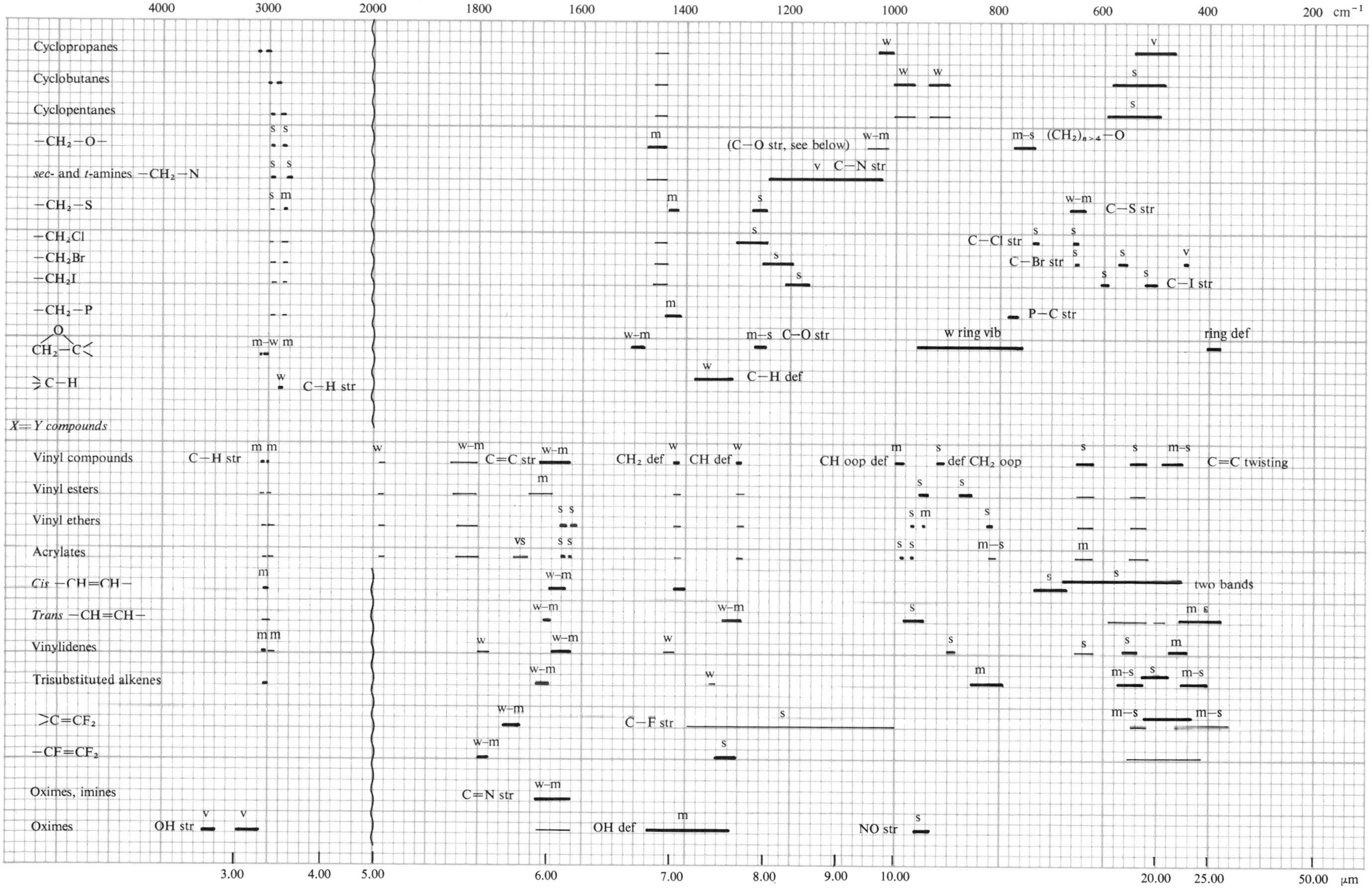
4000
3000
2000
1800
1600
1400
1200
1000
800
600
400
200 cm⁻¹
Cyclopropanes
Cyclobutanes
Cyclopentanes
−CH₂−O−
sec- and t-amines −CH₂−N
−CH₂−S
−CH₂Cl
−CH₂Br
−CH₂I
−CH₂−P
CH₂−C
≥C−H
X=Y compounds
Vinyl compounds
Vinyl esters
Vinyl ethers
Acrylates
Cis −CH=CH−
Trans −CH=CH−
Vinylidenes
Trisubstituted alkenes
>C=CF₂
−CF=CF₂
Oximes, imines
Oximes
C−H str
(C−O str, see below)
v C−N str
(CH₂)ₙ>₄−O
C−S str
C−Cl str
C−Br str
C−I str
P−C str
m−s C−O str
w ring vib
ring def
C−H def
C=C str
CH₂ def
CH def
CH oop def
def CH₂ oop
C=C twisting
two bands
C−F str
C=N str
OH str
OH def
NO str
3.00
4.00
5.00
6.00
7.00
8.00
9.00
10.00
20.00
25.00
50.00 μm

Chart 3 (*continued*)

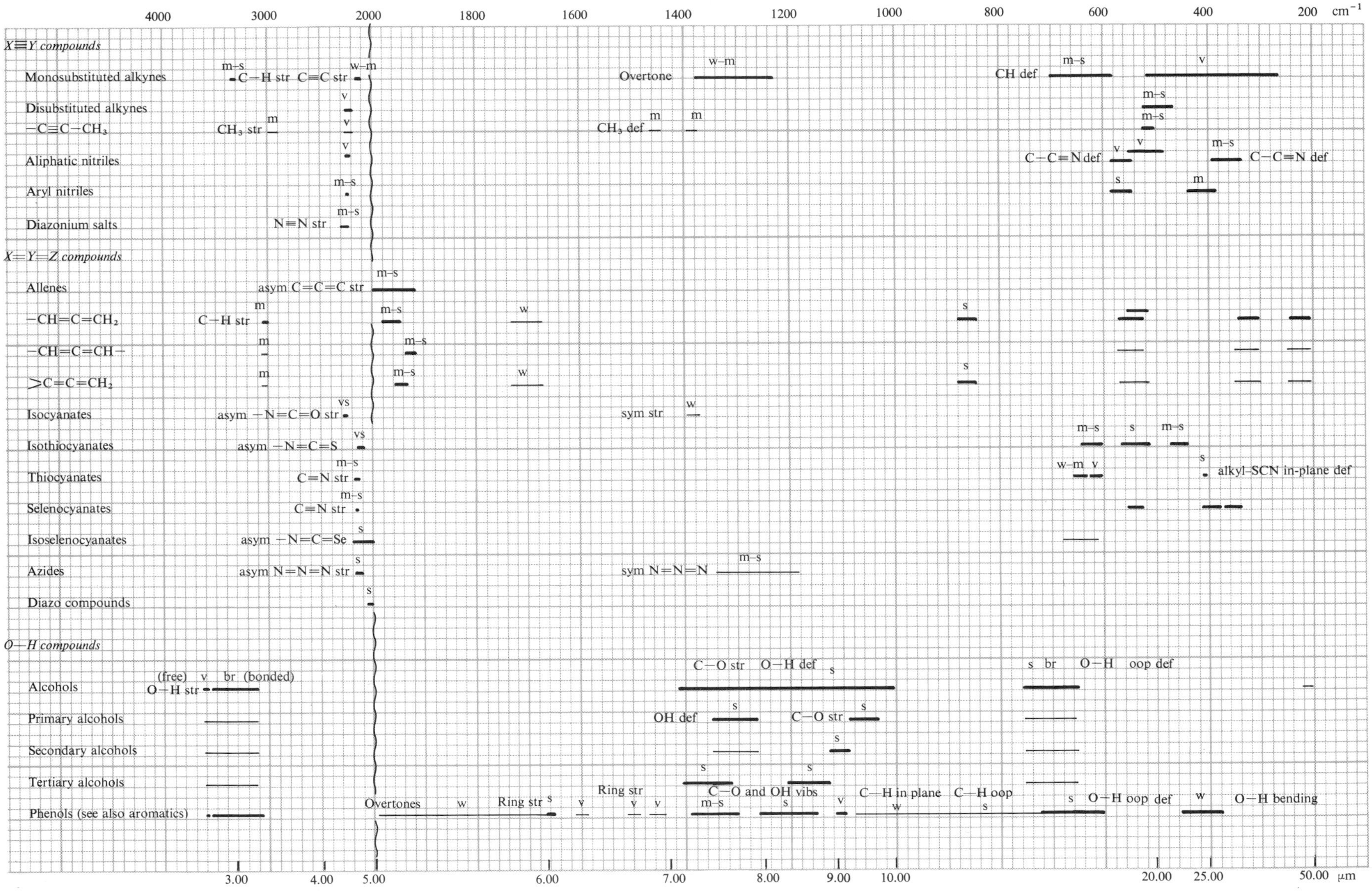

4000
3000
2000
1800
1600
1400
1200
1000
800
600
400
200
cm⁻¹
X≡Y compounds
Monosubstituted alkynes
m–s
C–H str
C≡C str
w–m
Overtone
w–m
CH def
m–s
v
Disubstituted alkynes
v
m–s
–C≡C–CH₃
CH₃ str
m
v
CH₃ def
m
m
m–s
Aliphatic nitriles
v
C–C≡N def
v
v
m–s
C–C≡N def
Aryl nitriles
m–s
s
m
Diazonium salts
m–s
N≡N str
X=Y=Z compounds
Allenes
asym C=C=C str
m–s
–CH=C=CH₂
C–H str
m
m–s
w
s
–CH=C=CH–
m
m–s
>C=C=CH₂
m
m–s
w
s
Isocyanates
asym –N=C=O str
vs
sym str
w
Isothiocyanates
asym –N=C=S
vs
m–s
s
m–s
Thiocyanates
C≡N str
m–s
w–m
v
s
alkyl–SCN in-plane def
Selenocyanates
C≡N str
m–s
Isoselenocyanates
asym –N=C=Se
s
Azides
asym N=N=N str
s
sym N=N=N
m–s
Diazo compounds
s
O–H compounds
Alcohols
(free)
v
br
(bonded)
O–H str
C–O str
O–H def
s
s
br
O–H
oop def
Primary alcohols
OH def
s
C–O str
s
Secondary alcohols
s
Tertiary alcohols
s
s
Phenols (see also aromatics)
Overtones
w
Ring str
s
v
Ring str
v
v
C–O and OH vibs
m–s
s
v
C–H in plane
w
C–H oop
s
s
O–H oop def
w
O–H bending
3.00
4.00
5.00
6.00
7.00
8.00
9.00
10.00
20.00
25.00
50.00
μm

Chart 3 (*continued*)

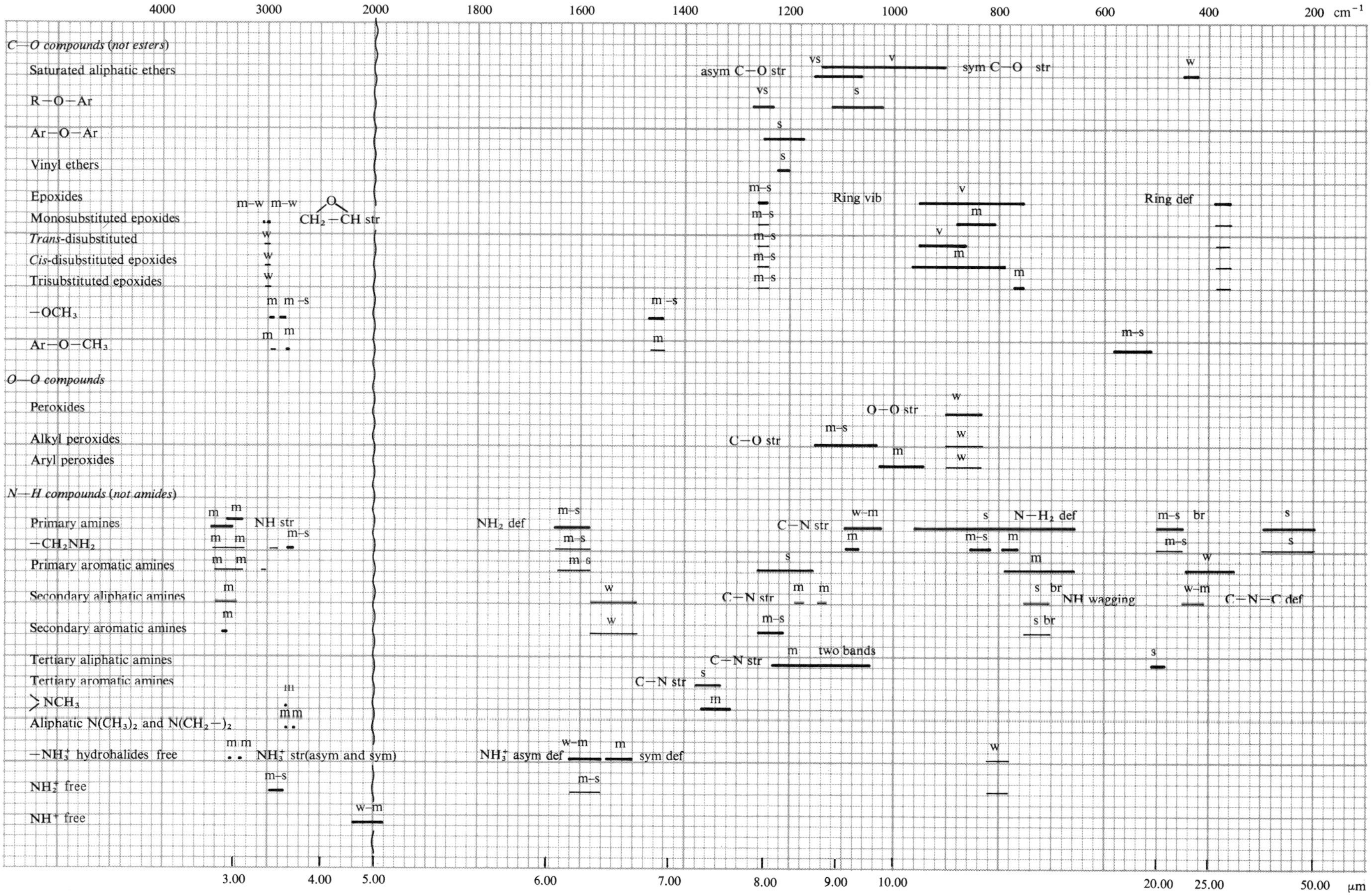
4000
3000
2000
1800
1600
1400
1200
1000
800
600
400
200 cm⁻¹
C—O compounds (not esters)
Saturated aliphatic ethers
R—O—Ar
Ar—O—Ar
Vinyl ethers
Epoxides
Monosubstituted epoxides
Trans-disubstituted
Cis-disubstituted epoxides
Trisubstituted epoxides
—OCH₃
Ar—O—CH₃
O—O compounds
Peroxides
Alkyl peroxides
Aryl peroxides
N—H compounds (not amides)
Primary amines
—CH₂NH₂
Primary aromatic amines
Secondary aliphatic amines
Secondary aromatic amines
Tertiary aliphatic amines
Tertiary aromatic amines
>NCH₃
Aliphatic N(CH₃)₂ and N(CH₂—)₂
—NH₃⁺ hydrohalides free
NH₂⁺ free
NH⁺ free
asym C—O str
sym C—O str
CH₂—CH str
Ring vib
Ring def
O—O str
C—O str
NH str
NH₂ def
C—N str
N—H₂ def
NH wagging
C—N—C def
two bands
NH₃⁺ str(asym and sym)
NH₃⁺ asym def
sym def
3.00
4.00
5.00
6.00
7.00
8.00
9.00
10.00
20.00
25.00
50.00 μm

Chart 3 (*continued*)

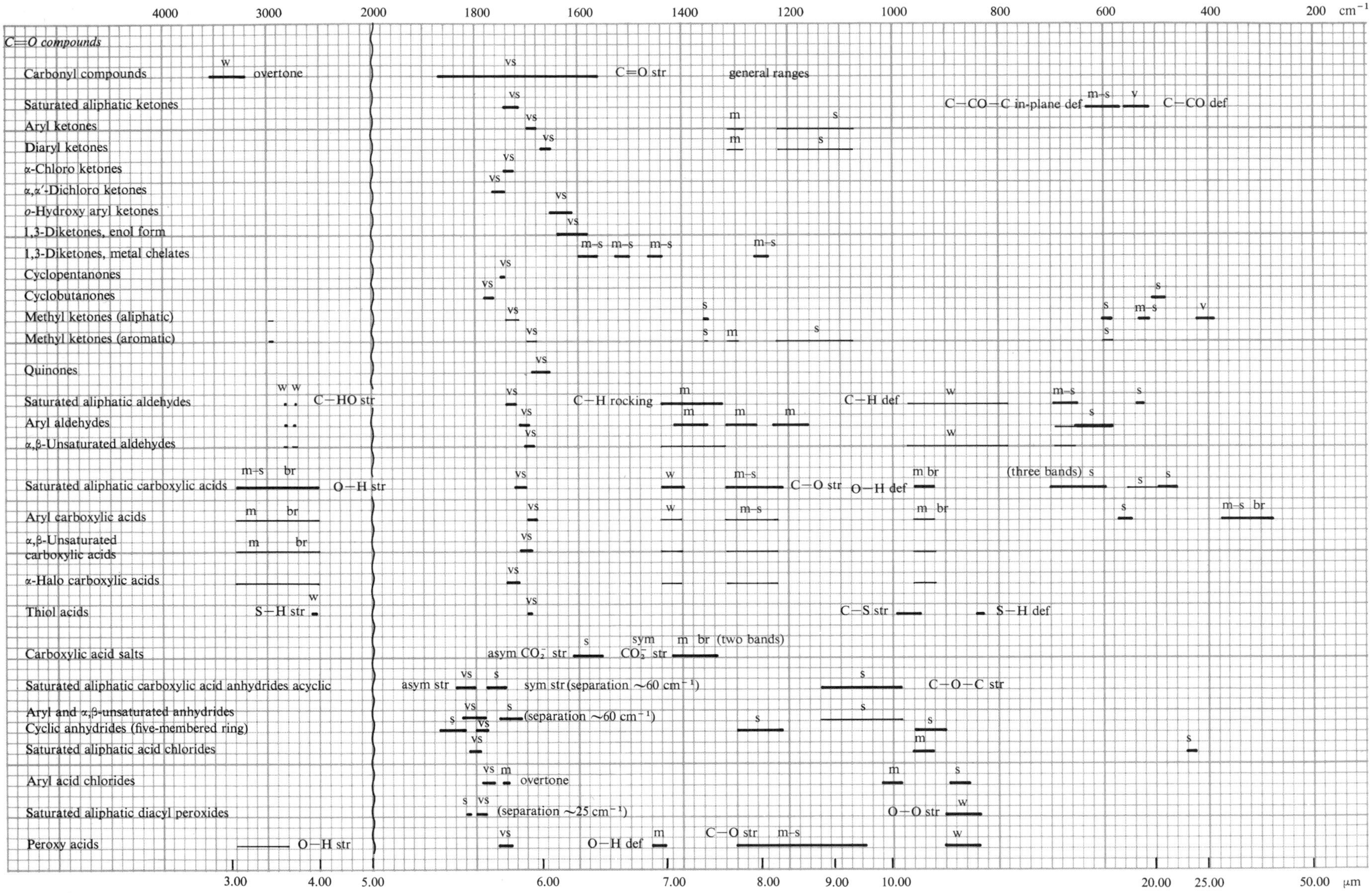
4000 3000 2000 1800 1600 1400 1200 1000 800 600 400 200 cm⁻¹
C=O compounds
Carbonyl compounds
w overtone
vs C=O str
general ranges
Saturated aliphatic ketones
C—CO—C in-plane def
m–s v C—CO def
Aryl ketones
Diaryl ketones
α-Chloro ketones
α,α′-Dichloro ketones
o-Hydroxy aryl ketones
1,3-Diketones, enol form
1,3-Diketones, metal chelates
Cyclopentanones
Cyclobutanones
Methyl ketones (aliphatic)
Methyl ketones (aromatic)
Quinones
Saturated aliphatic aldehydes
C—HO str
C—H rocking
C—H def
Aryl aldehydes
α,β-Unsaturated aldehydes
Saturated aliphatic carboxylic acids
O—H str
C—O str
O—H def
(three bands)
Aryl carboxylic acids
α,β-Unsaturated carboxylic acids
α-Halo carboxylic acids
Thiol acids
S—H str
C—S str
S—H def
Carboxylic acid salts
asym CO₂⁻ str
sym CO₂⁻ str
m br (two bands)
Saturated aliphatic carboxylic acid anhydrides acyclic
asym str
sym str (separation ~60 cm⁻¹)
C—O—C str
Aryl and α,β-unsaturated anhydrides
(separation ~60 cm⁻¹)
Cyclic anhydrides (five-membered ring)
Saturated aliphatic acid chlorides
Aryl acid chlorides
overtone
Saturated aliphatic diacyl peroxides
(separation ~25 cm⁻¹)
O—O str
Peroxy acids
O—H str
O—H def
C—O str
3.00 4.00 5.00 6.00 7.00 8.00 9.00 10.00 20.00 25.00 50.00 μm

Chart 3 (*continued*)

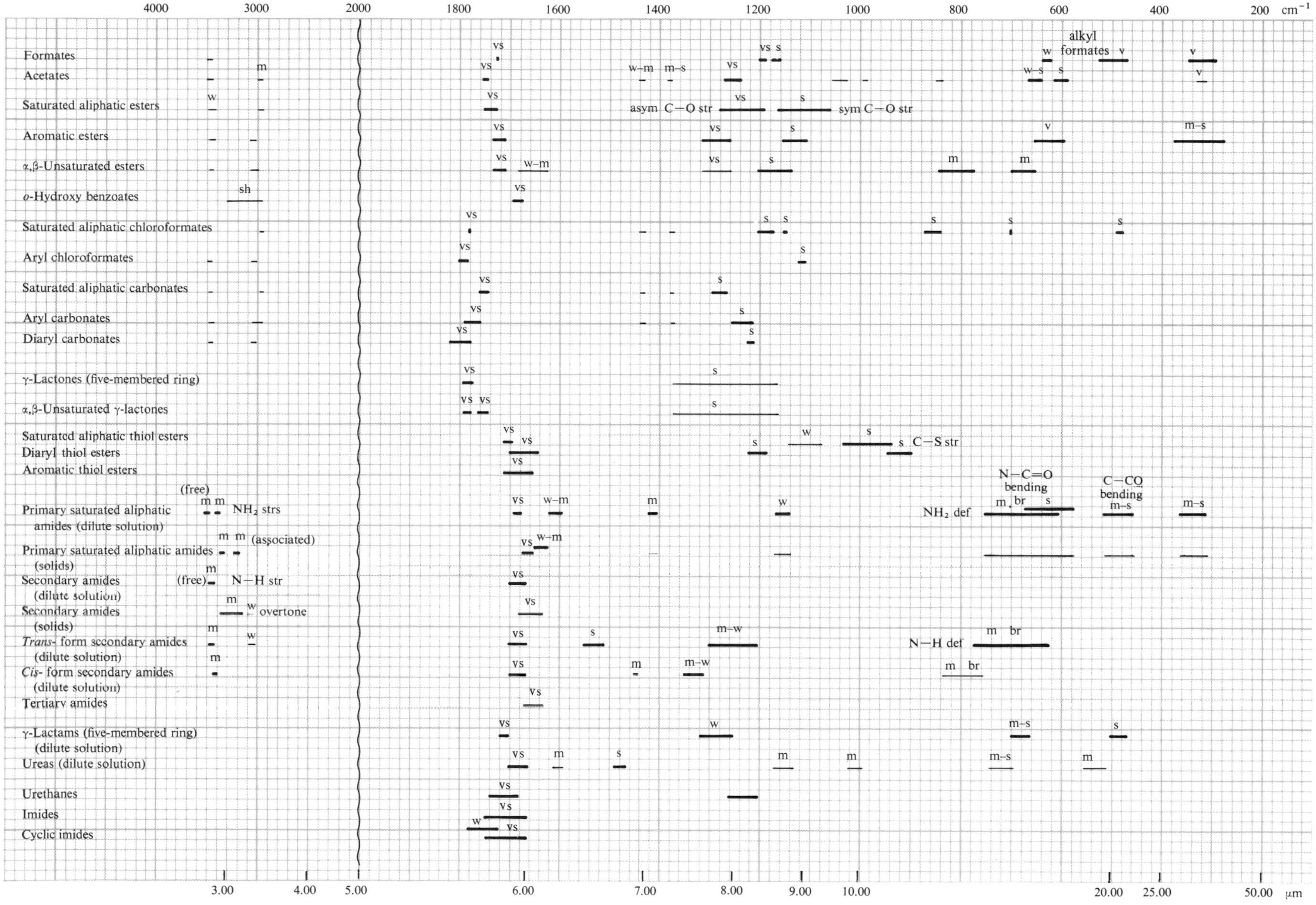
4000
3000
2000
1800
1600
1400
1200
1000
800
600
400
200
cm⁻¹
Formates
Acetates
Saturated aliphatic esters
Aromatic esters
α,β-Unsaturated esters
o-Hydroxy benzoates
Saturated aliphatic chloroformates
Aryl chloroformates
Saturated aliphatic carbonates
Aryl carbonates
Diaryl carbonates
γ-Lactones (five-membered ring)
α,β-Unsaturated γ-lactones
Saturated aliphatic thiol esters
Diaryl thiol esters
Aromatic thiol esters
Primary saturated aliphatic amides (dilute solution)
Primary saturated aliphatic amides (solids)
Secondary amides (dilute solution)
Secondary amides (solids)
Trans- form secondary amides (dilute solution)
Cis- form secondary amides (dilute solution)
Tertiary amides
γ-Lactams (five-membered ring) (dilute solution)
Ureas (dilute solution)
Urethanes
Imides
Cyclic imides
alkyl formates
asym C—O str
sym C—O str
C—S str
(free)
NH₂ strs
(associated)
N—H str
overtone
N—C=O bending
NH₂ def
C—CO bending
N—H def
m br
3.00
4.00
5.00
6.00
7.00
8.00
9.00
10.00
20.00
25.00
50.00
μm

Chart 3 (*continued*)

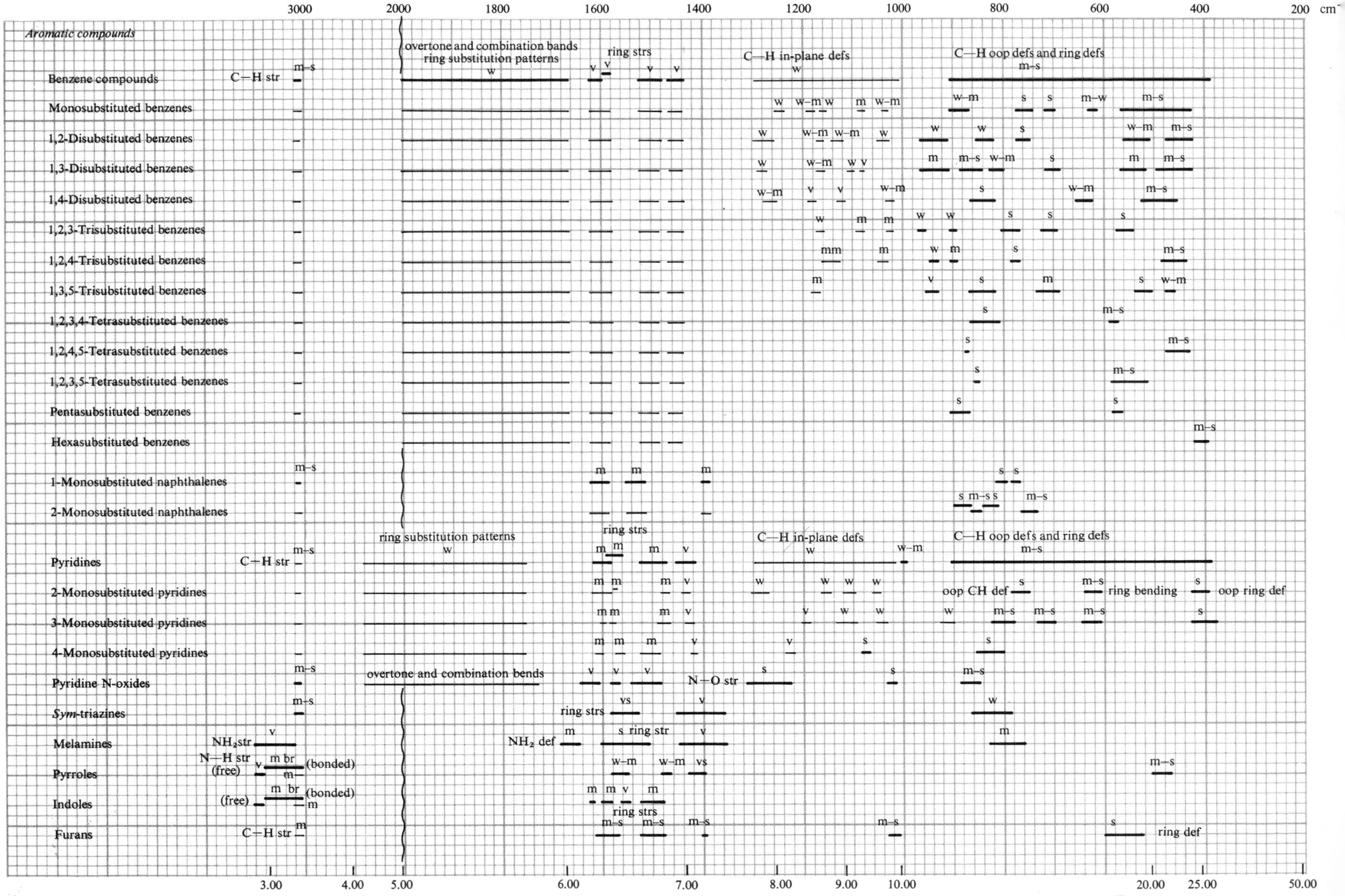
3000
2000
1800
1600
1400
1200
1000
800
600
400
200 cm⁻¹
Aromatic compounds
overtone and combination bands
ring substitution patterns
ring strs
C—H in-plane defs
C—H oop defs and ring defs
C—H str
Benzene compounds
Monosubstituted benzenes
1,2-Disubstituted benzenes
1,3-Disubstituted benzenes
1,4-Disubstituted benzenes
1,2,3-Trisubstituted benzenes
1,2,4-Trisubstituted benzenes
1,3,5-Trisubstituted benzenes
1,2,3,4-Tetrasubstituted benzenes
1,2,4,5-Tetrasubstituted benzenes
1,2,3,5-Tetrasubstituted benzenes
Pentasubstituted benzenes
Hexasubstituted benzenes
1-Monosubstituted naphthalenes
2-Monosubstituted naphthalenes
ring substitution patterns
ring strs
C—H in-plane defs
C—H oop defs and ring defs
Pyridines
2-Monosubstituted pyridines
3-Monosubstituted pyridines
4-Monosubstituted pyridines
Pyridine N-oxides
Sym-triazines
Melamines
Pyrroles
Indoles
Furans
oop CH def
ring bending
oop ring def
overtone and combination bends
N—O str
ring strs
NH₂ def
NH₂str
s ring str
N—H str
(free)
m br
(bonded)
ring def
3.00
4.00
5.00
6.00
7.00
8.00
9.00
10.00
20.00
25.00
50.00

Chart 3 (*continued*)

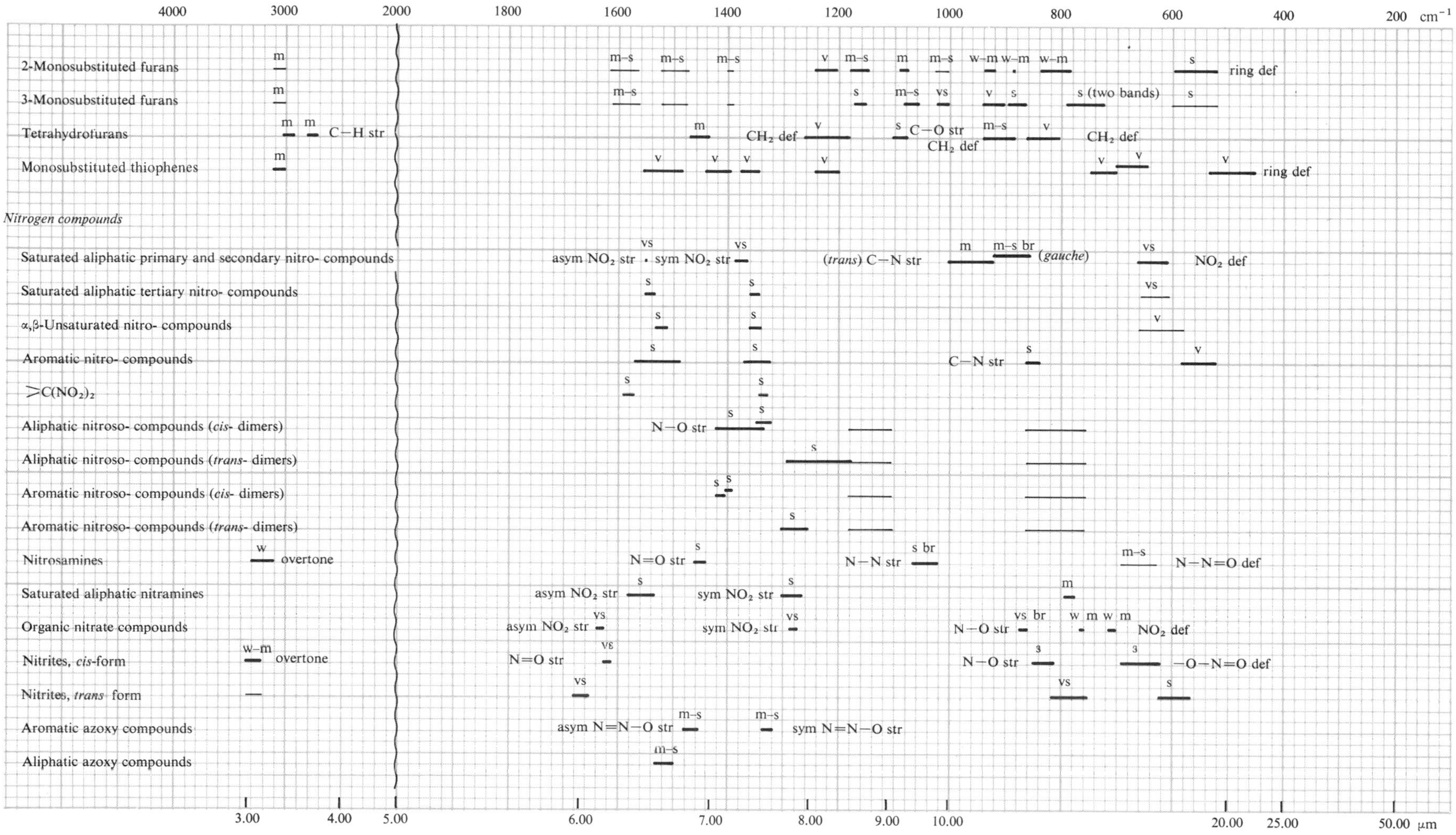

4000
3000
2000
1800
1600
1400
1200
1000
800
600
400
200 cm⁻¹
2-Monosubstituted furans
3-Monosubstituted furans
Tetrahydrofurans
Monosubstituted thiophenes
Nitrogen compounds
Saturated aliphatic primary and secondary nitro- compounds
Saturated aliphatic tertiary nitro- compounds
α,β-Unsaturated nitro- compounds
Aromatic nitro- compounds
>C(NO₂)₂
Aliphatic nitroso- compounds (cis- dimers)
Aliphatic nitroso- compounds (trans- dimers)
Aromatic nitroso- compounds (cis- dimers)
Aromatic nitroso- compounds (trans- dimers)
Nitrosamines
Saturated aliphatic nitramines
Organic nitrate compounds
Nitrites, cis-form
Nitrites, trans form
Aromatic azoxy compounds
Aliphatic azoxy compounds
ring def
s (two bands)
C—H str
CH₂ def
C—O str
CH₂ def
CH₂ def
ring def
asym NO₂ str
sym NO₂ str
(trans) C—N str
(gauche)
NO₂ def
C—N str
N—O str
overtone
N=O str
N—N str
N—N=O def
asym NO₂ str
sym NO₂ str
asym NO₂ str
sym NO₂ str
N—O str
NO₂ def
overtone
N=O str
N—O str
—O—N=O def
asym N=N—O str
sym N=N—O str
3.00
4.00
5.00
6.00
7.00
8.00
9.00
10.00
20.00
25.00
50.00 μm

Chart 3 (*continued*)

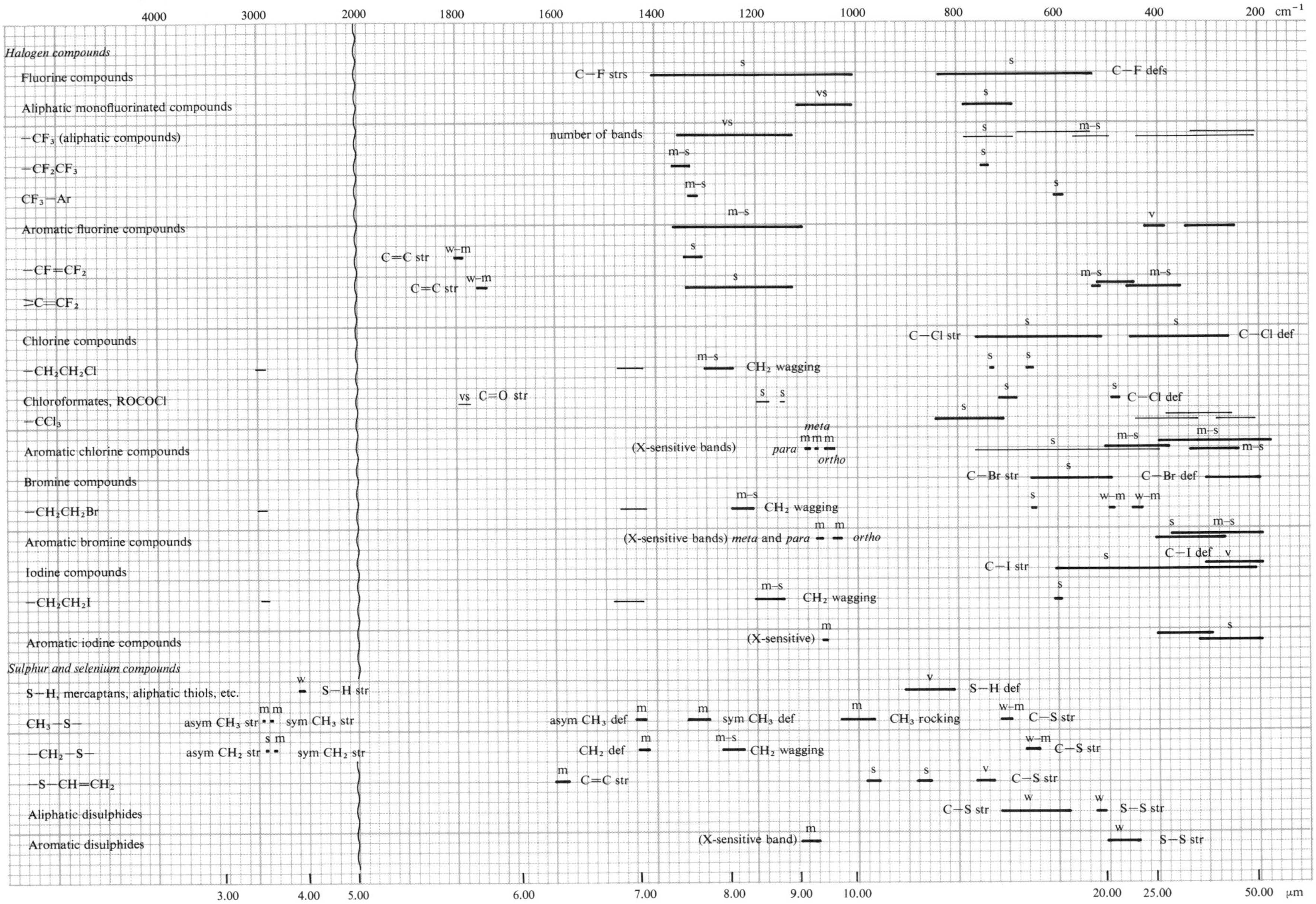
4000
3000
2000
1800
1600
1400
1200
1000
800
600
400
200 cm⁻¹
Halogen compounds
Fluorine compounds
C—F strs
s
C—F defs
Aliphatic monofluorinated compounds
vs
−CF₃ (aliphatic compounds)
number of bands
m–s
−CF₂CF₃
CF₃—Ar
Aromatic fluorine compounds
v
−CF=CF₂
C=C str
w–m
>C=CF₂
Chlorine compounds
C—Cl str
C—Cl def
−CH₂CH₂Cl
CH₂ wagging
Chloroformates, ROCOCl
vs C=O str
−CCl₃
Aromatic chlorine compounds
(X-sensitive bands)
meta
m m m
para
ortho
Bromine compounds
C—Br str
C—Br def
−CH₂CH₂Br
w–m
Aromatic bromine compounds
(X-sensitive bands) meta and para
m m
Iodine compounds
C—I str
C—I def
−CH₂CH₂I
Aromatic iodine compounds
(X-sensitive)
m
Sulphur and selenium compounds
S—H, mercaptans, aliphatic thiols, etc.
w
S—H str
S—H def
CH₃—S—
asym CH₃ str
sym CH₃ str
asym CH₃ def
sym CH₃ def
CH₃ rocking
C—S str
−CH₂—S—
asym CH₂ str
sym CH₂ str
s m
CH₂ def
−S—CH=CH₂
Aliphatic disulphides
S—S str
Aromatic disulphides
(X-sensitive band)
3.00
4.00
5.00
6.00
7.00
8.00
9.00
10.00
20.00
25.00
50.00 μm

Chart 3 (continued)

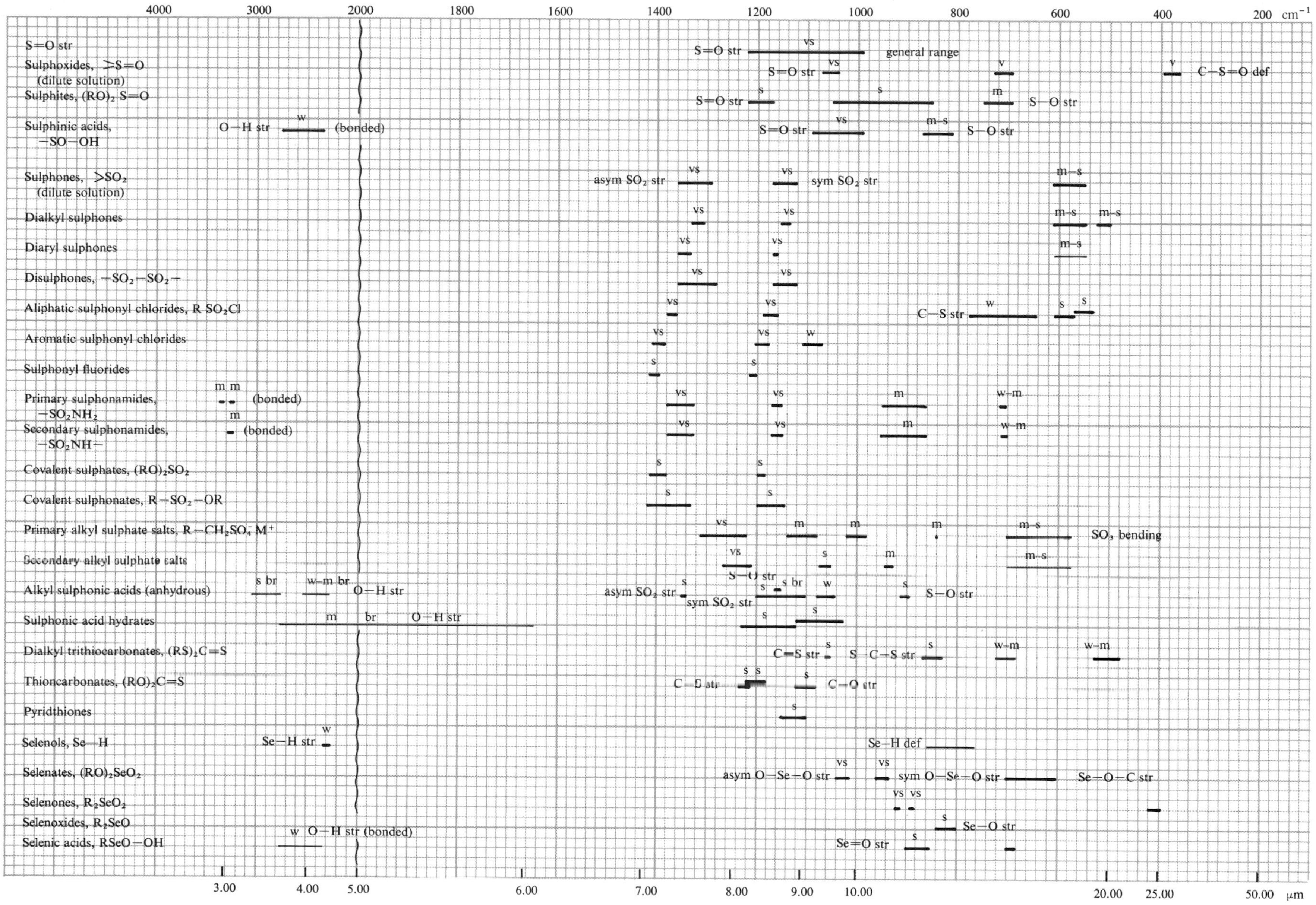

4000 3000 2000 1800 1600 1400 1200 1000 800 600 400 200 cm⁻¹
S=O str
Sulphoxides, >S=O (dilute solution)
Sulphites, (RO)₂ S=O
Sulphinic acids, —SO—OH
Sulphones, >SO₂ (dilute solution)
Dialkyl sulphones
Diaryl sulphones
Disulphones, —SO₂—SO₂—
Aliphatic sulphonyl chlorides, R SO₂Cl
Aromatic sulphonyl chlorides
Sulphonyl fluorides
Primary sulphonamides, —SO₂NH₂
Secondary sulphonamides, —SO₂NH—
Covalent sulphates, (RO)₂SO₂
Covalent sulphonates, R—SO₂—OR
Primary alkyl sulphate salts, R—CH₂SO₄⁻ M⁺
Secondary alkyl sulphate salts
Alkyl sulphonic acids (anhydrous)
Sulphonic acid hydrates
Dialkyl trithiocarbonates, (RS)₂C=S
Thioncarbonates, (RO)₂C=S
Pyridthiones
Selenols, Se—H
Selenates, (RO)₂SeO₂
Selenones, R₂SeO₂
Selenoxides, R₂SeO
Selenic acids, RSeO—OH
S=O str
general range
C—S=O def
S—O str
O—H str
(bonded)
asym SO₂ str
sym SO₂ str
C—S str
SO₃ bending
sym SO₂ str
C=S str
S—C—S str
C—O str
Se—H str
Se—H def
asym O—Se—O str
sym O—Se—O str
Se—O—C str
Se—O str
Se=O str
O—H str (bonded)
3.00 4.00 5.00 6.00 7.00 8.00 9.00 10.00 20.00 25.00 50.00 μm

Chart 3 (*continued*)

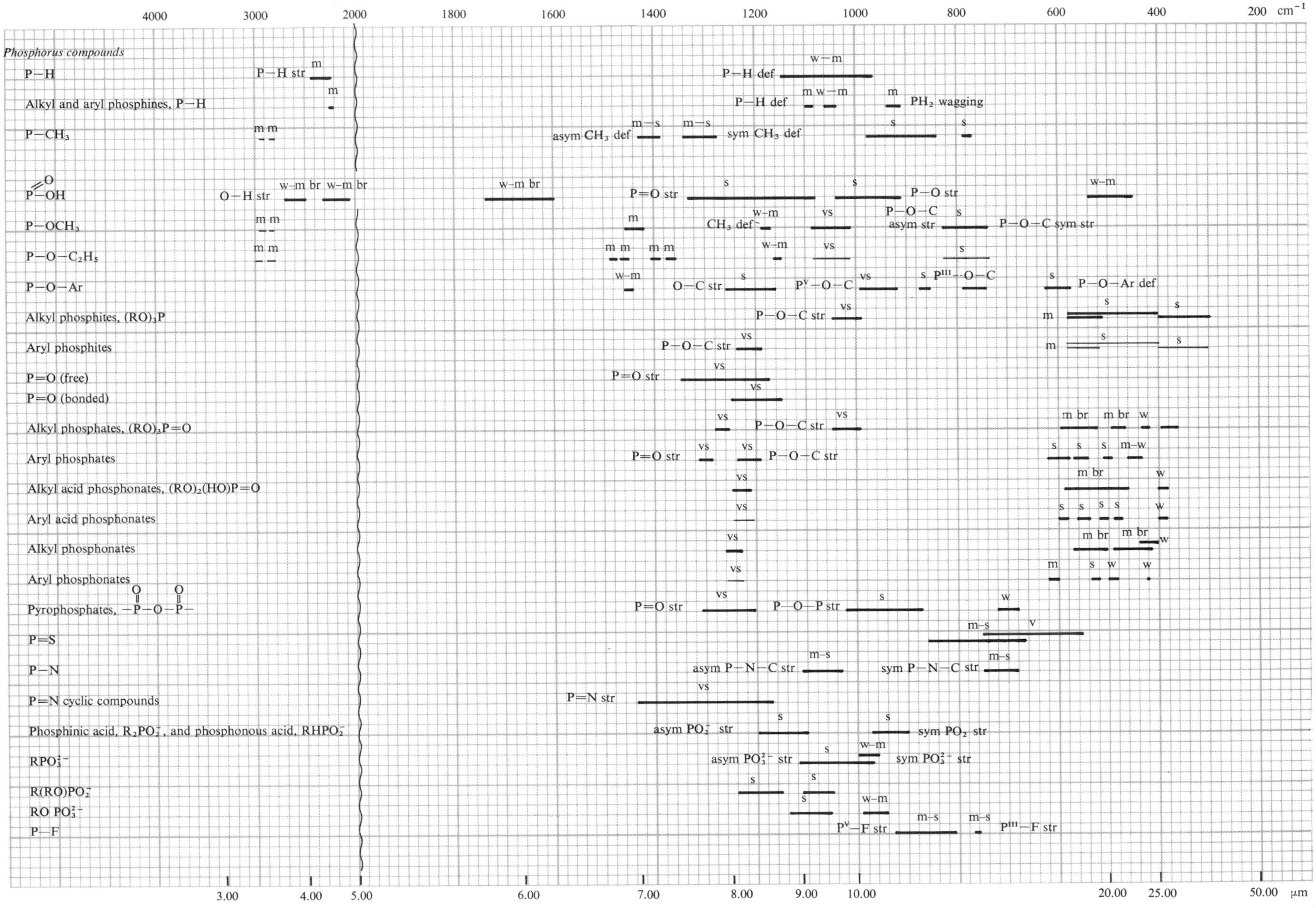
4000 3000 2000 1800 1600 1400 1200 1000 800 600 400 200 cm⁻¹
Phosphorus compounds
P—H
P—H str
P—H def
Alkyl and aryl phosphines, P—H
PH₂ wagging
P—CH₃
asym CH₃ def
sym CH₃ def
P(=O)—OH
O—H str
w–m br
P=O str
P—O str
P—OCH₃
CH₃ def
P—O—C asym str
P—O—C sym str
P—O—C₂H₅
P—O—Ar
O—C str
Pᵛ—O—C
Pᴵᴵᴵ—O—C
P—O—Ar def
Alkyl phosphites, (RO)₃P
P—O—C str
Aryl phosphites
P=O (free)
P=O (bonded)
Alkyl phosphates, (RO)₃P=O
Aryl phosphates
Alkyl acid phosphonates, (RO)₂(HO)P=O
Aryl acid phosphonates
Alkyl phosphonates
Aryl phosphonates
Pyrophosphates, —P(=O)—O—P(=O)—
P—O—P str
P=S
P—N
asym P—N—C str
sym P—N—C str
P=N cyclic compounds
P=N str
Phosphinic acid, R₂PO₂⁻, and phosphonous acid, RHPO₂⁻
asym PO₂⁻ str
sym PO₂ str
RPO₃²⁻
asym PO₃²⁻ str
sym PO₃²⁻ str
R(RO)PO₂⁻
RO PO₃²⁻
P—F
Pᵛ—F str
Pᴵᴵᴵ—F str
3.00 4.00 5.00 6.00 7.00 8.00 9.00 10.00 20.00 25.00 50.00 μm

Chart 3 (*continued*)

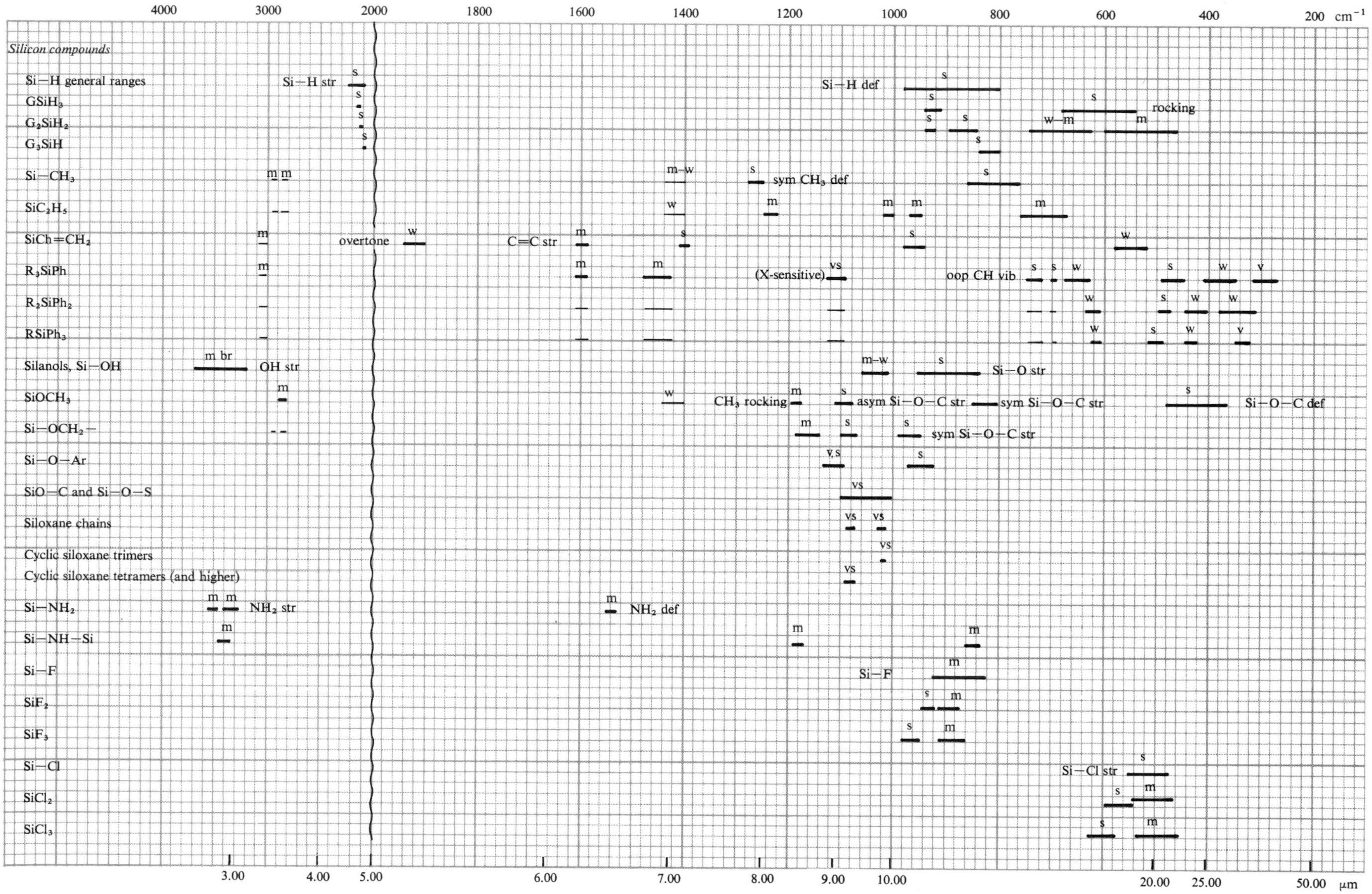
4000
3000
2000
1800
1600
1400
1200
1000
800
600
400
200 cm⁻¹
Silicon compounds
Si—H general ranges
GSiH₃
G₂SiH₂
G₃SiH
Si—CH₃
SiC₂H₅
SiCh=CH₂
R₃SiPh
R₂SiPh₂
RSiPh₃
Silanols, Si—OH
SiOCH₃
Si—OCH₂—
Si—O—Ar
SiO—C and Si—O—S
Siloxane chains
Cyclic siloxane trimers
Cyclic siloxane tetramers (and higher)
Si—NH₂
Si—NH—Si
Si—F
SiF₂
SiF₃
Si—Cl
SiCl₂
SiCl₃
Si—H str
Si—H def
rocking
sym CH₃ def
overtone
C=C str
(X-sensitive)
oop CH vib
m br
OH str
Si—O str
CH₃ rocking
asym Si—O—C str
sym Si—O—C str
Si—O—C def
NH₂ str
NH₂ def
Si—Cl str
3.00
4.00
5.00
6.00
7.00
8.00
9.00
10.00
20.00
25.00
50.00 μm

Chart 3 (*continued*)

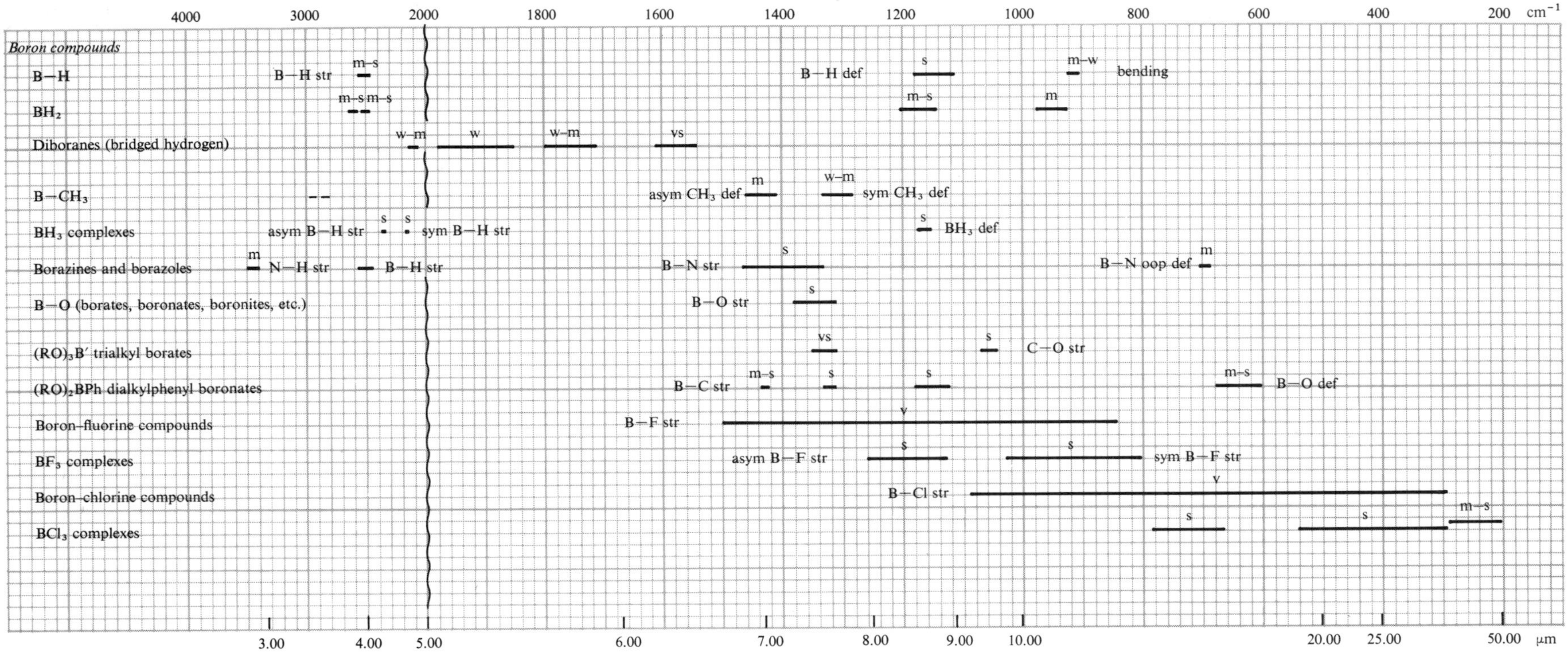

4000
3000
2000
1800
1600
1400
1200
1000
800
600
400
200
cm⁻¹
Boron compounds
B—H
B—H str
m–s
B—H def
s
m–w
bending
BH₂
m–s m–s
m–s
m
Diboranes (bridged hydrogen)
w–m
w
w–m
vs
B—CH₃
asym CH₃ def
m
w–m
sym CH₃ def
BH₃ complexes
asym B—H str
s
s
sym B—H str
s
BH₃ def
Borazines and borazoles
m
N—H str
B—H str
B—N str
s
B—N oop def
m
B—O (borates, boronates, boronites, etc.)
B—O str
s
(RO)₃B′ trialkyl borates
vs
s
C—O str
(RO)₂BPh dialkylphenyl boronates
B—C str
m–s
s
s
m–s
B—O def
Boron–fluorine compounds
B—F str
v
BF₃ complexes
asym B—F str
s
s
sym B—F str
Boron–chlorine compounds
B—Cl str
v
BCl₃ complexes
s
s
m–s
3.00
4.00
5.00
6.00
7.00
8.00
9.00
10.00
20.00
25.00
50.00
μm

Collections of Reference Spectra and Indexes

The most comprehensive collection of infrared spectra is that offered by Sadtler Research Laboratories.[1] It consists of many thousands of spectra covering a wide variety of compounds and new additions are made periodically. The spectra are all run under standard conditions on grating or prism spectrometers. Any given spectrum within the collection may be retrieved by the use of (a) an alphabetical index, (b) a molecular formula index, or (c) a chemical class index.

In addition, Sadtler also provide collections covering commercially available substances which deal with (a) agricultural chemicals, (b) dyes and pigments, (c) fats, waxes, and derivatives, (d) fibres, (e) intermediates, (f) lubricants, (g) monomers and polymers, (h) natural resins, (i) perfumes and flavours, (j) petroleum chemicals, (k) pharmaceuticals, (l) plasticizers, (m) polyols, (n) pyrolysates, (o) rubber chemicals, (p) solvents, (q) surface active agents, (r) textile chemicals. Many of the spectra in these collections are referred to by trade names.

An index which is useful in the identification of unknowns is the Sadtler Spec-Finder. It consists of a number of data sheets which have 13 vertical columns, each corresponding to 1 μm in the region 2–15 μm. The position of the most prominent band, if any, in each of these 13 regions is indicated in each column. Each row therefore represents a spectrum of a material given in the Sadtler collection, the serial number of the particular Sadtler spectrum card also being given. On denoting the spectrum of an unknown in a similar manner, this can then be compared with the rows of data until a good match is obtained. In order to avoid looking through the entire catalogue, the data are classified according to the position of the strongest band in the whole spectrum. One or more spectral cards may be found which provide a good approximation to the spectrum of the unknown. A computer search system is also now available, as is an index on IBM cards or magnetic tape. Sadtler also publish ultraviolet-visible, NMR, and DTA data for materials.

The Documentation of Molecular Spectroscopy (DMS)[2] system also provides a very large collection of spectra and relevant data which are collected from the literature and other sources to cover the range 3700–250 cm $^{-1}$. DMS also publish literature services for infrared, Raman, and microwave spectroscopy.

The American Petroleum Institute (API)[3] publish a large collection of spectra, mainly of hydrocarbons and compounds relevant to the petroleum industry.

The Infrared Data Committee of Japan (IRDC)[4] publish a large card system which is similar to that of DMS.

Mecke and Langenbucher[5] have published a small collection: *Infrared Spectra of Selected Chemical Compounds.*

The National Bureau of Standards[6] with the National Research Council have published infrared spectral and bibliographic punch cards which give spectra and a survey of literature.

Other sources of spectra, generally of a more specialized nature, are available,[7–21] as is useful information regarding band positions and assignments.[22]

Indexes of published spectra[23–27]

The American Society for Testing Materials (ASTM)[23] issue an IBM punched-card system, suitable for electronic sorting machines. The punched holes on the cards denote absorption bands at particular positions and other relevant physical data. This system provides an index of all infrared spectra available commercially or in the literature. ASTM provide computer retrieval programs on disc or tape. The index is also provided in book form.

A useful index of spectra published up to 1957 is *An Index of Published Infrared Spectra.*[24] An index covering the period 1945–1962 has been published by Hershenson.[25] An index covering inorganic and organometallic compounds is also available.[26]

Bibliography and References

There are a number of excellent books available from which an introduction to various aspects of infrared spectroscopy may be obtained. A few of these are given in the Appendix. Of course, there is some degree of overlap of subject matter but the titles of the books generally indicate their contents. References included in the Appendix are, in general, of the review type. At the end of each chapter are given references of a more specialized nature pertinent to that chapter.

It is intended that this book, rather than provide a complete bibliography or source of references, should act as a thorough guide to the newcomer to the field.

References

1. Sadtler Research Laboratories, 3314–20 Spring Garden Street, Philadelphia, Pa 19104, USA, and K. G. Heyden, 1 Vivien Avenue, Hendon, London NW4
2. DMS System, Butterworth Scientific Publications, 88 Kingsway, London WC2, and Verlag Chemie GmbH, Weiheim/Bergstrasse, West Germany
3. American Petroleum Institute Research Project 44, Chemistry Department, Agricultural and Mechanical College Texas, College Station, Tx 77843, USA
4. Sanyo Shuppan Boeki Co. Inc., Hoyn Building, 8,2-Chome, Takara-cho, Chuo-ku, Tokyo, Japan
5. R. Mecke and F. Langenbucher, *Infrared Spectra of Selected Chemical Compounds.* Heyden, London, 1966
6. Infrared Spectral and Bibliographic Data, National Bureau of Standards, Washington DC 20234, USA (or from Heyden, London)
7. Coblentz Society Spectra (available from Sadtler Research Laboratories or from Heyden)
8. Manufacturing Chemist Association Research Project, Chemistry Department, Agricultural and Mechanical College Texas, USA
9. C. J. Pouchert, *Aldrich Library of Infrared Spectra.* Aldrich Chemical Co. Inc., Milwaukee, Wisconsin, USA, 1975

10. A. F. Ardyukova, O. P. Shkurko, and V. F. Sedova, *Atlas of Spectra of Aromatic and Heterocyclic Compounds*. Nauka Sib. Otd., Novosibirsk, 1974
11. *Infrared and Ultraviolet Spectra of Some Compounds of Pharmaceutical Interest*, revised ed. Association of Official Analytical Chemists, Washington DC, 1972
12. R. L. Davidovich, T. A. Kaidolova, T. F. Levchishina, and V. I. Sergineko, *Atlas of Infrared Absorption Spectra and X-Ray Measurement Data for Complex Group IV and V Metal Fluorides*. Nauka, Moscow, 1972
13. K. Dobriner, E. R. Katzenellenbogen, and R. N. Jones, *Infrared Absorption Spectra of Steroids, An Atlas*, Vol. I. Interscience, New York, 1953
14. G. Roberts, B. S. Gallagher, and R. N. Jones, *Infrared Adsorption Spectra of Steroids, An Atlas*, Vol. II. Interscience, New York, 1958
15. D. Welti, *Infrared Vapour Spectra*. Heyden, London, 1970 (and with Sadtler Research Laboratories)
16. R. A. Nyquist and R. O. Kagel, *Infrared Spectra of Inorganic Compounds (3800–45 cm^{-1})*. Academic Press, New York, 1971
17. D. O. Hummel and F. Scholl, *Infrared Analysis of Polymers, Resins and Additives, An Atlas*, Vol. I, part 2. Wiley, New York, 1971
18. K. Yamaguchi, *Spectral Data of Natural Products*. Elsevier, Amsterdam, 1970. (Contains spectra and data on infrared, ultraviolet, NMR, mass spectroscopy, etc.)
19. F. F. Bentley, D. L. Smithson, and A. L. Rozek, *Infrared Spectra and Characteristic Frequencies $\sim$700–300 cm^{-1}*. Interscience, New York, 1968
20. L. Lang, S. Holly, and P. Sohar (eds), *Absorption Spectra in the Infrared Region*. Butterworth, London, 1974. (Contains about 300 spectra of mainly organic compounds.)
21. H. A. Szymanski, *Interpreted Infrared Spectra*. Plenum, New York, 1971
22. H. A. Szymanski, *Infrared Band Hand Book*. Vols I and II. Plenum, New York, 1965
23. American Society for Testing Materials (ASTM), 1916 Race Street, Philadelphia, Pa 19103, USA
24. *An Index of Published Infrared Spectra*. Ministry of Aviation, HMSO, London, 1960
25. H. M. Hershenson, *Infrared Absorption Spectra Index, 1945–1962*. Academic Press, New York, 1959 and 1964
26. N. N. Greenwood, E. J. F. Ross, and B. P. Straughan, *Index of Vibrational Spectra of Inorganic and Organometallic Compounds*, Vol. I. Butterworth, London, 1972; N. N. Greenwood and E. J. F. Ross, Vol. II, 1975
27. C. Clark and M. Chianta, *Ann. NY Acad. Sci.*, 1957, **69**, 205. (Bibliography of infrared spectra of biochemicals.)
28. C. N. R. Rao *et al.*, *Bibliography of Infrared Spectroscopy, 1960–*, Parts 1, 2, and 3. GPO, Washington DC, 1976

CHAPTER 2

Alkane Group Residues: C—H Group

Alkane Functional Groups

Residual alkane groups are found in a very large number of compounds and hence are an extremely important class.[1] Four types of vibration are normally observed, namely the stretching and the deformation of the C—H and the C—C bonds. The C—H vibration frequencies of the methyl and methylene groups fall in narrow ranges for saturated hydrocarbons. However, atoms directly attached to $—CH_3$ or $—CH_2—$ may result in relatively large shifts in the absorption frequencies. In general, the effect of electronegative groups or atoms is to increase the C—H absorption frequency.

Alkane C—H stretching vibrations

For aliphatic hydrocarbons, with the exception of small ring compounds, the C—H stretching vibrations occur in the region 2975–2840 cm^{-1} (3.36–3.52 μm).[2–4] In strained ring systems, the frequency of the methylene C—H stretching vibration is increased, e.g. cyclopropanes absorb near 3050 cm^{-1} (3.28 μm). The CH_3 asymmetric stretching vibration occurs at 2975–2950 cm^{-1} (3.36–3.39 μm) and may easily be distinguished from the nearby CH_2 absorption at about 2930 cm^{-1} (3.41 μm). The symmetric CH_3 stretching absorption band occurs at 2885–2865 cm^{-1} (3.47–3.49 μm), and that of the methylene group at 2870–2840 cm^{-1} (3.49–3.52 μm).

The position of the CH_3 symmetric stretching vibration band may be altered due to an adjacent group, whereas the asymmetric stretching band is relatively insensitive, e.g. for the group $—O—CH_3$,[5,6] the CH_3 symmetric stretching band occurs at 2835–2815 cm^{-1} (3.53–3.55 μm) whereas the asymmetric stretching band occurs in the normal position (similarly for $>N—CH_3$).[6,7] Correlations involving C—H stretching vibrations have been studied.[2] Information has also been derived from the intensities of these bands.[8,9] In the presence of a double bond adjacent to a methyl or a methylene group, the symmetric stretching vibration band splits into two. In polar molecules, a series of bands is observed between 2980 and 2700 cm^{-1} (3.36–3.70 μm) due to interactions between the fundamental vibrations of the methyl group and the overtones of their deformation vibrations.

Alkane C—H deformation vibrations

The methyl groups of hydrocarbons give rise to two vibration bands, the asymmetric deformation band occurring at 1465–1440 cm^{-1} (6.83–6.94 μm) and the symmetric band at 1390–1370 cm^{-1} (7.19–7.30 μm). The former band is often overlapped by the $—CH_2—$ scissor vibration band occurring at 1480–1440 cm^{-1} (6.76–6.94 μm). The intensity of the methyl symmetric vibration band relative to the higher-frequency band (due to scissor $—CH_2—$ and/or asymmetric $—CH_3$ vibrations) may be used to indicate the relative number of methyl groups in the sample.

The presence of adjacent electronegative atoms or groups can alter the position of the methyl symmetric band significantly, its range being 1450–1260 cm^{-1} (6.90–7.94 μm), whereas the asymmetric band is far less sensitive, its range being 1470–1410 cm^{-1} (6.80–7.09 μm).

t-Butyl groups have a strong band near 1365 cm^{-1} (7.32 μm) and a slightly weaker band near 1390 cm^{-1} (7.19 μm). The band normally found near 1380 cm^{-1} (7.25 μm) is split into two by resonance which occurs when two or three methyl groups are attached to a single carbon atom. The presence of a tertiary butyl group may be confirmed by its skeletal vibration bands which occur near 1255 cm^{-1} (7.97 μm) and 1210 cm^{-1} (8.27 μm), whereas the corresponding bands for the isopropyl group are found near 1170 cm^{-1} (8.55 μm) and 1145 cm^{-1} (8.73 μm).

Methyl rocking vibrations[10] are generally weak and not very useful for assignment purposes even though they are mass sensitive. For *n*-alkanes, a band due to the CH_2 wagging vibration occurs near 1305 cm^{-1} (7.66 μm), the intensity of this band being less than the band at ~1460 cm^{-1} (6.85 μm) while being dependent on the number of CH_2 groups present.

As mentioned previously, in the spectra of hydrocarbons, the methylene deformation band is found in the region 1480–1440 cm^{-1} (6.76–6.94 μm), but in the presence of adjacent unsaturated groups this band is found near 1440 cm^{-1} (6.94 μm). With an adjacent chlorine, bromine, iodine, sulphur, or phosphorus atom, or a nitrile, nitro-, or carbonyl group, this band occurs at 1450–1405 cm^{-1} (6.90–7.12 μm).

Alkane C—C Vibrations: Skeletal Vibrations

Of the skeletal vibrations, the C—C stretching absorptions occur in the region 1260–700 cm^{-1} (7.94–14.29 μm) and are normally weak and of little use in assignments. Dimethyl quaternary carbon compounds have a characteristic absorption near 1180 cm^{-1} (8.48 μm). The C—C deformation bands occur below 600 cm^{-1} (16.67 μm)[11,17] and these also are weak. Straight-chain alkanes have two bands, one at 540–485 cm^{-1} (18.52–20.62 μm) and the other near 455 cm^{-1} (21.98 μm). The former band is usually slightly more intense than the second band and tends to the higher-frequency end of the range as the length of the chain increases. An exception is *n*-pentane which has only one band, near 470 cm^{-1} (21.28 μm). Branched alkanes not containing methyl or ethyl groups have at least one band in the region 570–445 cm^{-1} (17.54–22.47 μm). Alkanes with three or more branches absorb near 515 cm^{-1} (19.42 μm). Straight-chain paraffins have two characteristic bands at 1150–1130 cm^{-1} (8.70–8.85 μm) and 1090–1055 cm^{-1} (9.17–9.48 μm), both due to C—C stretching and CH_3 rocking vibrations.

Cyclopropane derivatives[12, 13] have a band of variable intensity at 540–500 cm^{-1} (18.52–20.00 μm). An exception is that of vinylcyclopropane which has a strong absorption at 455 cm^{-1} (21.98 μm), and other unsaturated cyclopropanes also have a medium-intensity absorption in this region. Saturated aliphatic cyclopropyl compounds have a medium-to-weak band at about 1045 cm^{-1} (9.57 μm), a medium-intensity band at about 1020 cm^{-1} (9.80 μm), and a strong band at 470–460 cm^{-1} (21.28–21.74 μm). Cyclopentanes absorb strongly at 580–490 cm^{-1} (17.24–20.41 μm), alkyl monosubstituted cyclopentanes absorbing in the higher-frequency half of this range, 580–530 cm^{-1} (17.24–18.87 μm). Cyclohexane derivatives have bands of variable intensity in the region 570–435 cm^{-1} (17.54–22.99 μm).

Methyl-substituted benzenes have an absorption band of medium intensity in the range 390–260 cm^{-1} (25.64–38.46 μm) which is due to the in-plane bending of the aromatic C—CH_3 bond. Ethyl-substituted benzenes have a medium-to-strong absorption at 565–540 cm^{-1} (17.70–18.52 μm) and isopropyl benzenes have a medium-intensity absorption band at 545–520 cm^{-1} (18.35–19.23 μm). Both these vibrations are due to the in-plane bending of the =C—C—C group.

For propyl and butyl benzenes, two bands of medium intensity close together, usually not completely resolved, are observed at 585–565 cm^{-1} (17.09–17.70 μm).

Table 2.1. Alkane C—H stretching vibrations for alkane functional groups as part of a residual saturated hydrocarbon portion of the molecule

Functional Groups	Region cm^{-1}	Region μm	Intensity	Comments
$-CH_3$	2975–2950	3.36–3.39	m–s	asym } Frequency rasied by electronegative substituents
	2885–2865	3.47–3.49	m	sym
$-CH_2-$ (acyclic)	2940–2915	3.40–3.45	m–s	asym } Frequency raised by electronegative substituents
	2870–2840	3.49–3.52	m	sym
⋧CH (acyclic)	2890–2880	3.46–3.47	w	
Ar—CH_3	2985–2965	3.35–3.37	m–s	asym str, see refs 14, 15
	2955–2935	3.38–3.41	m–s	asym str, lower part of range for *ortho*-substituted compounds
	2930–2920	3.41–4.23	m–s	sym str
	2870–2860	3.48–3.50	m	def overtones
	2830–2740	3.53–3.65	w–m	def overtones
Cyclopropanes, $-CH_2-$	3100–3070	3.23–3.26	m	asym str, see ref. 18
	3035–2995	3.30–3.34	m	sym str
Cyclobutanes, $-CH_2-$	3000–2975	3.33–3.36	m	asym str
	2925–2875	3.42–3.48	m	sym str
Cyclopentanes, $-CH_2-$	2960–2950	3.38–3.39	m	asym str
	2870–2850	3.48–3.51	m	sym str
Cyclohexanes, $-CH_2-$	(As for acyclic $-CH_2-$ groups, see ref. 19)			

Table 2.2. Alkane C—H deformation vibrations for alkane functional groups as part of a residual saturated hydrocarbon portion of molecule

Functional Groups	Region cm^{-1}	Region μm	Intensity	Comments
$-CH_3$	1465–1440	6.83–6.94	m	asym } Frequency raised by electronegative substituents
	1390–1370	7.19–7.30	m–s	sym (characteristic of C—CH_3)
$>C(CH_3)_2$	1385–1365	7.22–7.25	m–s	Two bands of almost equal intensity
$-C(CH_3)_3$	1395–1380	7.18–7.25	m	CH_3 sym bending
	1375–1365	7.27–7.33	m–s	CH_3 sym bending
$>CH_2$	1480–1440	6.76–6.94	m	Scissor vib
⋧CH	~1340	~7.46	w	

Table 2.3. Alkane C—C skeletal vibrations for alkane functional groups as part of a residual saturated hydrocarbon portion of the molecule

Functional Groups	Region cm^{-1}	μm	Intensity	Comments
$>C(CH_3)_2$	1175–1165	8.51–8.58	m	C—C str. If no hydrogen on central carbon then one band at ~1190 cm^{-1}
	1150–1130	8.90–8.85	m	
	840–790	11.90–12.66	w	
	495–490	20.20–20.41	w	
$—C(CH_3)_3$	1255–1245	7.98–8.03	m	
	1225–1165	8.17–8.58	m	
$—CH(C_2H_5)_2$	510–505	19.61–19.80	w	
Straight-chain alkanes	540–485	18.52–20.62	w	Not *n*-pentane
	~455	~21.98	w	
Branched alkanes	570–445	17.54–22.47	w	At least one band
Monobranched alkanes	560–540	17.86–18.52	w	
	470–445	21.28–22.47	w	
Dibranched alkanes not possessing CH_3 or C_2H_5	555–535	18.02–18.69	w	
3,3-Dibranched alkanes	~530	~18.87	w	
2,2-Dibranched alkanes	~490	~20.41	w	
Alkanes with three or more branches	~515	~19.42	w	
$—(CH_2)_n—$ ($n > 3$)	725–720	13.79–13.89	w–m	Rocking vib; splits into two components in the crystalline phase
$—(CH_2)_3—$	735–725	13.61–13.79	w–m	
$—(CH_2)_2—$	745–735	13.42–13.61	w–m	
$—CH_2—$	785–770	12.74–12.99	w–m	
Methyl benzenes	390–260	25.64–38.46	m	In-plane bending of aromatic C-CH_3 bond
Ethyl benzenes	565–540	17.70–18.52	m–s	In-plane bending of =C—C—C group
Isopropyl benzenes	545–520	18.35–19.23	m	In-plane bending of —C—C—C group
Propyl and butyl benzenes	585–565	17.09–17.70	m	Two bands
Cyclopropanes	1030–1000	9.71–10.00	w	Often ~1020 cm^{-1}
	870–850	11.49–11.76	v	Often absent
	540–500	18.52–20.00	v	
Saturated aliphatic cyclopropanes	470–460	21.28–21.74	s	
Cyclobutanes	1000–960	10.00–10.42	w	(CH_2 scissoring vib, ~1445 cm^{-1})
	930–890	10.75–11.24	w	See ref. 16
	580–490	17.24–20.41	s	
Alkyl cyclobutanes	580–530	17.24–18.87	s	
Cyclopentanes	1000–960	10.00–10.42	w	
	930–890	10.75–11.24	w	
	595–490	16.81–20.41	s	
Saturated aliphatic cyclopentanes	585–530	17.09–18.87	s	
Cyclohexanes	1055–1000	9.48–10.00	w	See ref. 19
	1015–950	9.86–10.53	w	
	570–435	17.54–22.99	v	

Table 2.4. Symmetric C—H stretching vibrations for alkane residues attached to atoms other than saturated carbon atoms (excluding olefines)

Functional Groups	Region cm^{-1}	μm	Intensity	Comments
Methyl groups				
$—O—CH_3$ (ethers)	2995–2955	3.34–3.38	s	asym CH_3 str
	2900–2865	3.45–3.49	s	sym CH_3 str
	2835–2815	3.53–3.55	s	sym CH_3 str (overtone, see refs 5, 6
$RSCH_3$	2995–2955	3.34–3.38	m	asym CH_3 str
	2900–2865	3.45–3.49	m	sym CH_3 str
$N—CH_3$ (amines and imines)	2820–2760	3.55–3.62	s	sym CH_3 str, general range, see refs 6, 7
$N—CH_3$ (aliphatic amines)	2805–2780	3.56–3.60	s	sym CH_3 str, $>NCH_2—$ band may also occur in this region
$N—CH_3$ (aromatic amines)	2820–2810	3.55–3.56	s	sym CH_3 str
$—N(CH_3)_2$ (aliphatic)	2825–2810	3.54–3.56	s	sym CH_3 str, see ref. 7
	2775–2765	3.60–3.62	s	
$—N(CH_3)_2$ (aromatic)	2830–2800	3.53–3.57	s	sym CH_3 str
$CH_3—CO—$ (ketones)	3000–2900	3.33–3.45	w	sym CH_3 str
Methylene and other groups				
—CHO (aldehyde)	2900–2800	3.45–3.57	w	C—H str
	2775–2695	3.63–3.71	w	Probably overtone ?
$X—CH_2—$ (X = halogen)	~3050	~3.28	w	C—H str
$—O—CH_2—O—$	2820–2710	3.55–3.69	m	C—H str
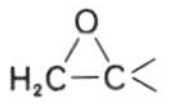 (epoxides)	3050–3030	3.28–3.03	w	asym C—H str, see ref. 18
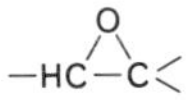 (epoxides)	3000–2990	3.33–3.34	w	C—H str
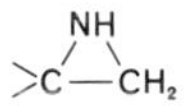	~3050	~3.28	m–s	asym C—H str

Table 2.5. C—H deformation vibrations for alkane residues attached to atoms other than saturated carbon atoms (excluding olefines)

Functional Groups	Region cm^{-1}	Region μm	Intensity	Comments
—O—CH_3	1470–1430	6.80–6.99	m	asym def
	1445–1430	6.92–6.99	s	sym def
	~1030	~9.71	w–m	C—C vib almost always observed
—OC$(CH_3)_3$	1200–1155	8.33–8.66	s	C—O str
	1040–1000	9.62–10.00	w–m	C—C vib almost always observed
	920–820	10.87–12.20	w–m	Skeletal vib
	770–720	13.00–13.89	w–m	*t*-Bu sym skeletal vib
—O—CH_2— (esters)	1475–1460	6.78–6.85	m–s	CH_2 sym def
	~1030	~9.71	w–m	Not always observed
Esters (acylcic)	1470–1435	6.80–6.97	m–s	CH_2 sym def
Esters (cyclic, small rings)	1500–1470	6.67–6.80	m	sym def, several bands
—O—CO—CH_3 (acetates)	1450–1400	6.90–7.14	s	asym def
	1385–1340	7.22–7.46	s	sym def
—CO—CH_3 (ketones)	1450–1400	6.90–7.14	s	asym def
	1360–1355	7.35–7.38	s	sym def
—CO—CH_2— (small-ring ketones)	1475–1425	6.78–7.02	s	asym def, several bands
—CO—CH_2— (acyclic ketones)	1435–1405	6.97–7.12	s	asym def
—CH_2—COOH	~1200	~8.33	m	CH_2 def
Acetyl acetonates	1415–1380	7.07–7.25	s	asym def
	1360–1355	7.35–7.38	s	sym def
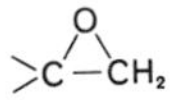 (epoxides)	~1500	~6.67	w–m	asym bending
—CHO (aldehydes)	1420–1370	7.04–7.30	m	CH def
>CHOH (secondary alcohols) (free)	1410–1350	7.09–7.41	w	CH def
	1300–1200	7.69–8.33	w	CH def
Secondary alcohols (bonded)	1440–1400	6.94–7.14	w	CH def
	1350–1285	7.41–7.78	w	CH def
—$(CH_2)_n$—O— ($n > 4$)	745–735	13.42–13.61	m–s	CH_2 def
>N—CH_3	1440–1390	6.94–7.19	m	sym def, usually moves to higher wavenumbers for hydrohalides
>N—CH_3 (amine hydrochlorides)	1475–1395	6.78–7.17	m	sym def
>N—CH_3 (amino acid hydrochlorides)	1490–1480	6.71–6.76	m	sym def
N—CH_3 (amides)	1420–1405	7.04–7.12	s	sym def (asym def 1500–1450 cm^{-1})
>N—CH_2— (amides)	~1440	~6.94	m	
>N—CH (amines) and groups with —O—CH such as acetals, orthoformates, and peroxides	1350–1315	7.41–7.61	w	CH def
N—CH_2— (ethylenediamine complexes)	1480–1450	6.76–6.90	s	sym def, two bands
	1400–1350	7.14–7.41	m–s	
—CH_2—NO_2	1425–1415	7.02–7.07	s	sym def
—CH_2—CN	1430–1420	6.99–7.04	s	sym def
—CH_2—C=C< ; —CH_2—C≡C—	1445–1430	6.92–6.99	m	Conjugation to CH_2 decreases (wavenumber)
X—CH_2— (X = halogen, X ≠ F)	1435–1385	6.97–7.22	m	(Strong band at 1300–1240 cm^{-1} due to CH_2 wagging)

Table 2.5 (*cont.*)

Functional Groups	Region cm^{-1}	Region μm	Intensity	Comments
F—CH_3	~1475	~6.78	m	sym def
Cl—CH_3	~1355	~7.38	m	sym def
Br—CH_3	~1305	~7.61	m	sym def
I—CH_3	~1250	~7.98	m	sym def
P—CH_3	1320–1280	7.58–7.81	s	sym bending
S—CH_3	1325–1300	7.55–7.69	m–w	sym def
S—CH_2—	1440–1415	6.04–7.07	s	sym def (strong band at 1270–1220 cm^{-1} due to CH_2 wagging)
Se—CH_3	~1280	~7.81	m	sym def
B—CH_3	1460–1405	6.85–7.12	m	asym def
	1320–1280	7.58–7.81	m	sym def
Si—CH_3	1265–1250	7.91–8.00	m–s	sym bending } CH_3—metal groups have strong band at 900–700 cm^{-1} due to CH_2 rocking
Sn—CH_3	1200–1180	8.33–8.48	m	sym bending
Pb—CH_3	1170–1155	8.55–8.66	m	sym bending
As—CH_3	1265–1240	7.91–8.07	m	sym def
Ge—CH_3	1240–1230	8.07–8.13	m	sym def
Sb—CH_3	1215–1195	8.23–8.37	m	sym def
Bi—CH_3	1165–1145	8.58–8.73	m	sym def
—CH_2—SO_2—	~1250	~8.00	m	sym def
—CH_2—metal (metal = Cd, Hg, Zn, Sn)	1430–1415	6.99–7.07	m	CH_2 def

References

1. N. Sheppard and D. M. Simpson, *Quart. Rev.*, 1953, **7**, 19
2. H. J. Bernstein, *Spectrochim. Acta*, 1962, **18**, 161
3. D. C. McKean *et al.*, *Spectrochim. Acta*, 1973, **29A**, 1037
4. M. T. Forel *et al.*, *J. Opt. Soc. Amer.*, 1960, **50**, 1228
5. H. B. Henbest *et al.*, *J. Chem. Soc.*, 1957, 1462
6. F. Dalton *et al.*, *J. Chem. Soc.*, 1960, 2927
7. R. D. Hill and G. D. Meakins, *J. Chem. Soc.*, 1958, 761
8. A. S. Wexler, *App. Spectrosc. Rev.*, 1968, **1**, 29
9. S. Higuchi *et al.*, *Spectrochim. Acta*, 1972, **28A**, 1335
10. J. Van Schooten *et al.*, *Polymer*, 1961, **2**, 357
11. F. F. Bentley and E. F. Wolfarth, *Spectrochim. Acta*, 1959, **15**, 165
12. K. H. Ree and F. A. Miller, *Spectrochim. Acta*, 1971, **27A**, 1
13. N. C. Craig *et al.*, *Spectrochim. Acta*, 1972, **28A**, 1175
14. G. M. Badger and A. G. Moritz, *Spectrochim. Acta*, 1959, **15**, 672
15. A. B. Dempster *et al.*, *Spectrochim. Acta*, 1972, **28A**, 373
16. H. E. Ulery and J. R. McClenon, *Tetrahedron*, 1963, **19**, 749
17. A. S. Gilbert *et al.*, *Spectrochim. Acta*, 1976, **32A**, 931
18. C. J. Wurrey and A. B. Nease, *Vib. Spectra Struct.*, 1978, **7**, 1
19. T. C. Rounds and H. L. Strauss, *Vib. Spectra Struct.*, 1978, **7**, 237

CHAPTER 3

Alkenes, Oximes, Imines, Amidines, Azo Compounds: C=C, C=N, N=N Groups

Alkene Functional Group, >C=C<

The most useful bands are those resulting from the C=C stretching and the C—H out-of-plane deformation vibrations, the latter bands being the strongest observed in the spectra of alkenes.[1]

Alkene >C=C< stretching vibrations

Non-conjugated alkenes have a weak C=C stretching absorption band in the range 1680–1620 cm^{-1} (5.95–6.17 μm). This band is absent for symmetrical molecules. Therefore, it is not surprising to find that olefines which have terminal double bonds have the most intense absorptions. Vinyl, vinylidene, and *cis*-disubstituted olefines tend to absorb at the lower end of the range given, below 1665 cm^{-1} (above 6.01 μm), whereas *trans*-disubstituted, tri-, and tetra-substituted olefines absorb at the higher wavenumbers.

In conjugated systems, the C=C stretching vibration frequency is lower than that of an isolated C=C group.[2–4] Often there is the same number of bands as there is of double bonds, e.g. with two double bonds, two bands of different intensities are observed due to the C=C—C=C symmetric and asymmetric stretching. For conjugated systems with a centre of symmetry, two absorption bands are normally observed, one at about 1650 cm^{-1} (6.06 μm) and another less intense band near 1600 cm^{-1} (6.25 μm). The presence of this latter band may be used to confirm the presence of conjugation. Alkenes conjugated to aromatic rings exhibit a strong absorption near 1625 cm^{-1} (6.15 μm).[3] In this case, the aromatic C=C absorption is at about 1590 cm^{-1} (6.28 μm). In poly-conjugated systems, a series of weak bands is observed at 2000–1660 cm^{-1} (5.00–6.02 μm), similar to that of aromatic compounds.

The effect of electronegative substituents such as chlorine, etc., attached directly to alkene groups, is generally to lower the C=C stretching vibration frequency. Fluorine, on the other hand, increases this frequency. In alkene strained-ring compounds, the frequency of the C=C stretching vibration is decreased[5–10]—the smaller the ring, the lower the frequency. Information on the integrated intensity of the band due to the C=C stretching vibration is also available.[15, 22]

Alkene C—H stretching vibrations

In general, bands due to both alkene and aromatic C—H stretching occur above 3000 cm^{-1} (below 3.33 μm). Although alkane C—H stretching vibrations generally occur below 3000 cm^{-1}, it must be noted that small-ring paraffins and alkanes substituted with electronegative atoms or groups also absorb above 3000 cm^{-1}. The $=CH_2$ stretching vibration of vinyl and vinylidine groups occurs at 3095–3075 cm^{-1} (3.24–3.25 μm) and the =CH stretching vibration at 3050–3000 cm^{-1} (3.28–3.33 μm), whilst their symmetric stretching vibration occurs near 2975 cm^{-1} (3.36 μm), although this is unfortunately often over-lapped by alkane absorptions.

Alkene C—H deformation vibrations

The deformation vibrations of C—H may be either perpendicular to or in the same plane as that containing the carbon–carbon double bonds and the other bonds:

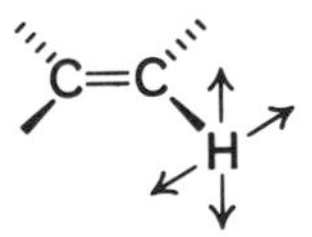

C=C
H

The arrows indicate the vibrational motions of a single C—H

The absorption bands due to the out-of-plane vibrations occur mainly at 1000–800 cm^{-1} (10.00–12.50 μm) and have strong-to-medium intensities. These bands are important in the characterization of alkenes,[11–13] e.g. for hydrocarbons:

(a) Vinyl groups, $—CH=CH_2$, absorb strongly[14] in the regions 995–980 cm^{-1} (10.05–10.20 μm) and 915–905 cm^{-1} (10.93–11.05 μm). For the nitrile compound, the first band occurs at 960 cm^{-1} (10.42 μm) and for the corresponding isothiocyanate and thiocyanate this band occurs near 940 cm^{-1} (10.64 μm).

(b) Vinylidene groups, $>C=CH_2$, absorb strongly at 895–885 cm^{-1} (11.17–11.30 μm).

(c) *Trans*-disubstituted alkenes, —CH=CH—, absorb strongly at 980–955 cm^{-1} (10.20–10.47 μm).

(d) *Cis*-disubstituted alkenes, —CH=CH—, absorb strongly at 730–665 cm^{-1} (13.70–15.04 μm).

(e) Trisubstituted alkenes, >C=CH—, absorb at 850–790 cm^{-1} (11.76–12.66 μm).

The $=CH_2$ out-of-plane deformation vibration is not mass sensitive for non-hydrocarbon olefines but it is sensitive to electronic changes. Groups that withdraw electrons mesomerically from the $=CH_2$ group, e.g.

$$-CO-\overset{\overset{\displaystyle O}{\|}}{C}-CH{=}CH_2 \quad \text{and} \quad CNCH{=}CH_2,$$

tend to raise the frequency and those which donate electrons mesomerically lower the frequency relative to that of the hydrocarbon olefine.

For vinylidene compounds[14, 15] with halogens directly bonded to the $>C{=}CH_2$ group, the out-of-plane deformation vibration frequency is decreased. This shift in frequency becomes greater with increase in the electronegativity of the halogen atom and appears to have an approximately additive effect. Oxygen atoms directly bonded to the vinylidene group also tend to decrease the $=CH_2$ out-of-plane vibration frequency.

Alkene Skeletal Vibrations[15–19]

For unbranched 1-alkenes, strong bands are observed near 635 cm^{-1} (15.75 μm) and 550 cm^{-1} (18.18 μm) and these have been assigned to ethylenic twisting vibrations. All *cis*-alkenes have two, well-separated, strong bands in the region 670–455 cm^{-1} (14.93–21.98 μm) and in general have weak bands or no bands in the region 455–370 cm^{-1} (21.98–27.03 μm), whereas all *trans*-alkenes have medium-to-strong absorption bands, usually only one, in this latter region. For example, unbranched *cis*-2-alkenes absorb in the regions 590–570 cm^{-1} (16.95–17.54 μm) and 490–465 cm^{-1} (20.41–21.51 μm) whereas unbranched *trans*-2-alkenes have absorptions at 420–385 cm^{-1} (23.81–25.97 μm) and 325–285 cm^{-1} (30.77–35.09 μm). For C=C conjugated to an aromatic group, an absorption band near 550 cm^{-1} (18.18 μm) is observed.

Table 3.1. Alkene C=C stretching vibrations

Functional Groups	Region cm^{-1}	Region μm	Intensity	Comments
Isolated C=C	1680–1620	5.95–6.17	w–m	May be absent for sym compounds
C=C conjugated with aryl	1640–1610	6.10–6.21	m	
C=C conjugated with C=C or C=O	1660–1580	6.02–6.33	s	See ref. 10
Conjugated, $CH_2{=}CH-C{\equiv}C-$	1620–1610	6.17–6.21	s	Conjugated with C≡C, see ref. 21
Vinyls				
Vinyl group, $-CH{=}CH_2$	1645–1640	6.08–6.10	w–m	Hydrocarbons
Halo- or cyano-vinyls	1620–1595	6.17–6.27	s	Fluoro- ~1650 cm^{-1}
Vinyl ether, $-O-CH{=}CH_2$	1640–1630	6.49–6.54	s	} Usually a doublet in region 1640–1610 cm^{-1}, see ref. 13
	1620–1610	6.17–6.21	s	
Vinyl ketone, $-CO-CH{=}CH_2$	1620–1615	6.17–6.19	s	(For dichlorovinyl ketones, see ref. 36)
Vinyl ester, $CH_2{=}CHOCOR$	1700–1645	5.88–6.08	s	
Acrylates, $CH_2{=}CHCOOR$	1640–1635	6.10–6.12	s	
	1625–1620	6.16–6.17	s	
Vinylenes				
cis $-CH{=}CH-$	1665–1635	6.01–6.12	w–m	Hydrocarbons
trans $-CH{=}CH-$	1675–1665	5.97–6.02	w–m	Hydrocarbons
Vinylidenes				
$>C{=}CH_2$	1660–1640	6.02–6.10	w–m	
Halo- and cyano-substituted $>C{=}CH_2$	1630–1620	6.14–6.17	v	Difluoro-substituted ~1730 cm^{-1}
$-CO-C{=}CH_2$, ketones	~1630	~6.14	m–s	
$-CO-O-C{=}CH_2$, esters	1675–1670	5.97–5.99	s	
α,β-unsaturated amines, $CH_2{=}CN<$	1700–1660	5.88–6.02	m	
Trisubstituted alkenes				
$>C{=}CH-$	1690–1665	5.92–6.01	w–m	Adjacent C=O decreases frequency and increases intensity
$CH_2{=}CF-$	1650–1645	6.06–6.08	m	
CF_2-CF-	1800–1780	5.56–5.62	m	See ref. 19
$>C{=}CF_2$	1755–1735	5.70–5.76	m	
$>C{=}C-N<$	1680–1630	5.95–6.13	m–s	More intense than normal C=C str band
Tetrasubstituted alkenes				
$>C{=}C<$	1690–1670	5.92–5.99	w	May be absent for symmetrical compounds
Internal double bonds				
Cyclopropene	~1655	~6.04	w–m	(Polyfluorinated compound ~1945 cm^{-1})

Table 3.1 (*cont.*)

Functional Groups	Region cm^{-1}	Region μm	Intensity	Comments
Cyclobutene	~1565	~6.39	w–m	See ref. 9 (polyfluorinated compound ~1800 cm^{-1})
Cyclopentene	~1610	~6.21	w–m	See ref. 18 (polyfluorinated compound ~1770 cm^{-1})
Cyclohexene	~1645	~6.08	w–m	(Polyfluorinated compound ~1745 cm^{-1})
Cycloheptene	~1650	~6.06	w–m	
1,2-Dialkylcyclopropenes	1900–1865	5.26–5.36	w–m	
1,2-Dialkylcyclobutenes	~1685	~5.93	w–m	
1,2-Dialkylcyclopentenes	1685–1670	5.93–5.65	w–m	
1,2-Dialkylcyclohexenes	1685–1675	5.93–5.63	w–m	
Exocyclic double bonds: 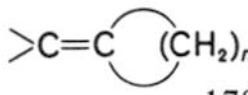				
$n = 2$ —	1780–1730	5.62–5.78	m	Shift to lower frequency as ring size increases
$n = 3$	~1680	~5.95	m	
$n = 4$	~1655	~6.04	m	
$n = 5$	~1650	~6.06	m	
Alkyl-substituted fulvenes, 	~1645	~6.08	m	Aromatic groups on the exo-double bond lower frequency to ~1600 cm^{-1}
benzofulvenes	~1630	~6.13	m	
A—CH=CH$_2$	1650–1580	6.23–6.33	v	A = heavy element, or group involving heavy element, directly attached to C=C, see ref. 26
C=C π-interaction with metal	1530–1500	6.54–6.67		e.g. Pt(C$_2$H$_4$), see refs 23–25

Table 3.2. Alkene C—H vibrations

Functional Groups	Region cm^{-1}	Region μm	Intensity	Comments
Vinyls				
Vinyl hydrocarbon compounds, —CH=CH$_2$	3095–3075	3.23–3.25	m	CH str of CH$_2$
	3030–2995	3.30–3.34	m	CH str of CH
	1985–1970	5.04–5.08	w	Overtone
	1850–1800	5.41–5.56	w	Overtone
	1420–1410	7.04–7.09	w	CH$_2$ in-plane def, scissoring
	1300–1290	7.69–7.75	w	CH in-plane def
	995–980	10.05–10.20	m	CH out-of-plane def
	915–905	10.93–11.05	s	CH$_2$ out-of-plane def, insensitive to conjugation, see ref. 21
Vinyl halogen compounds	945–925	10.58–10.83	m	CH out-of-plane def (nitrile substituted compound, 960 cm^{-1})
	905–865	11.05–11.56	s	CH$_2$ our-of-plane def (nitrile substituted compound, 960 cm^{-1})
Vinyl ethers, —O—CH=CH$_2$	965–960	10.36–10.42	s	CH out-of-plane def, see ref. 13
	945–940	10.58–10.64	m	CH out-of-plane def
	820–810	12.20–12.35	s	CH$_2$ out-of-plane def
Vinyl ketones, —COCH=CH$_2$	995–980	10.05–10.20	s	CH out-of-plane def
	965–955	10.36–10.47	m	CH$_2$ out-of-plane def
Vinyl esters, CH$_2$=CHOCOR	950–935	10.53–10.70	s	CH out-of-plane def
	870–850	11.49–11.76	s	CH$_2$ out-of-plane def
Acrylates, CH$_2$=CHCOOR	985–980	10.15–10.20	s	Out-of-plane def
	965–960	10.36–10.42	s	Out-of-plane def
Vinylidenes				
Hydrocarbons, >C=CH$_2$	3095–3075	2.53–2.67	m	CH asym str
	2985–2970	3.35–3.37	m	CH sym str
	1800–1780	5.56–5.62	w	Overtone
	1420–1410	7.04–7.09	w	CH$_2$ in-plane def, scissoring
	1320–1290	7.58–7.75	w	CH$_2$ in-plane def
	895–885	11.17–11.30	s	CH$_2$ out-of-plane def
Mono- and dihalogen-substituted >C=CH$_2$	880–865	11.36–11.56	s	CH$_2$ out-of-plane def (difluoro- at ~805 cm^{-1})
Cyano-substituted >C=CH$_2$	930–895	10.75–11.17	s	CH$_2$ out-of-plane def
—CO—C=CH$_2$ (ketones and esters)	~930	~11.07	s	CH$_2$ out-of-plane def
—CO—O—C=CH$_2$ (esters)	880–865	11.36–11.56	s	CH$_2$ out-of-plane def
Vinylenes				
cis —CH=CH— (hydrocarbons)	3040–3010	3.29–3.32	m	CH str
	1420–1400	7.04–7.14	w	CH in-plane def
	730–665	13.70–15.04	s	CH out-of-plane def, conjugation increases frequency range to 820 cm^{-1}

Table 3.2 (*cont.*)

Functional Groups	Region cm⁻¹	μm	Intensity	Comments
Halogen-substituted *cis* —CH=CH—	780–770	12.82–12.99	s	
trans —CH=CH— (hydrocarbons)	3040–3010	3.29–3.32	m	CH str
	1325–1290	7.55–7.75	w	CH in-plane def, sometimes absent
	980–955	10.20–10.47	s	CH out-of-plane def (usually ~965 cm^{-1}), conjugation increases frequency slightly and polar groups decrease it significantly (e.g. for *trans–trans* system, may be ~1000 cm^{-1})
Halogen-substituted *trans* —CH=CH—	~930	~10.75	s	CH out-of-plane def
trans —CH=CH— conjugated with C=C or C=O	~990	~10.10	s	CH out-of-plane def
trans —CH=CH—O— (ethers)	940–920	10.64–10.87	s	
Trisubstituted alkenes				
>C=CH— (hydrocarbons)	3040–3010	3.29–3.32	m	CH str
	1680–1600	5.95–6.25	w	Overtone
	1350–1340	7.41–7.46	m	CH in-plane def
	850–790	11.76–12.66	m	CH out-of-plane def, electronegative groups at lower end of frequency range
Cyclic alkenes (internal double bond)	3060–2995	3.27–3.34	m	=C—H str, ring-strain dependent—highest frequencies for smallest rings
	780–665	12.82–15.04	m	CH out-of-plane def
CH_2=CH—M (M = metal)	1410–1390	7.09–7.19	w	CH_2 def, see ref. 26
	1265–1245	7.91–8.03	w–m	CH rocking vib
	1010–985	9.90–10.15	m	CH out-of-plane vib
	960–940	10.42–10.64	s	CH_2 out-of-plane vib
Fulvenes	~765	~13.07	s	CH out-of-plane def (also, strong band at 1370–1340 cm^{-1} characteristic of unsaturated five-membered ring)
Benzofulvenes	~790	~12.66	s	CH out-of-plane def

Table 3.3. Alkene skeletal vibrations

Functional Groups	Region cm⁻¹	μm	Intensity	Comments
R—CH=CH_2	~635	~15.75	s	Ethylenic twisting vib, see ref. 20 (exception is propene ~578 cm^{-1})
	~550	~18.18	s	Ethylenic twisting vib
	485–445	20.62–22.47	m–s	
cis-Alkenes	670–455	14.93–21.98	s	Two bands
trans-Alkenes	455–370	21.98–27.03	m–s	Usually one band
Unbranched *cis*- R—CH=CH—CH_3	590–570	16.95–17.54	s	
	490–465	20.41–21.51	s	
Unbranched *trans*- R—CH=CH—CH_3	420–385	23.81–25.97	s	
	325–285	30.77–35.09	s	
cis-R_1CH=CHR_2	630–575	15.87–17.39	s	
	500–475	20.00–21.05	s	
trans-R_1CH=CHR_2	580–515	17.24–19.42	m–s	
	500–480	20.00–20.83	m–s	
	455–370	21.98–27.03	m–s	
R_1R_2C=CH_2	560–530	17.86–18.87	s	
	470–435	21.28–22.99	m	
R_1R_2C=CHR_3	570–515	17.54–19.42	s	Rocking motion, may have medium intensity
	525–470	19.05–21.28	s	Probably out-of-plane bending vib
	450–395	22.22–25.32	m–s	
Aryl olefines	~550	~18.18	m	

Oximes, >C=N—OH, Imines, >C=N—, Amidines, >N—C=N—, etc.

The N—H stretching vibration of the group C=N—H occurs in the region 3400–3300 cm^{-1} (2.94–3.03 μm). The frequency of the vibration is decreased in the presence of hydrogen bonding. For oximes and imines,[27,28,30,31] the C=N stretching vibration occurs in the region 1690–1620 cm^{-1} (5.92–6.17 μm), the band being weak in the case of aliphatic oximes and occurring at the higher-frequency end of the range given. For α,β-unsaturated and aromatic oximes[28,32] this band is of medium intensity and occurs in the lower-frequency half of the range. The closeness of this band to that due to the C=C stretching vibration often presents difficulties. Conjugated cyclic systems containing C=N have a band of variable intensity, due to the stretching vibration, in the region 1660–1480 cm^{-1} (6.02–6.76 μm), e.g. pyrrolines absorb at 1660–1560 cm^{-1} (6.02–6.41 μm).

The O—H stretching vibration[30,31] for oximes in a dilute solution using non-polar solvents occurs in the region 3650–3570 cm^{-1} (2.78–2.79 μm), a strong absorption being observed. If hydrogen bonding occurs, this band appears at 3300–3130 cm^{-1} (3.03–3.20 μm). In general, oximes have a strong band near 930 cm^{-1} (10.75 μm) due to the stretching vibration of the N—O bond.

Amidines[29] absorb strongly at 1685–1580 cm^{-1} (5.93–6.33 μm), due to the C=N stretching vibration, the band being found as low as 1515 cm^{-1} (6.60 μm) for amidines in solution. *N*-Unsubstituted amidine hydrochlorides have a strong band at 1710–1675 cm^{-1} (5.85–5.97 μm) and a weak band at 1530–1500 cm^{-1} (6.54–6.67 μm). *N*,*N*-Disubstituted amidine hydrochlorides have a medium-intensity band at 1590–1530 cm^{-1} (6.29–6.54 μm) due to the deformation vibration of the =NH_2 group. Substituted amidines absorb strongly at 1700–1600 cm^{-1} (5.88–6.25 μm).

Table 3.4. Oximes, imines, amidines, etc.: C=N stretching vibrations

Functional Groups	Region cm^{-1}	Region μm	Intensity	Comments
Aliphatic oximes and imines, >C=N—	1690–1650	5.92–6.06	w	
α,β-Unsaturated and aromatic oximes and imines	1650–1620	6.06–6.17	m	
Conjugated cyclic systems (oximes and imines)	1660–1480	6.02–6.76	v	
R_1R_2>C=N—H	1650–1640	6.06–6.10	s	sh
Ar(R)>C=N—H	1635–1620	6.12–6.17	m	
R—CH=N—R_2	1690–1630	5.92–6.13	v	(Schiff bases)
Ar—CH=N—Ar	1645–1605	6.08–6.23	v	Often two bands, see ref. 32
Guanidine, >N—C(=N)=N	1725–1625	5.80–6.15	s	
Azines, >C=N—N=C<	1670–1600	5.99–6.25		
Benzamidines, 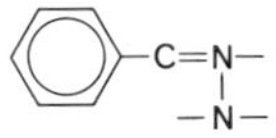	1630–1590	6.14–6.29		
Hydrazoketones, —CO—C=N—N	1600–1530	6.25–6.54	vs	
Amidines and guanidines >N—C=N—	1685–1580	5.93–6.33	v	
Imino ethers, —O—C=N—	1690–1645	5.92–6.08	v	Usually strong doublet due to rotational isomerism
—S—C=N—	1640–1605	6.10–6.23	v	
Imine oxides, C=N$^+$—O$^-$	1620–1550	6.17–6.45	s	

Table 3.5. Oximes, imines, amidines, etc.: other bands

Functional Groups	Region cm^{-1}	Region μm	Intensity	Comments
Oximes	3650–3500	2.74–2.86	v	Free O—H str, dilute solution
	3300–3130	3.03–3.20	v	Associated O—H str
	1475–1315	6.78–7.60	m	O—H def
	960–930	10.42–10.75	s	N—O str
Quinone oximes, 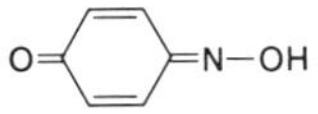	3540–2700	2.82–3.70	s	br, associated O—H str
	1670–1620	5.99–6.17	s	C=O str
	1560–1520	6.37–6.58	s	C=N str
Imines	3400–3300	2.94–3.03	v	Free N—H str
	3400–3100	2.94–3.23	m	Associated N—H str
—N—D	2600–2400	3.85–4.15	w–m	Free N—D str

Azo Compounds, —N=N—

Azo compounds[33–35] are difficult to identify by infrared spectroscopy because no significant bands are observed for them, the azo group being non-polar in nature. In addition, the weak absorption of the azo group occurs in the same region as the absorptions of aromatic compounds, the *cis-* form having slightly stronger bands normally than the *trans-* form.

Aromatic azo compounds in the *trans-* form absorb at 1440–1410 cm^{-1} (6.94–7.09 μm) and in the *cis-* form, near 1510 cm^{-1} (6.62 μm). Aromatic compounds which are in the *trans-* form absorb at the lower-frequency end of the range given if they are substituted with strong electron donors.

Table 3.6. Azo compounds

Functional Groups	Region cm^{-1}	Region μm	Intensity	Comments
Alkyl azo compounds	1575–1555	6.35–6.43	v	N=N str
α,β-Unsaturated azo compounds	~1500	~6.67	v	
Trans- aromatic azo compounds	1440–1410	6.94–7.09	w	N=N str
Cis- aromatic azo compounds	~1510	~6.62	w	N=N str
Aliphatic azoxy compounds, —N=N⁺—O⁻	1530–1495	6.54–6.69	m–s	Electron-withdrawing group on N—O nitrogen increases frequency
	1345–1285	7.43–7.78	m–s	
Aromatic azoxy compounds, —N=N⁺—O⁻	1480–1450	6.76–6.90	m–s	asym N=N—O str
	1340–1315	7.46–7.60	m–s	sym N=N—O str
Azothio compounds, —N=N⁺—S⁻	1465–1445	6.83–6.92	w	N=N str
	1070–1055	9.35–9.48	w	N—S str
Diazirines, 	~1620	~6.17	w	N=N str

References

1. N. Sheppard and D. M. Simpson, *Quart. Rev.*, 1952, **6,** 1
2. J. L. H. Allan *et al.*, *J. Chem. Soc.*, 1955, 1874
3. J. H. Wotiz *et al.*, *J. Amer. Chem. Soc.*, 1950, **72,** 5055
4. A. A. Petrov and G. I. Semenov, *J. Gen. Chem. Moscow*, 1959, **29,** 3689
5. K. B. Wiberg and B. J. Nist, *J. Amer. Chem. Soc.*, 1961, **83,** 1226
6. S. Pinchas *et al.*, *Spectrochim. Acta*, 1965, **25,** 783
7. J. Shabati *et al.*, *J. Inst. Petroleum*, 1962, **48,** 13
8. J. B. Miller, *J. Org. Chem.*, 1960, **25,** 1279
9. E. M. Suzuki and J. W. Nibler, *Spectrochim. Acta*, 1974, **30A,** 15
10. K. Noack, *Spectrochim. Acta*, 1962, **18,** 697 and 1625
11. W. J. Potts and R. A. Nyquist, *Spectrochim. Acta*, 1959, **15,** 679
12. E. M. Popov and G. I. Kajan, *Optics Spectrosc.*, 1962, **12,** 102
13. E. M. Popov *et al.*, *Optics Spectrosc.*, 1962, **12,** 17
14. J. Overend and J. R. Scherer, *J. Chem. Phys.*, 1960, **32,** 1720
15. G. P. Ford *et al.*, *J. Chem. Soc. Perkin Trans.*, 1974, 1569
16. F. F. Bentley and E. F. Wolfarth, *Spectrochim. Acta*, 1959, **15,** 165
17. P. M. Sverdlov, *Akad. Nauk SSSR Doklady*, 1957, **112,** 706
18. J. L. Lauer *et al.*, *J. Chem. Phys.*, 1959, **30,** 1489
19. D. E. Mann *et al.*, *J. Chem. Phys.*, 1957, **27,** 51
20. J. R. Scherer and W. J. Potts, *J. Chem. Phys.*, 1959, **30,** 1527
21. A. A. Petrov and G. I. Semenov, *J. Gen. Chem. Moscow*, 1957, **27,** 2974; 1958, **28,** 73
22. A. S. Wexler, *App. Spectrosc. Rev.*, 1968, **1,** 29
23. D. M. Adams and J. Chatt, *Chem. Ind.*, **1960,** 149
24. D. B. Powell and N. Sheppard, *J. Chem. Soc.*, **1960,** 2519
25. D. W. Wertz *et al.*, *Spectrochim. Acta*, 1973, **29A,** 1439
26. D. B. Powell *et al.*, *Spectrochim. Acta*, 1974, **30A,** 15
27. J. Fabian *et al.*, *Bull. Soc. Chim. France*, **1956,** 1499
28. D. Hadzi, *J. Chem. Soc.*, **1956,** 2725.
29. J. Fabian *et al.*, *Bull. Soc. Chim. France*, **1956,** 287
30. D. Hadzi and L. Premru, *Spectrochim. Acta*, 1967, **23A,** 35
31. M. St C. Flett, *Spectrochim. Acta*, 1957, **10,** 21
32. J. D. Margerum and J. A. Sousa, *App. Spectrosc.*, 1965, **19,** 91
33. L. E. Clougherty *et al.*, *J. Org. Chem.*, 1957, **22,** 462
34. R. Von Kübler, *Zeit. Electrochem.*, 1960, **64,** 650
35. K. J. Morgan, *J. Chem. Soc.*, **1961,** 2151
36. G. A. Gavrilova *et al.*, *Izv. Acad. Nauk SSSR Ser. Khim.*, 1978, **1,** 84

CHAPTER 4

Triple-bond Compounds: —C≡C—, —C≡N, —N≡C, —N≡N Groups

Alkyne Functional Group, —C≡C—

Two bands due to stretching vibrations may be observed, one due to the —C≡C— group and the other to the ≡C—H group.[1,2] Information is also available on band intensities.[3,4]

Alkyne C≡C stretching vibrations

This band is weak,[3,4] for monosubstituted alkynes[2,5] occurring in the region 2140–2100 cm^{-1} (4.67–4.76 μm) and for disubstituted alkynes[1,3] in the region 2260–2190 cm^{-1} (4.43–4.57 μm). For disubstituted alkynes, two bands are often observed due to Fermi resonance in the region 2310–2190 cm^{-1} (4.33–4.57 μm). The C≡C band is completely absent for some simple acetylenes because of symmetry considerations. Hence, as with alkenes, alkynes with a terminal triple bond have the most intense band due to C≡C stretching vibrations and as the triple bond is moved to an internal position its intensity becomes less. Conjugation[6–11] increases both the intensity[3,4] and the frequency of the C≡C stretching vibration. Information on cyclic acetylenes is also available.[14–17]

Alkyne C—H vibrations

For monosubstituted alkynes,[2,18] strong bands are observed at 3340–3300 cm^{-1} (2.99–3.03 μm) due to the C—H stretching vibration, and at 695–575 cm^{-1} (14.39–17.39 μm) due to the C—H deformation vibration. Care must be taken since the C—H stretching absorption occurs in the same region as those for N—H, which fortunately are usually much broader. The position of the band due to the ≡C—H stretching vibration is generally not sensitive to molecular structure changes, exceptions being acetylenes with halogen atoms directly bonded to the triple bond. Phase changes alter the position of the ≡C—H stretching vibration band significantly, in solid-phase spectra the band being up to 50 cm^{-1} lower (0.05 μm higher) than in dilute solution in inert solvents. An increase in wavenumber of similar magnitude is observed for vapour-phase spectra as compared with liquid-phase spectra.

A band of variable intensity and uncertain origin is sometimes observed in the region 1740–1630 cm^{-1} (5.75–6.14 μm). The hydrogen bonding of acetylenes[19] and their formation of complexes with nitrogen-containing compounds[20] have been studied.

Alkyne Skeletal Vibrations

All alkyl monosubstituted acetylenes have an absorption of variable intensity in the region 355–335 cm^{-1} (28.17–29.85 μm) due to the skeletal deformations of the C—C≡CH group. Monosubstituted acetylenes in which the substituent is not an alkyl group absorb in the region 510–260 cm^{-1} (19.61–38.46 μm) as a result of deformation vibrations. Methyl- and ethyl-substituted acetylenes absorb strongly at 520–495 cm^{-1} (19.23–20.20 μm) and 495–480 cm^{-1} (20.20–20.83 μm) respectively. Benzenes substituted with —C≡C—[12] absorb at about 550 cm^{-1} (18.18 μm).

All acetylenic compounds absorb at 960–900 cm^{-1} (10.42–11.11 μm) due to the ≡C—C stretching vibration.

Table 4.1. Alkyne C≡C stretching vibrations

Functional Groups	Region cm^{-1}	Region μm	Intensity	Comments
Monosubstituted alkynes, —C≡CH	2140–2100	4.67–4.76	w–m	See ref. 2
Disubstituted alkynes	2260–2190	4.43–4.57	v	Intensity decreases as symmetry of molecule increases (dialkyl acetylenes, 2210–2190 cm^{-1})
Conjugated alkynes	2270–2220	4.41–4.51	m	Conjugated with C=C
Conjugated alkynes	~2250	~4.43	s	Conjugated with COOH or COOR

Table 4.2. Alkynes: other bands

Functional Groups	Region cm⁻¹	μm	Intensity	Comments
Monosubstituted alkynes, —C≡CH	3340–3300	2.99–3.03	m	sh, CH str
	1375–1225	7.27–8.17	w–m	CH wagging overtone
	695–575	14.39–17.39	m–s	CH def, two bands if molecule has axial symmetry (fluoro ~580 cm^{-1})
Alkyl monosubstituted acetylenes	~630	~15.87	s	C≡C—H bending vib
	355–335	28.17–29.85	v	C—C≡CH def
—C≡CH	510–260	19.61–38.46	v	Non-alkyl substituent
R—C≡C—CH_3	520–495	19.23–20.20	m–s	
R—C≡C—C_2H_5	495–480	20.20–20.83	s	br
R—C≡C—$(CH_2)_2CH_3$	475–465	21.05–21.51	m	
—C≡C— substituted benzenes	~550	~18.18	m	
	~350	~28.57	v	
C≡C—X (X = Cl, Br, or I)	185–160	54.05–62.50	v	C≡C—X bending vib

Nitriles, —C≡N

Nitrile-containing compounds normally have an absorption in the region 2260–2200 cm^{-1} (4.43–4.55 μm). Care must be taken since acetylenic derivatives also absorb in this general region (due to the C≡C stretching vibration), as do compounds with cumulative double bonds.

For saturated aliphatic nitriles,[21,22] the band due to the stretching vibration of the —C≡N group occurs near 2250 cm^{-1} (4.44 μm) and for aryl and conjugated nitriles near 2230 cm^{-1} (4.48 μm).[23,24] The intensity of this band varies considerably. For example, oxygen atoms on neighbouring carbon atoms, —O—C—C≡N, tend to reduce the intensity of the band, for instance, cyanohydrins, >C(OH)CN, have no observable C≡N absorption whereas conjugation to the C≡N group appears to increase the intensity of the band. The intensity is reduced by electron-withdrawing atoms or groups, e.g. oxygen or chlorine atoms. Normally, medium-to-strong bands are observed for relatively small molecules not containing oxygen atoms. Aromatic nitriles with electron-donating substituents on the ring tend to have a more intense C≡N stretching band than those with electron-accepting groups. Solvents may also affect the intensity of this band.[31] The position of the band is about the same for dimers as for monomers.

In general, all aliphatic nitriles have a strong band at 390–350 cm^{-1} (25.64–28.57 μm) due to the C—C≡N deformation.[27] Saturated primary aliphatic nitriles have medium-to-strong bands at 580–555 cm^{-1} (17.24–18.02 μm) and 560–525 cm^{-1} (17.86–19.05 μm) due to the C—C—CN in-plane deformation vibration. These two bands may be assigned to rotational isomers, the first band to the isomer where the C≡N group is *trans*- to a carbon atom and the second band to that where the C≡N group is *trans*- to a hydrogen atom.

Aromatic nitriles have two bands, one strong at 580–540 cm^{-1} (17.24–18.52 μm) and one of medium intensity at 430–380 cm^{-1} (23.26–26.32 μm). The former band is due to the combination of the out-of-plane aromatic ring-deformation vibration and the in-plane deformation vibration of the —C≡N group. The latter band is due to the in-plane bending of the aromatic ring C—CN bond.

Inorganic cyanides[28] in the solid phase absorb over a wide range, 2250–2000 cm^{-1} (4.44–5.00 μm), as do coordination complexes: 2150–1980 cm^{-1} (4.65–5.05 μm).[28–30] For nitrile complexes with iodine monochloride, the band due to the C≡N stretching vibration is slightly higher by about 10 cm^{-1} (lower by 0.02 μm) than for the corresponding normal nitrile compound, the bands being broader and slightly stronger than usually observed. On the other hand, the coordination of nitriles to metal ions (R—C≡N → M) results in the band due to the C≡N stretching vibration being of greater intensity and occurring at a higher wavenumber, 2325–2265 cm^{-1} (4.30–4.42 μm), than for the uncoordinated nitrile compound. The cyanide ion absorbs at 2080–2070 cm^{-1} (4.81–4.83 μm).

Isonitriles, —N≡C

Alkyl and aryl isonitriles have strong absorptions in the regions 2175–2150 cm^{-1} (4.60–4.72 μm) and 2150–2115 cm^{-1} (4.72–4.74 μm) respectively.[33,34] The intensity of the band is very sensitive to changes in the substituent. Isonitriles have a characteristic band, not found for nitriles, at 1595 cm^{-1} (6.25 μm).

Nitrile N-oxides, —C≡N→O

Aryl nitrile *N*-oxides absorb strongly at 2305–2285 cm^{-1} (4.34–4.38 μm), due to the C≡N stretching vibration, and at 1395–1365 cm^{-1} (7.17–7.33 μm) due to the N—O stretching vibration.

Cyanamides, >N—C≡N

Cyanamides absorb more strongly at lower frequencies than might be expected for the C≡N stretching mode. This is due to the presence of the resonance >N—C≡N ↔ >N^+=C=N^- which reduces the force constant. The range is 2225–2210 cm^{-1} (4.49–4.53 μm). The same resonance effect is found for cyanoguanidines

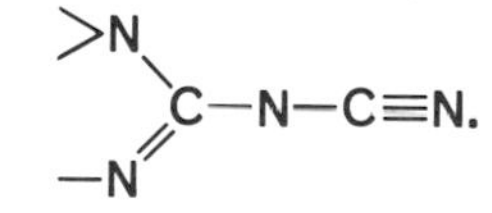

Table 4.3. Nitrile, isonitrile, nitrile *N*-oxide, and cyanamide C≡N stretching vibrations

Functional Groups	Region cm⁻¹	μm	Intensity	Comments
Saturated aliphatic nitriles	2260–2240	4.42–4.46	w–m	
α,β-Unsaturated nitriles	2250–2200	4.44–4.50	m–s	
Aryl nitriles	2240–2220	4.46–4.50	m–s	
α-Halogen substituted nitriles	2280–2240	4.39–4.46	w–m	
Aliphatic isonitriles	2175–2150	4.60–4.65	s	
Aryl isonitriles	2150–2115	4.65–4.73	s	
Aryl nitrile *N*-oxides	2305–2285	4.34–4.38	s	
Thiocyanates, S—C≡N	2175–2140	4.60–4.67	m–s	
Cyanamides	2225–2210	4.49–4.52	s	
Cyanoguanidines	2210–2175	4.52–4.60	s	Often multiple peaks
$—CF_2—C≡N$	2280–2270	4.39–4.41	m–s	

Table 4.4. Nitriles: other bands

Functional Groups	Region cm⁻¹	μm	Intensity	Comments
Aliphatic nitriles	390–350	25.64–28.57	m–s	C—C≡N def
Primary aliphatic nitriles	580–555	17.24–18.02	m–s	C—C≡N in-plane def, C≡N *trans-* to carbon atom
	560–525	17.86–19.05	m–s	C—C≡N in-plane def, C≡N *trans-* to hydrogen atom
Secondary aliphatic nitriles	580–550	17.24–18.18	v	C—C≡N in-plane def
	545–530	18.35–18.87	v	C—C≡N in-plane def, C≡N *trans-* to two hydrogen atoms
	565–535	17.70–18.69	v	C—C≡N in-plane def, C≡N *trans-* to one hydrogen atom
Tertiary aliphatic nitriles	~575	~17.39	m–s	C—C—CN in-plane def, C≡N *trans-* to three hydrogen atoms
	~595	~16.81	s	C—C—CN in-plane def, C≡N *trans-* to a carbon atom and two hydrogen atoms
Aromatic nitriles	580–540	17.24–18.52	s	Combination of C≡N in-plane bending and out-of-plane bending of aromatic ring
	430–380	23.26–26.32	m	In-plane bending of aromatic C—CN bond
Aryl nitrile *N*-oxides	1395–1365	7.17–7.33	s	N—O str

Diazonium Salts, Aryl—N≡N⁺X⁻

Diazonium salts[36–38] have a strong absorption in the region 2300–2230 cm⁻¹ (4.35–4.48 μm) which is due to the stretching vibration of the N≡N group. This band is dependent on the nature of the ring substituents but is less dependent on the nature of the anion, a shift of about 40 cm⁻¹ at most being observed for different anions. Aryl diazonium salts may be represented by the resonance structures

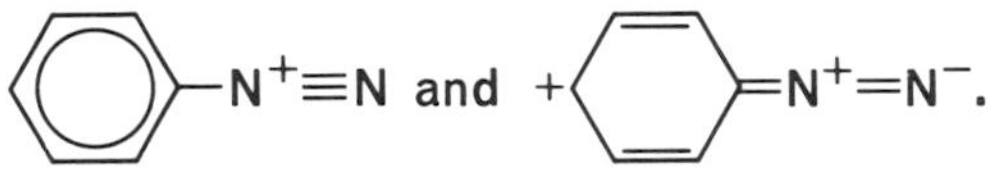
N⁺≡N and +
=N⁺=N⁻.

Electron-donating groups at *ortho* or *para* positions tend to increase the contribution of the second structure and hence tend to decrease the frequency of the N≡N stretching vibration, whereas electron-withdrawing groups have the opposite effect.

Table 4.5. Diazonium salts

Functional Groups	Region cm⁻¹	μm	Intensity	Comments
Diazonium salts	2300–2230	4.35–4.48	m–s	N≡N str

References

1. N. Sheppard and D. M. Simpson, *Quart. Rev.*, 1952, **6,** 1
2. E. A. Gastilovich and D. N. Shigorin, *Usp. Khim.*, 1973, **42,** 1358
3. A. S. Wexler, *App. Spectrosc. Rev.*, 1968, **1,** 29
4. T. L. Brown, *J. Chem. Phys.*, 1962, **38,** 1049
5. R. A. Nyquist and W. J. Potts, *Spectrochim. Acta*, 1960, **16,** 419
6. A. D. Allen and C. D. Cook, *Canad. J. Chem.*, 1963, **41,** 1084
7. A. A. Petrov *et al.*, *J. Gen. Chem. Moscow*, 1957, **27,** 2081
8. A. A. Petrov and G. I. Semenov, *J. Gen. Chem. Moscow*, 1957, **27,** 2974
9. A. A. Petrov and G. I. Semenov, *J. Gen. Chem. Moscow*, 1958, **28,** 73
10. T. V. Yakovlera *et al.*, *Optics Spectrosc.*, 1962, **12,** 106
11. J. L. H. Allan *et al.*, *J. Chem. Soc.*, **1955,** 1874
12. J. C. Evans and R. A. Nyquist, *Spectrochim. Acta*, 1960, **16,** 918
13. P. N. Daykin *et al.*, *J. Chem. Phys.*, 1962, **37,** 1087
14. F. Sondheimer *et al.*, *J. Amer. Chem. Soc.*, 1962, **84,** 270
15. F. Sondheimer and R. Wolovsky, *J. Amer. Chem. Soc.*, 1962, **84,** 260
16. N. A. Domnin ahd R. C. Kolinsky, *J. Gen. Chem. Moscow*, 1961, **33,** 1682
17. G. Eglington and A. R. Galbraith, *J. Chem. Soc.*, **1959,** 889
18. J. C. D. Brand *et al.*, *J. Chem. Soc.*, **1960,** 2526
19. R. West and C. S. Kraihanel, *J. Amer. Chem. Soc.*, 1961, **83,** 765
20. E. A. Gastilovich *et al.*, *Optics Spectrosc.*, 1961, **10,** 595
21. A. Hidalgo, *Anales Real Soc. Espãnola Fís. Chem. Madrid*, 1962, **58A,** 71
22. J. P. Jesson and H. W. Thompson, *Spectrochim. Acta*, 1958, **13,** 217

23. R. Heilmann and J. Bonner, *Compt. Rend.*, 1959, **248,** 2595
24. D. A. Long and W. O. George, *Spectrochim. Acta*, 1964, **20,** 1799
25. T. L. Brown, *Chem. Rev.*, 1958, **58,** 581
26. H. W. Thompson and G. Steel, *Trans. Faraday Soc.*, 1956, **52,** 1451
27. J. Hidalgo, *Compt. Rend.*, 1959, **249,** 395
28. K. Nakamoto, *Infrared Spectra of Inorganic and Coordination Compounds.* Wiley, New York, **1963**
29. J. P. Fackler, *J. Chem. Soc.*, **1962,** 1957
30. H. A. Brune and W. Zeil, *Zeit. Naturf.*, 1961, **16A,** 1251
31. G. L. Cadlow *et al.*, *Proc. Roy. Soc.*, 1960, **A254,** 17
32. L. van Haverbeke and M. A. Herman, *Spectrochim. Acta*, 1975, **31A,** 959
33. I. Ugi and R. Meyer, *Chem. Ber.*, 1960, **93,** 239
34. I. Ugi and C. Steinbruckner, *Chem. Ber.*, 1961, **94,** 2797 and 2802
35. S. Califano *et al.*, *J. Chem. Phys.*, 1957, **26,** 1777
36. L. A. Kazitsyna *et al.*, *Doklady Akad. Nauk SSSR*, 1963, **151,** 573
37. L. A. Katitsyna *et al.*, *J. Phys. Chem. Moscow*, 1960, **34,** 404
38. R. H. Nuttall *et al.*, *Spectrochim. Acta*, 1961, **17,** 947

CHAPTER 5

Cumulated Double-bond Compounds: X=Y=Z Group

Often resonance hybrids are possible for compounds of this type: X=Y=Z, $X^+-Y^-\equiv Z$, etc. The asymmetric stretching vibration of the cumulated double-bond group X=Y=Z gives rise to a band in the range 2275–1900 cm^{-1} (4.40–5.26 μm) which is in approximately the same region as the band due to the triple bond X≡Y, 2300–2000 cm^{-1} (4.35–5.00 μm).

The symmetric stretching vibration is generally weak and not very useful. It occurs in the region 1400–1100 cm^{-1} (7.14–9.09 μm).

It can be seen that some compounds dealt with in this chapter could, in fact, be considered as triple-bond compounds (depending on the triple-bond character) and therefore could equally well have been dealt with in the previous chapter, e.g. thiocyanates.

Allenes, >C=C=C< [1,19–21]

Monosubstituted allenes have a medium-to-strong absorption in the region 1980–1945 cm^{-1} (5.05–5.14 μm) which is due to the asymmetric stretching vibration of the C=C=C group. For polar substituents, this band is in the higher-frequency portion of this range, and also, with strong polar groups such as carbonyls or nitriles, the band is observed to consist of two peaks. Asymmetrically- and symmetrically-disubstituted allenes absorb at 1955–1930 cm^{-1} (5.12–5.18 μm) and 1930–1915 cm^{-1} (5.18–5.22 μm) respectively.[20] Tri- and tetrasubstituted allenes absorb in the region 2000–1920 cm^{-1} (5.00–5.21 μm).

Mono- and asymmetrically-substituted allenes absorb strongly at 875–840 cm^{-1} (11.43–11.90 μm) due to the out-of-plane deformation vibrations of the $=CH_2$ group. The overtone of this band occurs near 1700 cm^{-1} (5.88 μm). Symmetrically-disubstituted allenes absorb near 870 cm^{-1} (11.49 μm) due to the CH deformation vibrations. Trisubstituted allenes absorb strongly at 880–840 cm^{-1} (11.36–11.90 μm).

The C=C=C symmetric stretching band is of medium or weak intensity, or absent. It occurs in the region 1075–1060 cm^{-1} (9.30–9.43 μm) and is not a useful band in making assignments.

Bands due to the C—H stretching vibrations of $C=C=CH_2$ occur near 3050 cm^{-1} (3.28 μm) and 2990 cm^{-1} (3.34 μm), the first band being at a slightly lower frequency than the corresponding band for vinyl and vinylidene groups.

Table 5.1. Allenes

Functional Groups	Region cm^{-1}	Region μm	Intensity	Comments
Allenes	2000–1915	5.00–5.22	m–s	asym C=C=C str
Monosubstituted allenes, $-HC=C=CH_2$	1980–1945	5.05–5.14	m–s	asym C=C=C str
	~1700	~5.88	w	$=CH_2$ wagging overtone
	875–840	11.43–11.90	s	$=CH_2$ out-of-plane def
Symmetrically-disubstituted allenes, —CH=C=CH—	1930–1915	5.18–5.22	m–s	asym C=C=C str
Asymmetrically-disubstituted allenes, $>C=C=CH_2$	1955–1930	5.12–5.18	m–s	asym C=C=C str
	~1700	~5.88	w	$=CH_2$ wagging overtone
	875–840	11.43–11.90	s	$=CH_2$ out-of-plane def
Tri- and tetrasubstituted allenes	2000–1920	5.00–5.21	m–s	asym C=C=C str
Methyl, ethyl, propyl, and butyl allenes	~555	~18.02	m	C=C=C bending vib
	550–520	18.18–19.23	w–m	C=CH bending vib
	355–305	28.17–32.79	w–m	C=C=C bending vib
	~200	~50.00	w–m	C=C=C bending vib
Cyclopropyl allenes, 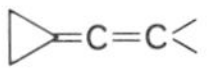	~2020	~4.95	m	asym C=C=C str

Isocyanates, —N=C=O

Due to the asymmetric stretching vibration of the —N=C=O group, a band, sometimes with shoulders, occurs at 2285–2250 cm^{-1} (4.38–4.44 μm)[2,7,18], except for methyl isocyanate which absorbs near 2230 cm^{-1} (4.48 μm). This band is slightly broader than the corresponding band observed for thiocyanates and is not affected by conjugation. However, for α,β-unsaturated compounds, the C=C stretching vibration is affected and moves from its normal position to 1670–1630 cm^{-1} (5.99–6.14 μm). The symmetric —N=C=O stretching vibration band occurs at 1460–1340 cm^{-1} (6.85–7.46 μm). It is weak and not usually of use for assignment purposes since it is often overlapped by aliphatic absorption bands which occur in the same region.

Aliphatic and aryl isocyanate trimers,[3] i.e. isocyanurates, have a strong band due to the carbonyl stretching vibration in the region 1715–1680 cm^{-1} (5.83–5.95 μm). Aromatic isocyanate dimers[3] have a strong similar band at 1785–1775 cm^{-1} (5.60–5.63 μm).

Isothiocyanates, —N=C=S[6–12]

Due to the asymmetric stretching vibration of the —N=C=S group, a very strong band in the region 2140–1990 cm^{-1} (4.67–5.03 μm) is observed. For aliphatic compounds, this band is usually a broad doublet, although it may sometimes have a shoulder. Alkyl isothiocyanates[11] absorb in the region 2140–2080 cm^{-1} (4.67–4.81 μm) whereas aryl derivatives[12, 35] tend to absorb in the region 2100–1990 cm^{-1} (4.76–5.03 μm). The symmetric stretching vibration gives rise to a band of variable intensity in the region 1250–1080 cm^{-1} (8.00–9.26 μm). Most alkyl isothiocyanates have absorptions at 640–600 cm^{-1} (15.63–16.67 μm) and at 565–510 cm^{-1} (17.70–19.61 μm) which are of strong-to-medium intensity. These bands have been assigned to the in-plane and out-of-plane deformation vibrations of the —NCS group. A medium-to-strong band is also usually observed at 470–440 cm^{-1} (21.28–22.73 μm).

Thiocyanates, —S—C≡N

(Rather than include this section in the previous chapter, it was felt that it would best be treated here together with isothiocyanate compounds.)

A band of medium-to-strong intensity is observed in the region 2175–2140 cm^{-1} (4.60–4.67 μm) due to the C≡N stretching vibration.[4–7, 9, 10, 13] The absorption due to aryl derivatives is found in the upper end of this frequency range while that for alkyl derivatives[4] is in the lower half of the range.

All aliphatic thiocyanates have a strong band at 405–400 cm^{-1} (24.69–25.00 μm) which is due to the in-plane deformation vibration of the —SCN group. Primary aliphatic thiocyanates have a weak-to-medium intensity band at 650–640 cm^{-1} (15.38–15.63 μm) due to the stretching vibration of the S—CN bond and a band of medium-to-strong intensity near 620 cm^{-1} (16.13 μm) due to the C—S stretching vibration (where the carbon is the α-carbon). Secondary aliphatic thiocyanates have a band of variable intensity at 610–600 cm^{-1} (16.39–16.67 μm) due to the S—CN bond stretching vibration and, in addition, as many as three bands may be observed due to different molecular configurations: one near 655 cm^{-1} (15.27 μm), another at 640–630 cm^{-1} (15.63–15.87 μm), and one near 575 cm^{-1} (17.39 μm). As with alkyl halides, cyanides, etc., different rotational isomers are possible.

Simple inorganic thiocyanates[14–16] absorb strongly near 2050 cm^{-1} (4.90 μm), this band usually being the predominant one in the region 5000–650 cm^{-1} (2–15 μm).

Selenocyanates and Isoselenocyanates[17]

Aromatic selenocyanates have a medium-to-strong sharp band near 2160 cm^{-1} (4.63 μm) whilst the corresponding isoselenocyanates have a strong, broad band, usually with two peaks, in the region 2200–2000 cm^{-1} (4.55–5.00 μm). The symmetric —N=C=Se stretching vibration band of isoselenocyanates occurs in the region 675–605 cm^{-1} (14.85–16.53 μm). Selenocyanates have a band at 520–545 cm^{-1} (18.35–19.23 μm) due to the stretching vibration of the Se—CN bond, another band at about 390 cm^{-1} (25.64 μm) due to the in-plane vibration of the Se—C≡N group, and a band at about 350 cm^{-1} (28.57 μm) due to the out-of-plane vibration of the group.

For isoselenocyanates where the nitrogen atom is not bound to a carbon atom (e.g. to an atom of Si, Sn, Ge, etc.), the asymmetric —N=C=Se stretching vibration band occurs at about 2140 cm^{-1} (4.67 μm), a single band being observed. For isoselenocyanatophosphates, $(RO)_2P$=ONCSe, and thiophosphates, $(RO)_2P$=SNCSe,[22] the N=C=Se asymmetric stretching vibration band occurs in the region 1975–1960 cm^{-1} (5.06–5.10 μm).

Isoselenocyanato- complexes have a strong band at 430–370 cm^{-1} (23.26–27.03 μm).

Azides, —N=N⁺=N⁻

Resonance is possible for these compounds:

$$—N=N^+=N^- \longleftrightarrow —N^-—N^+≡N.$$

Organic azides[23–27, 35] have a strong band in the region 2135–2080 cm^{-1} (4.68–4.81 μm) due to the asymmetric stretching vibration of the N=N=N group. This band is relatively insensitive to conjugation and to changes in the electronegativity of the adjacent group. A weak band at 1340–1180 cm^{-1} (7.46–8.48 μm) is also observed due to the symmetric stretching of the NNN group. This band is not observed for ionic azides,[28] which have their strong absorption in the range 2140–2030 cm^{-1} (4.67–4.93 μm). Information is also available for inorganic azides.[29, 30]

Diazo Compounds, >C=N⁺=N⁻

Diazo compounds may be represented by resonance hybrids:

$$C=N^+=N^- \longleftrightarrow C^-—N^+≡N.$$

Diazo compounds with the group —CH=N⁺=N⁻ have a strong absorption in the region 2050–2035 cm^{-1} (4.88–4.91 μm) and disubstituted compounds, >C=N⁺=N⁻, absorb at 2030–2000 cm^{-1} (4.93–5.00 μm). Diazoketones and diazoesters, >CO—C=N⁺=N⁻, have their carbonyl stretching frequencies slightly decreased from that expected for an ordinary ketone or ester. Similarly, the stretching vibration frequency of the C=N⁺=N⁻ group for these compounds is increased (probably due to coupling), indicating that there is an increase in the proportion of triple-bond character. Monosubstituted diazoketones, —CO—CH=N⁺=N⁻ absorb at 2100–2080 cm^{-1} (4.76–4.81 μm) and disubstituted diazoketones, —CO—C=N⁺=N⁻, absorb at 2075–2050 cm^{-1} (4.82–4.88 μm), the frequency of the carbonyl stretching absorption being lowered to 1650–1645 cm^{-1} (6.06–6.08 μm) for aliphatic compounds and to 1630–1605 cm^{-1} (6.14–6.23 μm) for aromatic compounds.

Table 5.2. X=Y=Z groups (except allenes)

Functional Groups	Region cm^{-1}	Region μm	Intensity	Comments
Isocyanates, —N=C=O	2285–2250	4.38–4.44	vs	asym NCO str, see ref. 35
	1460–1340	6.85–7.46	w	sym NCO str
Cyanate ion, NCO$^-$	2220–2130	4.51–4.70	s	asym NCO str
	1335–1290	7.49–7.75	w	sym NCO str
	1295–1200	7.72–8.33	w	Combination band
	640–605	15.63–16.53	s	Bending vib
Isothiocyanates, —N=C=S	2140–1990	4.67–5.03	vs	br, asym NCS str, usually a doublet
Alkyl isothiocyanates	640–600	15.63–16.67	m–s	br, —NCS in-plane def
	565–510	17.70–19.61	m–s	—NCS out-of-plane def
	470–440	21.28–22.73	m–s	
Thiocyanates, —SC≡N	2175–2140	4.60–4.67	m–s	Aryl at upper end of frequency range, alkyl at lower end
Alkyl thiocyanates	405–400	24.69–25.00	s	SCN group in-plane def
Primary aliphatic thiocyanates	650–640	15.38–15.63	w–m	S—CN str
	~620	~16.13	m–s	C_α—S str (absent for MeSCN)
Secondary aliphatic thiocyanates	~655	~15.27	w	C_α—S str
	640–630	15.63–15.87	w	C_α—S str
	610–600	16.39–16.00	v	C—SN str
	~575	~17.39	m	C_α—S str
Thiocyanate ion	2090–2020	4.79–4.95	s	asym N=C=S str
	~950	~10.53	w	Bending overtone
	~750	~13.33	w	sym str
	~470	~21.28	s	Bending
Coordinated thiocyanate ions, NCS—metal	720–690	13.89–14.49	m–s	See refs 15, 16, 34
Coordinated isothiocyanate ions, SCN—metal	860–780	11.63–12.82	m–s	
Aromatic selenocyanates, —SeCN	~2160	~4.63	s	sh
Aromatic isoselenocyanates	2200–2000	4.55–5.00	s	br, doublet
Azides, —N=N=N	2135–2080	4.68–4.81	s	asym str (sometimes a doublet) (—CO—N_3, ~2150 cm^{-1})
	1340–1180	7.46–8.48	m–s	sym str
Metal azides and azide ion	650–635	15.38–15.75		N=N=N bending vib
Acid azides and nitro-aromatic azides	2240–2180	4.46–4.59	s	asym N=N=N str
	2155–2140	4.64–4.67	s	asym N=N=N str
	1710–1690	5.85–5.92	s	C=O str for acid azides
	1260–1235	7.94–8.10	m	sym N=N=N str
Diazo compounds, >C=N$^+$=N$^-$	2050–2000	4.88–5.00	vs	br
Diazoketones and diazoesters, —CO—C=N$^+$=N$^-$	2075–2050	4.82–4.88	s	(Ketones: C=O str, 1650–1600 cm^{-1} and strong band, 1390–1330 cm^{-1}—may be doublet; alkyl ketones: C=O str, ~1645 cm^{-1})
Diazonium salts, Ar—N≡N$^+$X$^-$	2300–2230	4.35–4.48	m–s	N≡N str
Ketenes, >C=C=O	2200–2080	4.45–4.81	s	Often found near 2150 cm^{-1}
	~1130	~8.85	s	
Ketenimines, >C=C=N—	2050–2000	4.88–5.00	s	

Table 5.2 (*cont.*)

Functional Groups	Region cm^{-1}	Region μm	Intensity	Comments
Aliphatic carbodi-imines, R—N=C=N—R	2155–2130	4.64–4.70	vs	asym N=C=N str, see ref. 33
Aryl carbodi-imines, Ar—N=C=N—Ar	2145–2135	4.66–4.68	vs	C=N str doublet, due to Fermi resonance, band at ~2110 cm^{-1} usually being the stronger
	2115–2105	4.73–4.75	vs	
—N=S=O	1300–1230	7.69–8.13		
	1180–1100	8.48–9.09		

References

1. L. M. Sverdlov and M. G. Borisov, *Optics Spectrosc.*, 1960, **9**, 227
2. E. A. Nicol *et al.*, *Spectrochim. Acta*, 1974, **30**, 1717
3. B. Taub and C. E. McGinn, *Dyestuffs*, 1958, **42**, 263
4. R. P. Hirschmann *et al.*, *Spectrochim. Acta*, 1964, **20**, 809
5. K. Kottke *et al.*, *Pharmazie*, 1973, **28**, 736
6. N. S. Ham and J. B. Willis, *Spectrochim. Acta*, 1960, **16**, 279
7. G. D. Caldow and H. W. Thompson, *Spectrochim. Acta*, 1958, **13**, 212
8. E. Svatek *et al.*, *Acta Chem. Scand.*, 1959, **13**, 442
9. A. Foffani *et al.*, *R. C. Acad. Lincei*, 1960, **29**, 355
10. E. Lieber *et al.*, *Spectrochim. Acta*, 1959, **13**, 296
11. R. N. Knisely *et al.*, *Spectrochim. Acta*, 1967, **23A**, 109
12. P. Kristián *et al.*, *Coll. Czech. Chem. Comm.*, 1964, **29**, 2507
13. R. A. Cummins, *Austral. J. Chem.*, 1964, **17**, 838
14. C. Pecile *et al.*, *R. C. Acad. Lincie*, 1960, **28**, 189
15. A. Turco and C. Pecile, *Nature*, 1961, **191**, 66
16. A. Tramer, *J. Chim. Phys.*, 1962, **59**, 232
17. W. J. Franklin *et al.*, *Spectrochim. Acta*, 1974, **30A**, 1293
18. R. P. Hirschmann *et al.*, *Spectrochim. Acta*, 1965, **21**, 2125
19. J. H. Wotiz and D. E. Mancuso, *J. Org. Chem.*, 1957, **22**, 207
20. A. A. Petrov *et al.*, *Optics Spectrosc.*, 1959, **7**, 170
21. M. G. Borisov and L. M. Sverdlov, *Optics Spectrosc.*, 1963, **15**, 14
22. T. Gabrio and G. Barnikow, *Zeit. Chem.*, 1969, **9**, 183
23. E. Mantica and G. Zerbi, *Gazz. Chim. Ital.*, 1960, **90**, 53
24. E. Lieber and A. E. Thomas, *App. Spectrosc.*, 1961, **15**, 144
25. E. Lieber and E. Oftedahl, *J. Org. Chem.*, 1959, **24**, 1014
26. E. Leiber *et al.*, *Analyt. Chem.*, 1957, **29**, 916
27. W. R. Carpenter, *App. Spectrosc.*, 1963, **17**, 70
28. J. I. Bryant and G. C. Turrell, *J. Chem. Phys.*, 1962, **37**, 1069
29. H. A. Papazian, *J. Chem. Phys.*, 1961, **34**, 1614
30. W. Dobramsyl *et al.*, *Spectrochim. Acta*, 1975, **31A**, 905
31. L. A. Kazicyna *et al.*, *J. Phys. Chem. Moscow*, 1960, **34**, 850
32. L. A. Kazicyna *et al.*, *Doklady Akad. Nauk. SSSR*, 1963, **151**, 573
33. G. D. Meakins and R. J. Moss, *J. Chem. Soc.*, **1957**, 993
34. J. Lewis *et al.*, *J. Chem. Soc.*, **1961**, 4590
35. A. El Shahawy and R. Gaufres, *J. Chim. Phys. Phys.-Chim. Biol.*, 1978, **75**, 196

CHAPTER 6

Hydroxyl Group Compounds: O—H Group

Alcohols, R—OH

Bands due to O—H stretching and bending vibrations and C—O stretching vibrations are observed.

Alcohol O—H stretching vibrations

Unassociated hydroxyl groups absorb strongly in the region 3670–3580 cm^{-1} (2.73–2.80 μm).[1,2] However, free hydroxyl groups only occur in the vapour phase or in very dilute solutions in non-polar solvents. The band due to the free hydroxyl group is sharp and its relative intensity increases in the following order:

aromatic alcohols < tertiary alcohols < secondary alcohols < primary alcohols

In very dilute solution in non-polar solvents, the normal O—H absorptions of alcohols are:

primary aliphatic alcohols	3640–3630 cm^{-1}	(2.75–2.76 μm)
secondary aliphatic alcohols	3625–3620 cm^{-1}	(~2.76 μm)
tertiary aliphatic alcohols	3620–3610 cm^{-1}	(2.76–2.77 μm)
R—OH···O=C<	3600–3450 cm^{-1}	(2.78–2.90 μm)

The relative intensity of the band due to the hydroxyl stretching vibration decreases with increase in concentration, with additional broader bands appearing at lower frequencies 3580–3200 cm^{-1} (2.73–3.13 μm). These bands are the result of the presence of intermolecular bonding, the amount of which increases with concentration. The precise position of the O—H band is dependent on the strength of the hydrogen bond. In some samples, intramolecular hydrogen bonding[7,8] may occur, the resulting hydroxyl group band which appears at 3590–3400 cm^{-1} (2.79–2.94 μm) being sharp and unaffected by concentration changes.

For solids, liquids, and concentrated solutions, a broad band is normally observed at about 3300 cm^{-1} (3.00 μm). Polyhydric alcohols in dilute solution in non-polar solvents normally have a sharp band at about 3600 cm^{-1} (2.78 μm) and a broader band at 3550–3450 cm^{-1} (2.82–2.90 μm). Hydroxyl groups which are hydrogen-bonded to aromatic ring π-electron systems absorb at 3580–3480 cm^{-1} (2.79–2.87 μm).

Overtone bands of carbonyl stretching vibrations also occur in the region 3600–3200 cm^{-1} (2.78–3.13 μm) but are, of course, of weak intensity. Bands due to N—H stretching vibrations may also cause confusion. However, these bands are normally sharper than those due to intermolecularly hydrogen-bonded O—H groups.

Alcohol C—O stretching vibrations

The absorption region for the alcohol C—O group due to its stretching vibration is 1200–1000 cm^{-1} (8.33–10.00 μm). Hydrogen bonding has the effect of decreasing the frequency of this band slightly: saturated primary alcohols absorb strongly in the region 1085–1030 cm^{-1} (9.22–9.71 μm); secondary alcohols absorb at 1125–1085 cm^{-1} (8.90–9.22 μm); tertiary alcohols absorb strongly at 1205–1125 cm^{-1} (8.30–8.90 μm). These ranges, which are given for pure liquids, should be extended slightly for solution spectra. In general, the presence of unsaturation and chain branching both lower the C—O stretching vibration frequency. Care must be taken since esters, carboxylic acids, acid anhydrides, and ethers all absorb strongly in the general range 1300–1000 cm^{-1} (7.69–10.00 μm) due to the C—O stretching vibration.

Alcohol O—H deformation vibrations

The in-plane O—H deformation vibration gives rise to a strong band in the region 1440–1260 cm^{-1} (6.94–7.93 μm). In concentrated solutions, this band is very broad, extending over approximately 1500–1300 cm^{-1} (6.67–7.69 μm). On dilution, the band becomes weaker and is eventually replaced by a sharp, narrow band at about 1260 cm^{-1} (7.93 μm). In the presence of hydrogen bonding, the O—H deformation vibration is lowered in frequency. (Bands due to CH_3 deformation vibrations may also be present in this region.)

The O—H out-of-plane vibration gives a broad band in the region 700–600 cm^{-1} (14.29–16.67 μm). The position of this band is dependent on the strength of the hydrogen bond—the stronger the hydrogen bond, the higher the wavenumber.

In the far infrared spectroscopic region,[11–13] aliphatic alcohols in a cyclohexane solvent exhibit a characteristic strong band at 220–200 cm^{-1} (45.45–50.00 μm) due to the torsional motion of the O—H group.[13] This band is insensitive to steric effects but becomes broad with increase in concentration, eventually disappearing. In benzene solution, a band is observed at about 300 cm^{-1} (33.33 μm) which is believed to be due to an alcohol–benzene complex which is formed.

Table 6.1. Hydroxyl group O—H stretching vibrations

Functional Groups	Region cm^{-1}	Region μm	Inten-sity	Comments
Free O—H	3670–3580	2.73–2.80	v	sh, OH str
Hydrogen-bonded O—H (intermolecular), 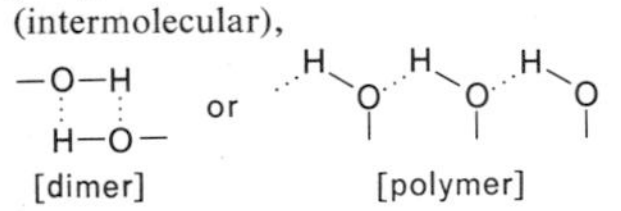	3550–3230	2.82–3.10	m–s	Usually broad but may be sharp, frequency is concentration-dependent
Hydrogen-bonded O—H (intramolecular), 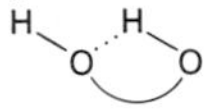	3590–3400	2.79–2.94	v	Usually sharp, frequency is concentration-independent
Chelated O—H, 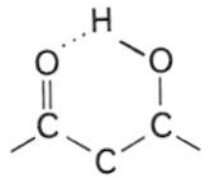	3200–2500	3.13–4.00	v	Usually broad, frequency concentration-independent
—OD	2780–2400	3.60–4.17	v	O—D str
OH of enol form of β-diketones	2700–2500	3.71–4.00	v	br, chelated OH
Intramolecular-bonded *ortho*-phenols	3200–2500	3.13–4.00	m	Free phenols ~3610 cm^{-1}
Carboxylic acids, —COOH	3300–2500	3.03–4.00	w–m	br, O—H str, hydrogen-bonded, sometimes a number of weak bands in region 2700–2500 cm^{-1}. Band is concentration-dependent
OH of water of crystallization	3600–3100	2.78–3.23	w	In solid-state spectra
	1630–1615	6.13–6.19	v	
OH of water in dilute solution	~3760	~2.66	w–m	In non-polar solvents
Free oximes, >C=N—OH	3600–3570	2.78–2.79	w–m	sh
Free hydroperoxides, —O—O—H	3560–3530	2.82–2.83	m	
Peracids, —CO—O—OH	~3280	~3.05	m	
Tropolones	~3100	~3.23	w–m	
Phosphorus acids, 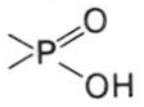	2700–2560	3.70–3.91	m	br

Table 6.2. Hydroxyl group O—H deformation vibrations

Functional Groups	Region cm^{-1}	Region μm	Inten-sity	Comments
Primary and secondary alcohols	1350–1260	7.41–7.94	s	In-plane def
Tertiary alcohols	1410–1310	7.09–7.63	s	In-plane def
Alcohols	700–600	14.29–16.67		br, out-of-plane def
Phenols	1410–1310	7.09–7.63	s	O—H def and C—O str combination
Carboxylic acids	955–915	10.47–10.93	s	Out-of-plane def, br diffuse band
Deuterated carboxylic acids	~675	~14.81	s	O—D in-plane def

Table 6.3. Alcohol C—O stretching vibrations

Functional Groups	Region cm^{-1}	μm	Intensity	Comments
Saturated primary alcohols, —CH_2—OH	1085–1030	9.22–9.71	s	
Saturated secondary alcohols, > CH—OH	1125–1085	8.90–9.22	s	
Saturated tertiary alcohols ⩾ C—OH	1205–1125	8.30–8.90	s	
Primary alcohols				
RCH_2CH_2OH	~1050	~9.52	s	Ethanol ~1065 cm^{-1}
$R_1R_2CHCH_2OH$	~1035	~9.66	s	
$R_1R_2R_3CCH_2OH$	~1020	~9.80	s	
(Unsat group) —CH_2CH_2OH	~1015	~9.85	s	Vinyl or aryl substituted
Secondary alcohols				
RH_2C(>CHOH)H_3C	~1085	~9.22	s	(Isopropyl alcohol ~1100 cm^{-1}.) Each additional alkyl group increases the wavenumber by ~15 cm^{-1}
(Unsat group) —$CH_2CH(OH)CH_3$	~1070	~9.35	s	
[(Unsat group)CH_2]$_2$-CHOH	~1010	~9.90	s	
(Aryl—CH_2)$_2$CHOH	~1050	~9.52	s	
Tertiary alcohols				
$RCH_2(CH_3)_2COH$	~1135	~8.81	s	(*t*-Butyl alcohol ~1150 cm^{-1})
$(R_1CH_2)(R_2CH_2)COH$	~1120	~8.93	s	Each additional alkyl group increases the wavenumber by ~15 cm^{-1}
(Unsat group) —$CH_2(CH_3)_2COH$	~1120	~8.93	s	
[(Unsat group)—CH_2]$_2$-CH_3COH	~1060	~9.43	s	
[(Unsat group)—CH_2]$_3$-COH	~1010	~9.90	s	
Aromatic and α-unsaturated secondary alcohols	1085–1030	9.22–9.71	s	
α Unsaturated and cyclic tertiary alcohols	1125–1085	8.90–9.22	s	See refs 9 and 10
Alicyclic secondary alcohols (three- or four-membered rings)	1060–1020	9.43–9.80	s	
Alicyclic secondary alcohols (five- or six-membered rings)	1085–1030	9.22–9.71	s	
Phenols	1260–1180	7.94–8.48	s	O—H def and C—O str combination

Phenols

In the absence of intramolecular hydrogen bonding and in the case of a dilute solution in a non-polar solvent[14–16] (i.e. in the additional absence of intermolecular hydrogen bonding), phenols have an absorption band at 3620–3590 cm^{-1} (2.76–2.79 μm) due to the O—H stretching vibration.[17, 18] If strong intramolecular hydrogen bonding does occur, for example, to a carbonyl group, then a relatively sharp band is found at about 3200 cm^{-1} (3.13 μm). If, on the other hand, hydrogen bonding is inhibited by the presence of large groups in the *ortho* positions,[19–21] the absorption occurs in the region 3650–3600 cm^{-1} (2.74–2.78 μm). Phenols without bulky *ortho* groups, whether in concentrated solutions or as solids or in the pure liquid phase, have a broad absorption at 3400–3230 cm^{-1} (2.94–3.10 μm).

Medium-to-strong bands are observed at 1255–1240 cm^{-1} (7.97–8.07 μm), 1175–1150 cm^{-1} (8.51–8.70 μm), and 835–745 cm^{-1} (11.98–13.42 μm) for alkyl phenols.[22, 23] In addition, *o*-phenols usually have a band near 1320 cm^{-1} (7.58 μm) and *m*-alkyl phenols one at 1185 cm^{-1} (8.44 μm).[24] The three main bands may be attributed to the C—O stretching and the O—H in-plane and out-of-plane deformation vibrations.

The C—O stretching vibration for *p*-monosubstituted phenols,[24] i.e. the strongest band in the region 1300–1200 cm^{-1} (7.69–8.33 μm), increases in frequency with the electron-withdrawing ability of the substituent.

In the solid phase, or in cases where strong hydrogen bonding may occur, a broad absorption at 720–600 cm^{-1} (13.89–16.67 μm) is observed due to the out-of-plane deformation of the O—H group. In dilute solution, i.e. in the unassociated state, this absorption occurs near 300 cm^{-1} (33.33 μm). In the presence of hydrogen bonding, a characteristic weak absorption, due to the in-plane bending of the ring C—OH bond, is observed at 450–375 cm^{-1} (22.22–26.67 μm).[25, 26] In the absence of hydrogen bonding, this band may be shifted by about 20–40 cm^{-1} (1.00–2.20 μm). For monosubstituted phenols,[25] the position of this weak band is influenced by the nature of the substituent. In the case of electron-accepting or almost neutral groups, such as alkyl groups,[25] the band is found above 400 cm^{-1} (below 25.00 μm) whereas with electron-donating substituents the band occurs below 400 cm^{-1} for solid samples.

Table 6.4. Phenols: O—H stretching vibrations

Functional Groups	Region cm⁻¹	μm	Intensity	Comments
Unassociated	3620–3590	2.76–2.79	m	(In dilute solution) sh
Associated	3250–3000	3.08–3.33	v	(In solution) br, concentration and solvent dependent
Ortho-substituted, 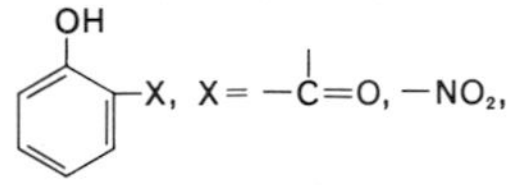	3200–2500	3.13–4.00	m	Intramolecular hydrogen bonded

OH
X, X = —C=O, —NO₂,

Table 6.5. Phenols: interaction of O—H deformation and C—O stretching vibrations

Functional Groups	Region cm⁻¹	μm	Intensity	Comments
Associated	1390–1330	7.19–7.52	m	
	1260–1180	7.94–8.48	s	
Unassociated (dilute solution)	1360–1300	7.35–7.69	m	
	1225–1150	8.17–8.70	s	
o-Alkyl phenols (solution)	~1320	~7.58	s	
	1255–1240	7.97–8.07	s	
	1175–1150	8.51–8.70	s	
m-Alkyl phenols (solution)	1285–1270	7.78–7.87	s	
	1190–1180	8.40–8.48	s	
	1160–1150	8.62–8.70	s	
p-Alkyl phenols (solution)	1260–1245	7.94–8.03	s	
	1175–1165	8.51–8.58	s	

Table 6.6. Phenols: other bands

Functional Groups	Region cm⁻¹	μm	Intensity	Comments
Phenols	~1660	~6.02	s	Usually a doublet, aromatic ring C=C str
	~1110	~9.01	v	Aromatic C—H deformation
	720–600	13.89–16.67	s	br, O—H out-of-plane bending (hydrogen bonding), see ref. 26
	450–375	22.22–26.67	w	In-plane bending of aromatic C—OH bond

References

1. I. Motoyama and C. H. Jarboe, *J. Phys. Chem.*, 1966, **70,** 3226
2. J. H. van der Maas and E. T. G. Lutz, *Spectrochim. Acta*, 1974, **30A,** 2005
3. J. S. Cook and I. H. Reece, *Austral. J. Chem.*, 1961, **14,** 211
4. U. Liddel, *Ann. NY Acad. Sci.*, 1957, **69,** 70
5. S. Siggia *et al.*, In *Chemistry of the Hydroxyl Group*, Part 1 (S. Patai, ed.). Interscience, London, 1971, p. 311
6. A. S. Wexler, *App. Spectrosc. Rev.*, 1968, **1,** 29
7. A. O. Diallo, *Spectrochim. Acta*, 1972, **28A,** 1765
8. L. T. Pitzner and R. H. Atalla, *Spectrochim. Acta*, 1975, **31A,** 911
9. A. A. Petrov and G. I. Semenov, *J. Gen. Chem. USSR AQOTU*, 1957, **27,** 2974
10. J. C. Richer and P. Bélanger, *Canad. J. Chem.*, 1966, **44,** 2057
11. J. E. Chamberlain *et al.*, *Nature*, 1975, **255,** 319
12. W. F. Passchier *et al.*, *Chem. Phys. Letters*, 1970, **4,** 485
13. S. M. Craven, *U.S. Nat. Tech. Inform. Service AD Rep.*, **1971,** No. 733 666
14. M. St C. Flett, *Spectrochim. Acta*, 1957, **10,** 21
15. N. A. Putnam, *J. Chem. Soc.*, **1960,** 5100
16. K. U. Ingold, *Canad. J. Chem.*, 1960, **38,** 1092
17. T. Cairns and G. Eglinton, *J. Chem. Soc.*, **1965,** 5906
18. Z. Yoshida and E. Osawa, *J. Amer. Chem. Soc.*, 1966, **88,** 4019
19. L. J. Bellamy *et al.*, *J. Chem. Soc.*, **1961,** 4762
20. A. W. Baker *et al.*, *Spectrochim. Acta*, 1964, **20,** 1467 and 1477
21. K. U. Ingold and D. R. Taylor, *Canad. J. Chem.*, 1961, **39,** 471 and 481
22. D. D. Shrewsbury, *Spectrochim. Acta*, 1960, **16,** 1294
23. J. H. S. Green *et al.*, *Spectrochim. Acta*, 1972, **28A,** 33
24. J. H. S. Green *et al.*, *Spectrochim. Acta*, 1971, **27A,** 2199
25. R. J. Jakobsen, *Wright Air Development Division Tech. Rep.*, **1960,** No. 60–204
26. V. Bekarek and K. Pragerova, *Coll. Czech. Chem. Comm.*, 1975, **40,** 1005

CHAPTER 7

Ethers: $G_1—O—G_2$ *Group*

The mass and bond strength for the C—O group is similar to that of C—C and therefore, as expected, there is a close similarity in their band positions. However, the change in dipole moment of the C—O group is much larger and therefore the intensity of the band due to the C—O stretching vibration is considerably greater.

Ethers have characteristic, strong absorption bands in the range 1270–1060 cm^{-1} (7.94–9.43 μm) which may be associated with the C—O—C asymmetric stretching vibration.[1] Carboxylic esters and lactones also absorb strongly in this region. For saturated aliphatic ethers, this band may be found at 1150–1060 cm^{-1} (8.70–9.43 μm), usually within the range 1140–1110 cm^{-1} (8.77–9.01 μm). In the case of branched-chain aliphatic ethers, two peaks may be observed. Benzyl ethers absorb at about 1090 cm^{-1} (9.17 μm) and cyclic ethers absorb at 1270–1030 cm^{-1} (7.87–9.71 μm). Aryl ethers absorb strongly in the region 1270–1230 cm^{-1} (7.87–8.13 μm). Alkyl aryl ethers have two strong absorptions, the most intense of which is at 1270–1230 cm^{-1} (7.87–8.13 μm), the other being at 1120–1020 cm^{-1} (8.93–9.80 μm), these bands being due to the asymmetric and symmetric vibrations of the group C—O—C respectively.

The asymmetric C—O—C stretching vibration frequency depends on the group directly bonded to the oxygen atom and decreases in the following order:

$$C_6H_5— > CH_2{=}CH— > R_3C— > RCH_3CH— > RCH_2— > C_6H_5CH_2—$$

For aliphatic ethers, a weak band is observed, usually in the region 930–900 cm^{-1} (10.75–11.11 μm) but sometimes found as high as 1140 cm^{-1} (8.77 μm) when it is usually strong. This band is due to the symmetric stretching vibration of the C—O—C group and may be absent for symmetric ethers due to symmetry factors (see below).

Vinyl ethers usually absorb very close to 1205 cm^{-1} (8.30 μm) in the range 1225–1200 cm^{-1} (8.16–8.33 μm) due to the asymmetric =C—O stretching vibration. The C=C stretching vibration results in a band which appears as a strong doublet, the stronger portion being at 1620–1610 cm^{-1} (6.17–6.21 μm), the other peak being near 1640 cm^{-1} (6.10 μm). The doubling is due to the presence of rotational isomerism which is the result of rotation being restricted about the =C—O bond. The stronger band is due to the more stable, planar, *trans-* form and the weaker band is due to the *gauche-* form, the *cis-* form being sterically inhibited.

As a rough approximation, the asymmetric stretching vibration of the C—O bond occurs at about 1130 cm^{-1} (8.85 μm) when the carbon is fully saturated and at about 1200 cm^{-1} (8.33 μm) when it is unsaturated. This may be either aromatic or olefinic unsaturation.

Often the symmetric and asymmetric C—O—C absorption bands are well separated, by about 200 cm^{-1} (1.7 μm). On simple theoretical grounds these bands would be expected to occur closer together. The large difference is due to coupling. When a central atom is attached to two groups of similar mass by bonds of similar order, coupling of vibrations may occur, e.g. coupling occurs for CH_3OCH_3 whereas there is no coupling for CH_3OH. In general, it has been found that the separation of coupled frequencies is a maximum if the bond angle between the central atom and the two attached groups is 180° and a minimum if it is 90°. For symmetrical ethers, e.g. diethyl ether, due to the presence of coupling, the C—O—C asymmetric stretching vibration band, which is of strong intensity, occurs at about 1110 cm^{-1} (9.01 μm) and the symmetric stretching vibration band is weak or absent.

In general, the asymmetric stretching frequency is lowered for molecules with electron-withdrawing groups since the electron density of the C—O bond is reduced. The opposite is true of electron-donating groups. Any group which increases the double-bond character of the C—O group tends to increase the stretching vibration frequency of this bond and this may be the result of either electronic induction or resonance.

As a result of resonance, aromatic ethers have a contribution from $=O^+—$, e.g.

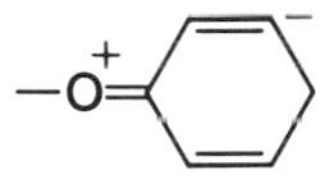

which tends to increase the force constant of the C—O (aromatic carbon–oxygen bond) and hence increases the C—O stretching vibration frequency as compared with aliphatic compounds. Electron-donating groups at *ortho-* or *para-* positions on the ring tend to reduce this frequency relative to a similar *meta*-substituted compound. The reverse is true of electron-attracting groups.

The CH_3—O group for aliphatic ethers may be distinguished from the group CH_3—C since the former absorbs at 1470–1440 cm^{-1} (6.80–6.94 μm) due to both the CH_3 symmetric and asymmetric deformation vibrations, whereas the latter group absorbs at 1385–1370 cm^{-1} (7.22–7.30 μm) due to the symmetric deformation vibration of the CH_3 group. The OCH_3 group can usually be distinguished by its CH_3 symmetric stretching vibration band which occurs in the region 2830–2815 cm^{-1} (3.53–3.55 μm). Aromatic compounds with methoxy groups have an absorption in the region 580–505 cm^{-1} (17.24–19.80 μm), of medium-to-strong intensity, due to the in-plane deformation vibration of the C—O—C groups.

Cyclic ethers often have several bands of medium intensity in the region 1050–800 cm^{-1} (9.50–12.50 μm).

Epoxides[2–4] absorb near 1250 cm^{-1} (8.00 μm) due to the C—O stretching vibration and near 370 cm^{-1} (27.03 μm) due to their ring deformations.

In the case of acetals and ketals, the C—O stretching vibration band is split into three: 1190–1040 cm^{-1} (8.40–9.62 μm), 1195–1125 cm^{-1} (8.37–8.89 μm), and 1100–1060 cm^{-1} (9.09–9.43 μm). The vibration modes may be considered as similar to the asymmetric C—O stretching vibration of ethers. A fourth band at 1060–1035 cm^{-1} (9.43–9.65 μm) which is due to the symmetric vibration may sometimes be observed. In addition, acetals have a characteristic, strong band in the region 1115–1105 cm^{-1} (8.96–9.02 μm) due to a C—H deformation vibration being perturbed by the neighbouring C—O groups. This band may be used to distinguish between acetals and ketals.

Table 7.1. Ether C—O stretching vibrations

Functional Groups	Region cm^{-1}	Region μm	Intensity	Comments
Saturated aliphatic ethers,	1150–1060	8.70–9.43	vs	asym C—O str
C—O—C	1140–900	8.77–11.11	v	sym C—O str
Alkyl-aryl ethers, =C—O—C	1270–1230	7.87–8.13	vs	asym =C—O str
	1120–1020	8.93–9.80	s	sym C—O str
Vinyl ethers, CH_2=CH—O—	1225–1200	8.16–8.33	s	asym C—O str, usually ~1205 cm^{-1}
Trimethylene oxides	~1030	~9.71	w	sym C—O str
(four-membered ring)	980–970	10.20–10.31	s	
Ar—O—CH_2—O—Ar	1265–1225	7.91–8.17	s	=C—O str
	1050–1025	9.52–9.76	s	
Cyclic ethers	1270–1030	7.87–9.71	s	
Acyclic diaryl ethers, =C—O—C=	1250–1170	8.00–8.55	s	
Ring, =C—O—C=	1200–1120	8.33–8.93	s	
	1100–1050	9.09–8.70	s	
Oxirane compounds:				
Epoxides, 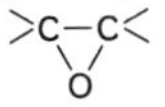	1260–1240	7.94–8.07	m–s	C—O str
Monosubstituted epoxides, 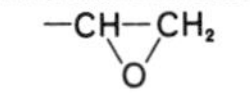	880–805	11.36–12.42	m	Ring vib
Trans-epoxides	950–860	10.53–11.63	v	Ring vib
Cis-epoxides	865–785	11.56–12.74	m	Ring vib
Trisubstituted epoxides	770–750	12.99–13.33	m	Ring vib
Ketals and acetals (see ref. 5), 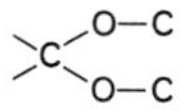	1190–1140	8.40–8.77	s	C—O—C—O—C vib
	1195–1125	8.37–8.89	s	C—O—C—O—C vib
	1100–1060	9.09–9.43	s	C—O—C—O—C vib, strongest band
	1060–1035	9.43–9.65	s	C—O—C—O—C vib, sometimes observed
Acetals	1115–1105	8.96–9.02	s	C—H def (perturbed by C—O group)
Phthalans	915–895	10.93–11.17	s	
Aromatic methylene dioxy compounds, 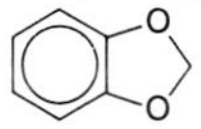	1265–1235	7.90–8.10	s	

Table 7.2. Ethers: other bands

Functional Groups	Region cm^{-1}	μm	Intensity	Comments
Aliphatic ethers, $-OCH_3$	2995–2955	3.34–3.38	m	asym $-CH_3$ str
	2895–2840	3.45–3.52	m	sym $-CH_3$ str
	1470–1440	6.80–6.94	m	asym and sym $-CH_3$ def
$-OCH_2-$	2955–2920	3.38–3.43	m	asym CH_2 str
	2880–2835	3.47–3.53	m	sym CH_2 str (almost equal in intensity to asym str)
	1475–1445	6.78–6.92	m	CH_2 def
$-O-CH_2-O-$	~2780	~3.60	m	CH str
Aliphatic ethers	~430	~23.26	w	$C-O-C$ def
Methyl aromatic ethers,	2830–2815	3.53–3.55	m	CH_3 str
$=C-O-CH_3$	580–505	17.24–19.80	m–s	$C-O-C$ def
$Ar-O-CH_2-O-Ar$	1375–1350	7.27–7.41	v	C—H def
	940–920	10.64–10.87	s	
Vinyl ethers	3150–3000	3.18–3.33	w	C—H str, a number of bands
	1660–1635	6.02–6.12	m	C=C str, *gauche-* form
	1620–1610	6.17–6.21	s	C=C str, *trans-* form
	~1320	~7.58	v	=CH rocking
	970–960	10.31–10.42	m	=CH wagging, *trans-* form
	820–810	12.20–12.35	s	$=CH_2$ wagging
Epoxides	3050–2990	3.28–3.34	w	C—H str, one or two bands, see ref. 6
	~370	~27.03		Ring def
Aromatic ethers	1310–1230	7.63–8.13	s	X-sensitive band
Phenoxy, Ph—O—	765–750	13.10–13.33	s	C—H out-of-plane def ring def
	695–690	14.39 14.49	s	
Benzyloxy, $Ph-CH_2-O-$	745–730	13.42–13.70	s	C—H out-of-plane def ring def
	700–695	14.29–14.39	s	
Acetals	2830–2820	3.53–3.55	w	
Aromatic methylene dioxy	2950–2750	3.40–3.64	m	C—H str, two bands
compounds,	1485–1350	6.73–7.60	v	Several bands
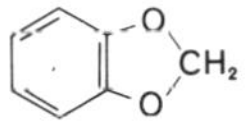				

O
CH₂
O

References

1. A. R. Katritzky and N. A. Coats, *J. Chem. Soc.*, **1959,** 2062
2. A. J. Durbetaki, *J. Org. Chem.*, 1961, **26,** 1017
3. H. von Hoppff and H. Keller, *Helv. Chim. Acta*, 1959, **42,** 2457
4. J. Bomstein, *Analyt. Chem.*, 1958, **30,** 544
5. B. Wladislaw *et al.*, *J. Chem. Soc. B*, **1966,** 586
6. C. J. Wurrey and A. B. Nease, *Vib. Spectra Struct.*, 1978, **7,** 1

CHAPTER 8

Peroxides and Hydroperoxides: —O—O— Group

Peroxides[1–3] and hydroperoxides[2–4] have two main structural units, the C—O and O—O groups. The band due to the C—O stretching vibration occurs in the region 1300–1000 cm^{-1} (7.69–10.00 μm). Electron-withdrawing substituents attached to the C—O group tend to reduce the frequency of this absorption band.

All peroxides have a band at 900–830 cm^{-1} (11.11–12.05 μm) which is usually weak. However, for symmetrical peroxides this O—O stretching vibration is infrared inactive, although as a result of environmental interaction it may still be observed. Tertiary peroxides and tertiary hydroperoxides have a strong band in the region 920–830 cm^{-1} (10.87–12.05 μm) which is believed to be due to the skeletal vibration of the group

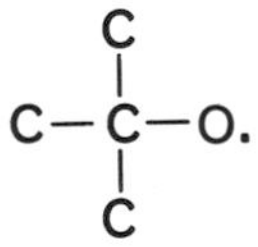

Symmetrical aliphatic diacyl peroxides, —CO—O—O—CO—, have two strong bands in the region 1820–1790 cm^{-1} (5.49–5.56 μm) due to the stretching vibrations of the C=O groups. Similarly, symmetrical aromatic diacyl peroxides have two strong bands in the region 1820–1760 cm^{-1} (5.50–5.88 μm), the position of these bands being dependent on the nature and position of the aromatic substituents.

Ozonides have a medium-intensity absorption at 1065–1040 cm^{-1} (9.39–9.62 μm) due to the stretching vibration of the C—O bond. This band is often not of use for assignment purposes since alcohols and ethers also absorb in this region.

Table 8.1. Peroxides and hydroperoxides

Functional Groups	Region cm^{-1}	Region μm	Intensity	Comments
Peroxides	900–830	11.11–12.05	w	O—O str
Alkyl peroxides	1150–1030	8.70–9.71	m–s	C—O str
Aryl peroxides	~1000	~10.00	m	C—O str
Peracids, peroxides of the type	~3450	~2.90	m	O—H str
G.CO.OO.H	1785–1755	5.60–5.70	vs	C=O str
	~1175	~8.51	m–s	C—O str
Aliphatic diacyl peroxides,	1820–1780	5.50–5.62	vs	C=O str, two bands
—CO—OO—CO—	1300–1050	7.69–9.52	m–s	C—O str
Aryl and unsaturated diacyl	1805–1755	5.54–5.70	vs	C=O str, two bands
peroxides	1300–1050	7.69—9.52	m–s	C—O str
Ozonides, 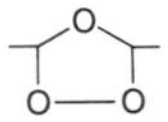	1065–1040	9.39–9.62	m	C—O str

References

1. D. Swern (ed.), *Organic Peroxides*, Vol. 2. Interscience, New York, 1971, pp. 683–697
2. H. A. Szymanski, *Prog. Infrared Spectrosc.*, 1967, **3**, 139
3. W. P. Keaveney *et al.*, *J. Org. Chem.*, 1967, **32**, 1537
4. M. A. Kovner *et al.*, *Optics Spectrosc.*, 1960, **8**, 64

CHAPTER 9

Amines, Imines, and Their Hydrohalides

Amine Functional Groups

Amine N—H stretching vibrations

As solids or liquids, in which hydrogen-bonding may occur, primary aliphatic amines[1–5] absorb in the region 3450–3250 cm^{-1} (2.90–3.08 μm). This is a broad band of medium intensity which may show structure depending on the hydrogen-bond polymers formed. In dilute solution in non-polar solvents, two bands are observed for primary amines due to the N—H asymmetric and symmetric vibrations. In the aliphatic case,[1,2] they are in the range 3550–3250 cm^{-1} (2.82–3.08 μm) whereas in the aromatic case[6–9] they are of medium intensity,[15] one at 3520–3450 cm^{-1} (2.84–2.90 μm) and the other at 3420–3350 cm^{-1} (2.92–2.99 μm). Various empirical relationships[3,10] between these bands have been proposed one of which is $\bar{\nu}_{sym} = 345.5 + 0.876\bar{\nu}_{asym}$ where the two N—H bonds of the primary amine are equivalent. For primary amines in the solid phase, the two bands are usually observed at approximately 100 cm^{-1} lower (0.09 μm higher) than for dilute non-polar solvent solutions.

Secondary amines[5,11,12] have only one N—H stretching band which is usually weak and occurs in the range 3500–3300 cm^{-1} (2.86–3.03 μm). In the solid and liquid phases, a band of medium intensity may be observed at 3400 3300 cm^{-1} (2.94–3.03 μm) for secondary aromatic amines.[12] As a result of hydrogen bonding, bands due to the N—H stretching vibrations may, in some solvents, be found as low as 3100 cm^{-1} (3.23 μm). In general, bands due to the N—H stretching vibration are sharper and weaker than, and do not occur in as wide a range as, those due to the O—H stretching vibration.

It is sometimes useful to convert tertiary amines into their hydrochlorides and then examine the resulting spectra for the presence of a band due to the N—H stretching vibration,[13,14] a technique which may also be found useful for distinguishing between imines and amines.

Amine N—H deformation vibrations

Primary amines have a medium-to-strong absorption band in the 1650–1580 cm^{-1} (6.06–6.33 μm) region and secondary amines have a weaker band at 1580–1490 cm^{-1} (6.33–6.71 μm). Primary aromatic amines normally absorb at 1615–1580 cm^{-1} (6.19–6.33 μm). Care must be taken since aromatic ring absorptions also occur in this general region.

Amines often exhibit a number of peaks when examined as pressed discs, due to a reaction with the dispersing agent and the formation of amine hydrohalides. Hydrogen bonding has the effect of moving the N—H deformation band to higher frequencies. This shift is dependent on the strength of the hydrogen bond. Primary amines have a broad absorption at 900–650 cm^{-1} (11.11–15.40 μm) which alters in shape and position depending on the amount of hydrogen bonding present.

Secondary aliphatic amines have an absorption in the range 750–700 cm^{-1} (13.33–14.29 μm).

Amine C—N stretching vibrations

The C—N stretching absorption of primary aliphatic amines is weak and occurs in the region 1090–1020 cm^{-1} (9.17–9.77 μm). Secondary aliphatic amines have two bands of medium intensity at 1190–1170 cm^{-1} (8.40–8.55 μm) and 1145–1130 cm^{-1} (8.73–8.85 μm).

For aromatic and unsaturated amines =C—N, two bands are observed at 1360–1250 cm^{-1} (7.36–8.00 μm) and 1280–1180 cm^{-1} (7.81–8.48 μm) due to conjugation of the electron pair of the nitrogen with the ring imparting double-bond character to the C—N bond, primary and secondary aromatic amines absorbing strongly in the first region. The C—N band for tertiary aromatic amines is found at 1380–1330 cm^{-1} (7.25–7.52 μm).

Imines with aliphatic groups attached to the nitrogen atom have a band near 1670 cm^{-1} (5.99 μm), with aromatic groups attached, this band is near 1640 cm^{-1} (6.10 μm) and, with extended conjugated groups, it is near 1620 cm^{-1} (6.17 μm).

Amine $>N—CH_3$ and $>N—CH_2—$ absorptions

A band of medium-to-strong intensity due to the stretching vibration of the C—H bond of the N—C—H group occurs near 2820 cm^{-1} (3.55 μm). This band is lower in frequency and more intense than ordinary alkyl bands and is therefore easily identified. Aliphatic amines with $—N(CH_3)_2$ have two bands, one near 2820 cm^{-1} (3.55 μm) and the other near 2770 cm^{-1} (3.61 μm).

Other amine bands

Primary aromatic amines (e.g. anilines) have a weak-to-medium intensity band at 445–345 cm^{-1} (22.47–28.99 μm) which is probably due to the in-plane deformation vibration of the aromatic ring–amine bond. For monosubstituted aminobenzenes with electron-donating substituents, this band is observed below

400 cm^{-1} (above 25.00 μm), whereas with electron-accepting or alkyl substituents in the ring, this band is above 400 cm^{-1}.

Primary alkyl amines have a strong absorption in the vicinity of 200 cm^{-1} (50.00 μm). It has been suggested that this band is due to torsional oscillations about the C—N bond and that the band which occurs at 495–445 cm^{-1} (20.20–22.47 μm) is an overtone of this band.

Table 9.1. Amine N—H stretching vibrations

Functional Groups	Region cm^{-1}	Region μm	Intensity	Comments
Primary amines, —NH_2	3550–3330	2.82–3.00	w–m	asym NH_2 str
(dilute solution spectra)	3450–3250	2.90–3.08	w–m	sym NH_2 str
Primary amines (condensed phase spectra)	3450–3250	2.90–3.08	w–m	br
Secondary aliphatic amines, >NH	3500–3300	2.86–3.03	w	
Secondary aromatic amines	3450–3400	2.90–2.94	m	Greater intensity than aliphatic compounds
N—D (free)	2600–2400	3.85–4.15	w	
Diamines (see ref. 16)	3360–3340	2.98–2.99	w–m	asym N—H str
(condensed phase)	3280–3270	3.05–3.06	w–m	sym N—H str
Imines, C=NH (see ref. 17)	3400–3300	2.94–3.03	m	

Table 9.2. Amine N—H deformation vibrations

Functional Groups	Region cm^{-1}	Region μm	Intensity	Comments
Primary amines	1650–1580	6.06–6.33	m–s	Aromatic amines at lower end of frequency range
	900–650	11.11–15.40	s	br, N—H out-of-plane bending, usually multiple bands
Primary aliphatic amines, R—CH_2NH_2 and	850–810	11.76–12.35	m–s	
$R_1R_2R_3CNH_2$	795–760	12.58–13.10	m	
Primary aliphatic amines,	~830	~12.05	s	
R_1R_2>CHNH$_2$	~795	~12.58	s	
Secondary amines	1580–1490	6.33–6.71	w	May be masked by an aromatic band at ~1580 cm^{-1}
	750–700	13.33–14.29	s	br, N—H wagging
Secondary aliphatic amines: R_1—CH_2—NH—CH_2—R_2 and $R_1R_2R_3$C—NH—C$R_4R_5R_6$	750–710	13.33–1408	s	
R_1R_2CH—NH—RHR_3R_4	735–700	13.61–14.29	s	
Imines, >C=N—H	1590–1500	6.29–6.67	m	N—H bending

Table 9.3. Amine C—N stretching vibrations

Functional Groups	Region cm^{-1}	Region μm	Intensity	Comments
Primary aliphatic amines	1090–1020	9.17–9.80	w–m	General range
Primary aliphatic amines, —CH_2—NH_2	1090–1065	9.17–9.39	m	
Primary aliphatic amines,	1140–1080	8.77–9.26	w–m	
>CH—NH_2	1045–1035	9.57–9.66	w	
Primary aliphatic amines,	1240–1170	8.07–8.55	w–m	
⩾CNH_2	1040–1020	9.62–9.80	w	
Secondary aliphatic amines	1190–1170	8.40–8.55	m	General range
	1145–1130	8.73–8.85	m	
Secondary aliphatic amines, —CH_2—NH—CH_2—	1145–1130	8.73–8.85	m–s	
Secondary aliphatic amines, >CH—NH—CH<	1190–1170	8.40–8.55	m	
Tertiary aliphatic amines	1230–1030	8.13–9.71	m	General range, doublet
Tertiary aliphatic amines,	1210–1150	8.25–8.70	m	
(—CH_2)$_3$N	1100–1030	9.10–9.70	m	
Tertiary dimethyl amines,	~1270	~7.87	m	
$(CH_3)_2$N—CH_2—	~1190	~8.40	m	
	~1040	~9.62	m	
Tertiary diethyl amines,	~1205	~8.30	m	
$(C_2H_5)_2$N—C⩽	~1070	~9.35	m	
Primary and secondary aromatic amines	1360–1250	7.36–8.00	s	X-sensitive band (see p. 138)
Tertiary aromatic amines, N⩽	1380–1330	7.25–7.52	s	
Imines, >C=N—	1690–1640	5.92–6.10	v	C=N str

Table 9.4. Amines: other vibrations

Functional Groups	Region cm^{-1}	Region μm	Intensity	Comments
>N—CH_3 and >N—CH_2	2820–2760	3.55–3.62	m–s	C—H str
CH_3—N<	1370–1310	7.30–7.64	m	
Primary aliphatic amines	495–445	20.20–22.47	m–s	Broad
	~290	~34.48	s	Broad
Primary aromatic amines	445–345	22.47–28.99	w	In-plane bending of aromatic —NH_2 bond
Secondary aliphatic amines	455–405	21.98–24.69	w–m	C—N—C def
Tertiary aliphatic amines	510–480	19.61–20.83	s	
>C=C(—)—N<	1680–1630	5.95–6.14	m–s	C=C str, usually more intense than normal C=C str

Amine Hydrohalides,[13, 14] $-NH_3^+$, $>NH_2^+$, $\geqslant NH^+$, and Imine Hydrohalides, $>C{=}NH^+-$

Amine hydrohalide N—H$^+$ stretching vibrations

In the solid phase, amine hydrohalides containing $-NH_3^+$ have an absorption of medium intensity at 3350–3100 cm^{-1} (2.99–3.23 μm) due to stretching vibrations. Depending on the amount of hydrogen bonding, a number of bands may appear in this region. Also in the solid phase, amines with $>NH_2^+$, $\geqslant NH^+$, and $C{=}NH^+-$ have a broad absorption of medium intensity at 2700–2250 cm^{-1} (3.70–4.44 μm). In dilute solution using non-polar solvents, the stretching vibrations of $-NH_3^+$ result in two bands, one near 3380 cm^{-1} (2.96 μm) and the other near 3280 cm^{-1} (3.05 μm), the stretching vibrations of $>NH_2^+$ result in strong bands at 3000–2700 cm^{-1} (3.33–3.70 μm), those of $\geqslant NH^+$ result in a weak-to-medium intensity band at 2200–1800 cm^{-1} (4.55–5.56 μm) while for $>C{=}NH^+-$, a strong absorption is the result at 2700–2330 cm^{-1} (3.70–4.29 μm). Quaternary salts have no characteristic absorption bands.

Amine hydrohalide N—H$^+$ deformation vibrations

Amine $-NH_3^+$ groups have medium-to-strong absorptions near 1600 cm^{-1} (6.25 μm) and 1500 cm^{-1} (6.67 μm) due to asymmetric and symmetric deformation vibrations. Secondary amine hydrohalides have only one band which is near to 1600 cm^{-1} (6.25 μm). Unfortunately, aromatic ring C=C stretching vibrations also give rise to bands in this general region so that care must be exercised in interpretations.

Amine and imine hydrohalides: other bands

Other relevant bands have, of course, been discussed in the previous section dealing with uncharged amines and this should be referred to.

Primary amine hydrohalides have a number of sharp bands in the region 2800–2400 cm^{-1} (3.57–4.15 μm), and a band around 2000 cm^{-1} (5.00 μm) which is believed to be a combination band involving NH_3^+ deformation vibrations. Secondary amine hydrohalides have two sharp bands, at about 2500 cm^{-1} (4.00 μm) and 2400 cm^{-1} (4.15 μm). Primary amine hydrohalides also absorb in the region 1280–1050 cm^{-1} (7.81–9.10 μm) due mainly to the rocking vibration of the NH_3^+ group. Most amine hydrohalides have in addition a medium-to-strong band at about 1120 cm^{-1} (8.93 μm).

Imine hydrohalides have one or more bands of medium intensity in the region 2200–1800 cm^{-1} (4.55–5.56 μm) which may be used to distinguish them clearly from amine hydrohalides.

Table 9.5. Amine and imine hydrohalide N—H$^+$ stretching vibrations

Functional Groups	Region cm^{-1}	Region μm	Intensity	Comments
$-NH_3^+$	3350–3100	2.99–3.23	m	br, solid phase spectra
$>NH_2^+$, $\geqslant NH^+$, $>C{=}NH^+-$	2700–2250	3.70–4.44	m	br, sometimes a group of sharp bands, solid phase spectra
$-NH_3^+$	~3380	~2.96	m	asym str, dilute solution spectra
	~3280	~3.05	m	sym str, dilute solution spectra
$>NH_2^+$	3000–2700	3.33–3.70	m–s	Dilute solution spectra, two bands
$\geqslant NH^+$	2200–1800	4.55–5.56	w–m	Dilute solution spectra
$>C{=}NH^+-$	2700–2330	3.70–4.29	m–s	Dilute solution spectra, overtone bands occur at 2500–2300 cm^{-1}
Ammonium salts, NH_4^+	3300–3030	3.03–3.30	s	br

Table 9.6. Amine and imine hydrohalide N—H$^+$ deformation and other vibrations

Functional Groups	Region cm^{-1}	Region μm	Intensity	Comments
$-NH_3^+$	~2500	~4.00	w	Overtone (sometimes absent)
	~2000	~5.00	w	Overtone (sometimes absent)
	1625–1560	6.15–6.41	m	asym NH_3^+ def
	1550–1500	6.45–6.67	m	sym NH_3^+ def
	~800	~12.50	w	NH_3^+ rocking vib
$>NH_2^+$	~2000	~5.00	w	Overtone (sometimes absent)
	1620–1560	6.17–6.41	m–s	
	~800	~12.50	w	NH_2^+ rocking vib
Imine, $>C{-}N^+{-}H$	2200–1800	4.55–5.56	m	One or more bands
	~1680	~5.95		C=N$^+$ str
Ammonium salts, NH_4^+	1430–1390	6.99–7.19	s	N—H def

References

1. H. J. Bernstein, *Spectrochim. Acta*, 1962, **18,** 161
2. W. J. Orville-Thomas *et al.*, *J. Chem. Soc.*, **1958,** 1047
3. L. J. Bellamy and R. L. Williams, *Spectrochim. Acta*, 1957, **9,** 341
4. E. V. Titov and M. V. Poddubnaya, *Teor. Eksp. Khim.*, 1972, **8,** 276
5. J. E. Stewart, *J. Chem. Phys.*, 1959, **30,** 1259
6. S. F. Mason, *J. Chem. Soc.*, **1958,** 3619
7. P. J. Kreuger and W. H. Thompson, *Proc. Roy. Soc.*, 1957, **A243,** 143
8. P. J. Kreuger, *Nature*, 1962, **194,** 1077
9. A. Bryson, *J. Amer. Chem. Soc.*, 1960, **82,** 4862
10. A. I. Finkelshtejn, *Optics Spectrosc.*, 1962, **12,** 454
11. L. K. Dyall and J. E. Kemp, *Spectrochim. Acta*, 1966, **22,** 467
12. A. G. Moritz, *Spectrochim. Acta*, 1960, **16,** 1176
13. B. Chenon and C. Sandorfy, *Canad. J. Chem.*, 1958, **36,** 1181
14. L. Segal and F. V. Eggeston, *App. Spectrosc.*, 1961, **15,** 112
15. A. S. Wexler, *App. Spectrosc. Rev.*, 1968, **1,** 29
16. M. E. Baldwin, *Spectrochim. Acta*, 1962, **18,** 1455
17. J. Fabian, *Bull. Soc. Chim. Fr.*, **1956,** 1499

CHAPTER 10

The Carbonyl Group: C=O

Introduction

The carbonyl group is contained in a large number of different classes of compounds, e.g. aldehydes, ketones, carboxylic acids, esters, amides, acid anhydrides, acid halides, lactones, urethanes, lactams, etc., for which a strong absorption band due to the C=O stretching vibration[1,2] is observed in the region 1850–1550 cm^{-1} (5.41–6.45 μm). Because of its high intensity[3,10] and the relatively interference-free region in which it occurs, this band is reasonably easy to recognize.

The frequency of this carbonyl stretching vibration is dependent on various factors:

1. The structural environment of the C=O group.
   (a) The more electronegative an atom or group directly attached to a carbonyl group, the greater is the frequency.
   (b) Unsaturation[11–17] in the α,β- position tends to decrease the frequency, except for amides which are little influenced, by 15–40 cm^{-1} from that expected without the conjugation, further conjugation having little effect on the frequency.
   (c) Hydrogen bonding[18–26] to the C=O results in a decrease in the frequency of 40–60 cm^{-1}, this being independent of whether the bonding is inter- or intramolecular.
   (d) In situations where ring strain occurs, the greater the strain, i.e. the smaller the ring, the greater is the frequency.[21,27–34]
2. The physical state of the sample. In the solid phase, the frequency of the vibration is slightly decreased compared with that in dilute non-polar solutions.[35,36] The presence of hydrogen bonding is an important contributing factor to this decrease in frequency.

In cases where more than one of these influences is present, the net shift in the position of the band due to the C=O stretching vibration appears to be the result of an approximately additive process, although this does not always hold in cases where hydrogen bonding to the C=O group is present.

If the double-bond character of the carbonyl group is increased (i.e. the force constant of the bond is increased) by its neighbouring groups, then the frequency of the stretching vibration is increased (i.e. the wavelength is decreased). If the presence of an adjacent group results in resonance hybrids, such as **II**(below), making a greater contribution, then this will tend to decrease the double-bond character of the carbonyl group and hence decrease the frequency of the carbonyl stretching vibration:

I **II**

On the other hand, an electron-accepting group tends, through the inductive effect, to increase the double-bond character and hence increases the frequency of the vibration:

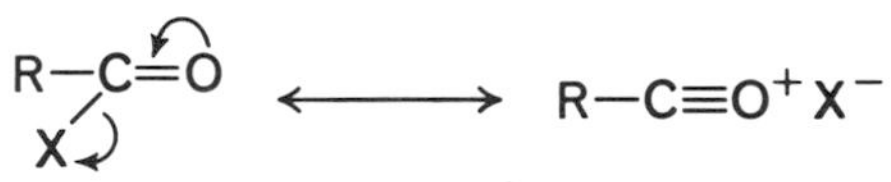

Hence, for a particular group, these two opposite effects determine the frequency of the vibration and it is therefore possible, in general, to give an approximate order for the C=O bond stretching vibration frequency for different groups:

$$RCOO^- < RCONH_2 < \underset{\text{dimer}}{(RCOOH)_2} < RCOR' < RCHO < RCOOR' < \underset{\text{monomer}}{RCOOH} < RCOOCOR'$$

Hydrogen bonding tends to decrease the double-bond character of the carbonyl group, therefore shifting the absorption band to lower frequency:

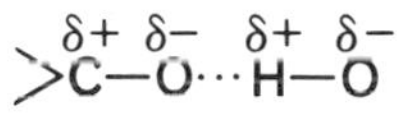

For example, the C=O stretching vibration band of aliphatic carboxylic acids as monomers appears near 1760 cm^{-1} (5.68 μm), but as dimers (which are predominant in liquid and solid samples) the band occurs near 1700 cm^{-1} (5.88 μm). For this hydrogen-bonded dimer, a characteristic broad band is observed at about 920 cm^{-1} (10.87 μm) due to its out-of-plane bending vibration.

Due to hydrogen bonding with solvents such as chloroform, the band due to the carbonyl stretching vibration of ketones may be split by about 5–10 cm^{-1} (0.02–0.04 μm).

If a carbonyl group is part of a conjugated system, then the frequency of the carbonyl stretching vibration decreases, the reason being that the double-bond character of the C=O group is less due to the π-electron system being delocalized.

For *meta*- and *para*-substituted aromatic carbonyl compounds, a linear relationship exists between the carbonyl absorption frequency and the Hammett reactivity constant.[2,5,18,37,38] A relationship between the carbonyl stretching vibration frequency of aromatic carbonyl compounds and the pK[39,40] of the

Chart 4 The band positions of compounds containing the carbonyl group (together with carboxylic acids and their salts, etc.) are given. All these bands are of very strong intensity.

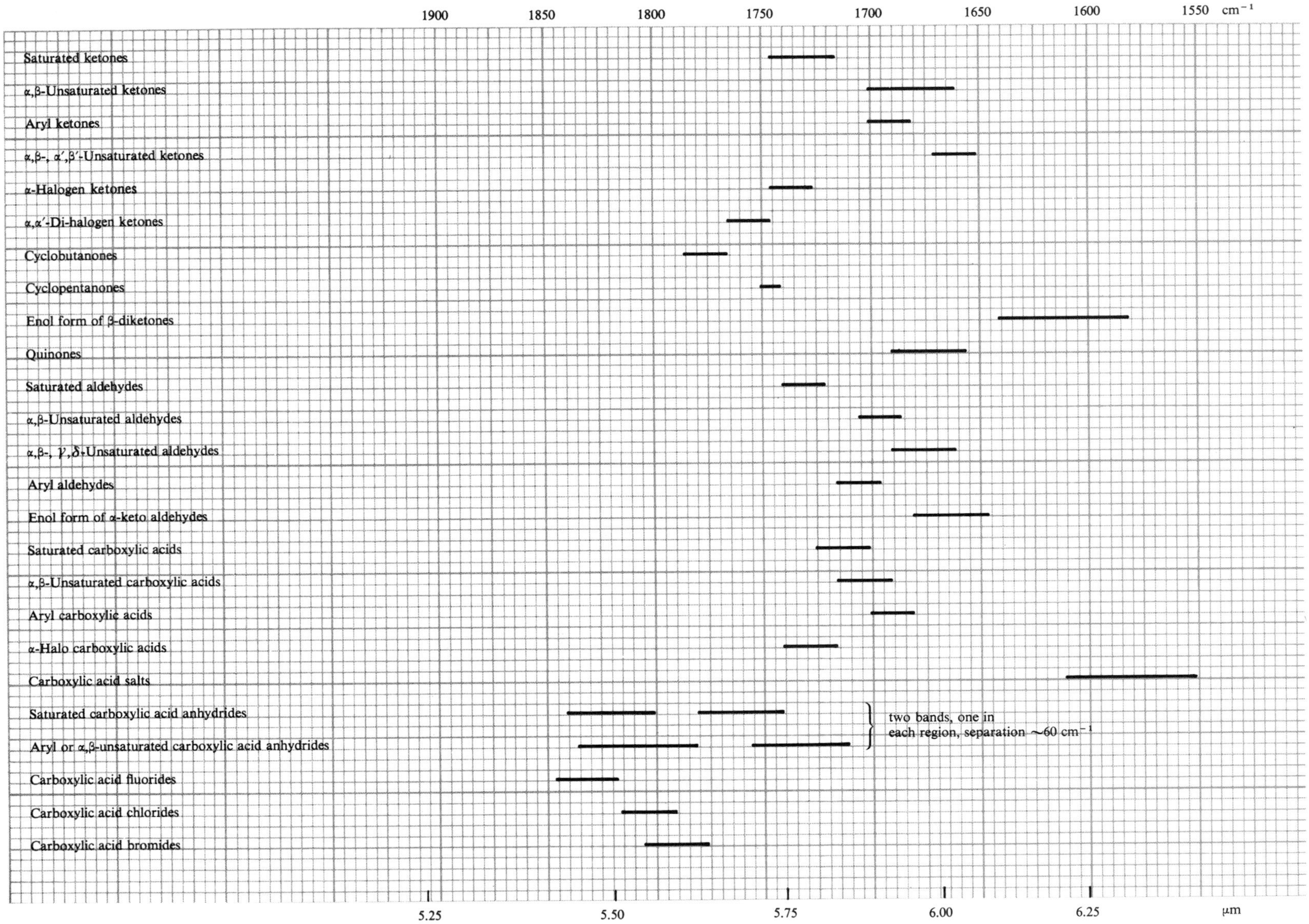
1900
1850
1800
1750
1700
1650
1600
1550 cm⁻¹
Saturated ketones
α,β-Unsaturated ketones
Aryl ketones
α,β-, α′,β′-Unsaturated ketones
α-Halogen ketones
α,α′-Di-halogen ketones
Cyclobutanones
Cyclopentanones
Enol form of β-diketones
Quinones
Saturated aldehydes
α,β-Unsaturated aldehydes
α,β-, γ,δ-Unsaturated aldehydes
Aryl aldehydes
Enol form of α-keto aldehydes
Saturated carboxylic acids
α,β-Unsaturated carboxylic acids
Aryl carboxylic acids
α-Halo carboxylic acids
Carboxylic acid salts
Saturated carboxylic acid anhydrides
Aryl or α,β-unsaturated carboxylic acid anhydrides
two bands, one in each region, separation ~60 cm⁻¹
Carboxylic acid fluorides
Carboxylic acid chlorides
Carboxylic acid bromides
5.25
5.50
5.75
6.00
6.25
μm

Chart 4 (*continued*)

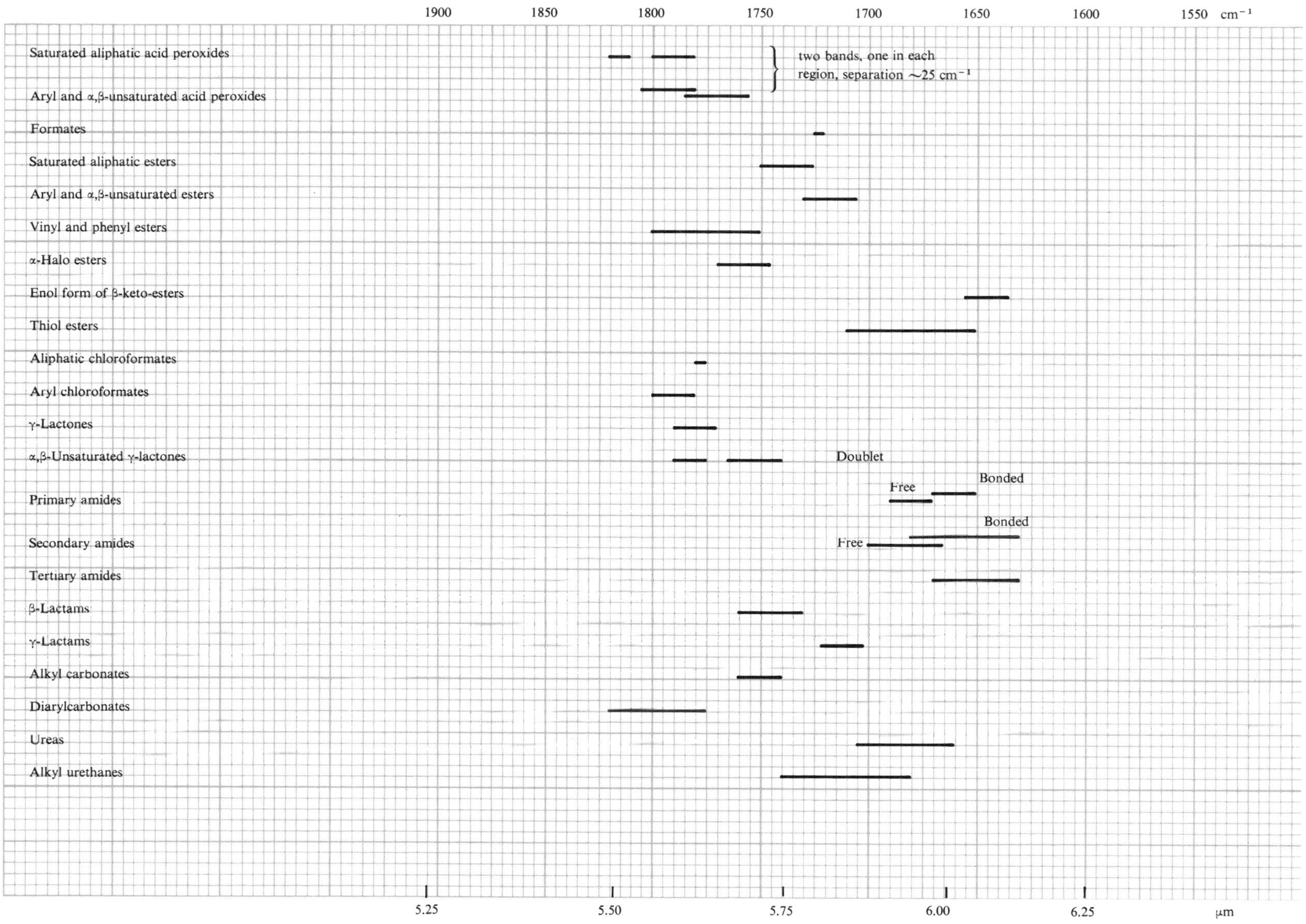
1900
1850
1800
1750
1700
1650
1600
1550 cm⁻¹
Saturated aliphatic acid peroxides
two bands, one in each region, separation ~25 cm⁻¹
Aryl and α,β-unsaturated acid peroxides
Formates
Saturated aliphatic esters
Aryl and α,β-unsaturated esters
Vinyl and phenyl esters
α-Halo esters
Enol form of β-keto-esters
Thiol esters
Aliphatic chloroformates
Aryl chloroformates
γ-Lactones
α,β-Unsaturated γ-lactones
Doublet
Primary amides
Free
Bonded
Secondary amides
Free
Bonded
Tertiary amides
β-Lactams
γ-Lactams
Alkyl carbonates
Diarylcarbonates
Ureas
Alkyl urethanes
5.25
5.50
5.75
6.00
6.25
μm

corresponding aromatic carboxylic acid has been demonstrated. Correlations with other parameters, such as electronegativities,[41] ionization potentials, Taft σ^* values, half-wave potentials, etc., have also been made.[42,43] For aromatic compounds with *ortho*- substituents, a combination of factors may be important, such as chelation, steric effects, and field effects (dipole interactions through space).

In detailed spectral studies of carbonyl compounds in which conjugation with an olefinic group occurs, geometrical isomerism must be taken into account.[13,16,17] α,β-Unsaturated carbonyl compounds have a contribution from the $\overset{\beta}{C}^{+}{-}\overset{\alpha}{C}{=}C{-}O^{-}$ form in addition to the form $\overset{\beta}{C}{=}\overset{\alpha}{C}{-}C{=}O$. Some partial double-bond character exists between the C=O and the α,β-unsaturated C=C bond. Hence, geometrical isomerism about this 'single' bond is possible, resulting in s-*trans* and s-*cis* forms, where the s indicates restricted rotation about a single bond:

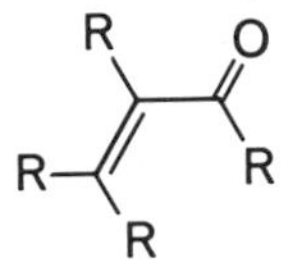
s-*trans*

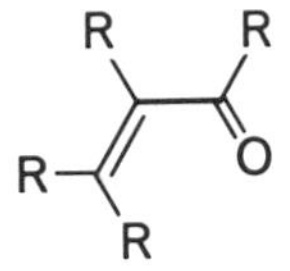
s-*cis*

If the R groups are different, then various s-*trans* and s-*cis* forms may exist, e.g.

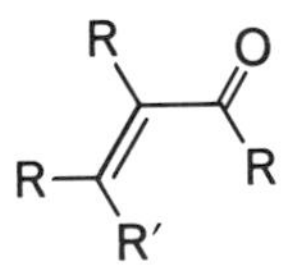

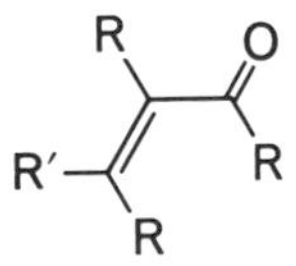

Different s-*trans* forms

In the case of two olefinic groups conjugated to a carbonyl group, configurations are possible such as

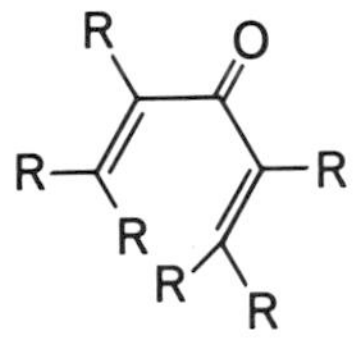
s-*trans*, s-*trans*

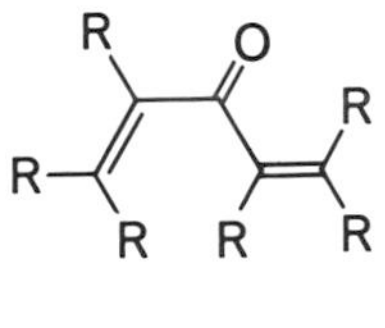
s-*trans*, s-*cis*

α-Dicarbonyl[44,45] compounds may exist in two configurations, cisoid and transoid:

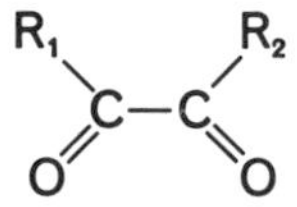
cisoid

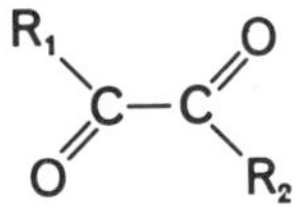
transoid

In the cisoid conformation, a degree of interaction between the dipoles of the two carbonyl groups would be expected which would result in an increase in the carbonyl character or possibly result in enolization. However, for acyclic α-dicarbonyl compounds, no such interaction is observed: the carbonyl stretching vibration frequency is virtually the same as for the equivalent monocarbonyl compound. This can be explained if the α-dicarbonyl substances exist in the more energetically-favoured transoid conformation. For the symmetric stretching vibration, the dipole interactions of the two carbonyls would be cancelled. The symmetric stretching mode is infrared inactive since there would be no net change in the dipole moment during vibration. In the case of cyclic α-dicarbonyls, the two C=O groups are held, depending on the ring size, more-or-less rigidly in the cisoid conformation. This results in these cyclic compounds with smaller rings having a marked tendency to enolization.

The great difference between the spectra of a carboxylic acid and its salt may be useful when doubt exists as to whether or not a C=O band should be attributed to a carboxyl group.

In general, for the C=O stretching vibration band, acids absorb more strongly than ketones, aldehydes, or amides. The intensity of the C=O absorptions of ketones and aldehydes is approximately the same, whereas that of amides may vary greatly.

A relatively small number of compounds containing only one carbonyl group has more than one band due to the carbonyl stretching vibration, examples being benzoyl chloride,[46] cyclopentanone,[47–49] cyclopent-2-enone, ethylene carbonate and certain α,β-unsaturated lactones (five- and six-membered rings)[50,51] and lactams. It would seem that Fermi resonance is responsible for this doubling of the carbonyl band.[52,72,170]

Fermi resonance occurs if the energy associated with a combination or an overtone band coincides approximately with that for a fundamental energy level of a different vibration. This may be thought of as a to-and-fro transfer of energy between the two levels. Fermi resonance results in two bands of similar intensity almost equidistant from the position at which the fundamental and combination bands would have occurred. These doublets are, of course, concentration-independent but may depend on temperature and solvent polarity.

With the exception of thioacids, the carbonyl stretching vibration frequency of thiol compounds[53,54] is found approximately 40 cm^{-1} lower (0.15 μm higher) than that of the corresponding oxygen compound. Similarly, dithiol carbonates have bands which are about 80 cm^{-1} lower (0.35 μm higher) than for the corresponding —O—CO—O— compound.

Ketones >C=O

Ketone C=O stretching vibrations

Ketones and aldehydes have almost identical carbonyl absorption frequencies. Aldehydes usually absorb at about 10 cm^{-1} higher (0.03 μm lower) than the corresponding ketone.

Saturated aliphatic ketones[43] and cyclic ketones (six-membered rings and greater) in the pure liquid and solid phases absorb strongly in the range 1725–1705 cm^{-1} (5.80–5.86 μm). In dilute solution in non-polar solvents, the absorption occurs at 1745–1715 cm^{-1} (5.73–5.83 μm). Therefore, in general, in the solid phase, the frequency of the C=O stretching vibration is 10–20 cm^{-1} lower than that observed in dilute solutions using non-polar solvents. In non-polar solvents, aryl ketones[42, 55] absorb at 1700–1680 cm^{-1} (5.88–5.95 μm), diaryl ketones at 1670–1650 cm^{-1} (5.99–6.06 μm), α,β-unsaturated ketones ($>\overset{\beta}{C}{=}\overset{\alpha}{C}{-}C{=}O$) at 1700–1660 cm^{-1} (5.88–6.02 μm), α-halo-ketones at 1750–1725 cm^{-1} (5.71–5.80 μm),[4, 56, 57] and α,α′-dihalo-ketones at 1765–1745 cm^{-1} (5.66–5.73 μm).[58–60] α-Chloro-ketones absorb at the higher frequencies if the chlorine atom is near the oxygen and at the lower values if away from it.[4] In the case of α,β-unsaturation, the C=C stretching vibration frequency is also reduced. The aromatic band near 1600 cm^{-1} (6.25 μm) usually appears as a doublet and the band near 1500 cm^{-1} (6.67 μm) can be very weak.

Enolized β-diketones[44, 61, 63, 67] have a very strong band in the region of 1610 cm^{-1} (6.21 μm) (the band due to the C=C stretching vibration being at 1520–1500 cm^{-1}). For α-diketones, a single band is observed at a slightly higher frequency than that expected for the single ketone. Unsymmetrical *para*-substituted benzils have two bands at 1690–1660 cm^{-1} (5.92–6.02 μm). *Ortho*-hydroxy or *ortho*-amino-aryl ketones[64] exhibit a strong band in the region 1655–1610 cm^{-1} (6.04–6.21 μm) due to the carbonyl stretching vibration. The presence of intramolecular hydrogen bonding causes this frequency to be lower than might otherwise be expected.

As mentioned previously, the band due to the C=O stretching vibration is shifted from its expected position by a number of parameters, these influences being approximately additive in their effect. The approximate magnitude of these shifts is given below.

Table 10.1. Influence on C=O stretching vibration for ketones and aldehydes

	Wavenumber shift/cm^{-1}	Wavelength shift/μm
α,β-Unsaturation	−30	+0.11
α-Halogen	+20	−0.07
α,α′-Dihalogen	+40	−0.15
α,α-Dihalogen	+20	−0.07
Solid phase	−20	+0.07

Methyl and methylene deformation vibrations in ketones

For the group $-CH_2-CO-$, the methylene scissoring vibration occurs in the range 1435–1405 cm^{-1} (6.97–7.12 μm).[65] This is lower than that for CH_2 in aliphatic hydrocarbons which occurs in the range 1480–1440 cm^{-1} (6.76–6.94 μm). For methyl groups adjacent to carbonyl groups, the symmetrical C—H bending vibration has a lower frequency, 1360–1355 cm^{-1} (7.35–7.38 μm), than that for aliphatic hydrocarbons, 1390–1370 cm^{-1} (7.19–7.30 μm). Ketones with the structure $-CH_2-CO-CH_2-$ have a medium-intensity band at 1230–1100 cm^{-1} (8.13–9.09 μm) due to the asymmetric stretching vibration of the backbone. For methyl ketones, this band is near 1170 cm^{-1} (8.55 μm).

Ketone skeletal and other vibrations

A band of medium-to-strong intensity due to the C—C stretching vibration may be found at 1325–1115 cm^{-1} (7.55–8.95 μm) for aliphatic ketones[65] and at 1225–1075 cm^{-1} (8.16–9.30 μm) for aromatic ketones. However, this band is not normally used for assignment purposes.

Due to the in-plane deformation of the C—CO—C group, aliphatic ketones have a strong absorption at 630–620 cm^{-1} (15.87–16.13 μm)[66] which is shifted to lower frequencies, 580–565 cm^{-1} (17.24–17.70 μm), if α-branching occurs.

Aliphatic ketones have an absorption at 540–510 cm^{-1} (18.52–19.61 μm) which is due to C—C=O deformation. This band is shifted to 560–550 cm^{-1} (17.86–18.18 μm) if α-branching occurs. Small-ring cyclic ketones absorb strongly at 505–480 cm^{-1} (19.80–20.83 μm).[66]

With the exception of acetone and α-branched compounds, methyl ketones have prominent bands at about 1355 cm^{-1} (7.38 μm) and at about 1170 cm^{-1} (8.55 μm), the former band being due to CH_3 deformation vibrations.[65] Methyl ketones generally (including aromatic methyl ketones) have a strong absorption at 600–580 cm^{-1} (16.67–17.24 μm) which is due to the in-plane deformation vibration of the C—CO—C group. Other aromatic ketones also exhibit this absorption band.

Table 10.2. Ketone C=O stretching vibrations

Functional Groups	Region cm^{-1}	Region μm	Intensity	Comments
Saturated aliphatic ketones	1745–1715	5.73–5.83	vs	
Aryl ketones	1700–1680	5.88–5.95	vs	
Diaryl ketones	1670–1650	5.99–6.06	vs	See ref. 55
α,β-Unsaturated ketones,	1700–1660	5.88–6.02	vs	General range
>C=C—C=O (β, α)	1690–1660	5.92–6.02	vs	s-*trans* form (C=C str, 1645–1615 cm^{-1})
	1700–1685	5.88–5.93	vs	s-*cis* form (C=C str, 1625–1615 cm^{-1})
α,β-unsaturated, β-amino ketones, *cis* form	1640–1600	6.10–6.25	s	Intramolecular hydrogen bonding occurs. *Trans* form has no hydrogen bonding and carbonyl band occurs in normal range
α,β-, α′,β′-Di-unsaturated ketones, >C=C—CO—C=C<	1670–1650	5.99–6.06	vs	
α-Halo-ketones	1750–1725	5.71–5.80	vs	
α,α′-Dihalo-ketones	1765–1745	5.66–5.73	vs	
Keto form of β-diketones, —CO—C—CO—	~1700	~5.88	vs	
Enol form of β-diketones, 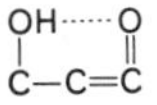	1640–1580	6.10–6.33	vs	br, extremely strong band (other bands at ~1500 cm^{-1}, ~1450 cm^{-1}, ~1260 cm^{-1}; O—H str, 3000–2700 cm^{-1})
Cyclic β-ketones, enol form	1630–1610	6.14–6.21	s	
α-Diketones, —CO—CO—	1730–1705	5.78–5.86	vs	
—CO.O.CH$_2$CO—	1745–1725	5.73–5.80	vs	
Cyclopentanone derivatives	1750–1740	5.71–5.75	vs	Fermi resonance doublet
Cyclobutanone derivatives	1790–1765	5.59–5.67	vs	
Cyclopropenones	1870–1845	5.35–5.42	vs	C=C and C=O in-phase and out-of-phase str; as mass of substituents increases, band at ~1475 cm^{-1}, mainly due to C=C str, disappears and strong band at 1655–1620 cm^{-1} appears instead
	1655–1620	6.04–6.17	s	
Cyclopropanones	~1820	~5.49	vs	Liquid phase (vapour phase, ~1905 cm^{-1})
o-Hydroxy-, and *o*-amino-aryl ketones	1655–1610	6.04–6.21	vs	Intramolecular hydrogen bonding occurs
β-Diketones, metal chelates	1600–1560	6.25–6.41	m–s	position dependent on complex stability; All four bands due to C⋯O and C⋯C stretching vibrations
	1530–1500	6.54–6.67	m–s	
	1450–1370	6.90–7.30	m–s	
	~1250	~8.00	m–s	
Flavones, 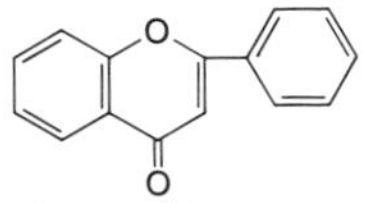	1670–1625	5.99–6.15	vs	
Cyclopropyl ketones, 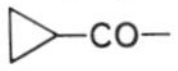	1705–1685	5.86–5.94	vs	
Aliphatic silyl ketones, R—CO—Si≤	1645–1635	6.08–6.12	vs	
Ar—CO—Si≤	~1620	~6.17	vs	

Table 10.3. Ketones: other bands

Functional Groups	Region cm^{-1}	Region μm	Intensity	Comments
>C=O	3550–3200	2.82–3.13	w	C=O stretching vibration overtone
Aliphatic ketones (straight chain)	630–620	15.87–16.13	s	C—CO—C in-plane def
	540–510	18.52–19.61	s	C—CO in-plane def
α-Branched aliphatic ketones	580–565	17.24–17.70	m–s	C—CO—C in-plane def
	560–550	17.86–18.18	v	C—CO in-plane def
Small-ring cyclic ketones	505–480	19.80–20.83	s	C—CO in-plane def
Methyl ketones	1360–1355	7.35–7.38	s	CH$_3$ def
	600–580	16.67–17.24	s	C—CO—C in-plane def
Aliphatic methyl ketones	530–510	18.87–19.61	m–s	C—CO in-plane def
	420–385	23.81–25.97	v	C—CO out-of-plane def
Aromatic methyl ketones	600–580	16.67–17.24	s	C—CO—C in-plane def
—CH$_2$CO—	1435–1405	6.97–7.12	s	CH$_2$ def
Alkyl ketones	1325–1215	7.55–8.23	s	
	~1100	~9.09	m	Generally several bands
Aryl ketones	1225–1075	8.16–9.30	s	Phenyl–carbon str
	~1300	~7.69	m	C—C—C bending and C—CO—C bending, generally several bands

Quinones 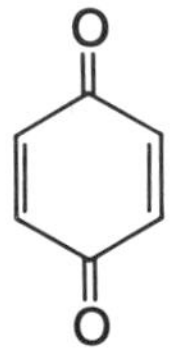 and 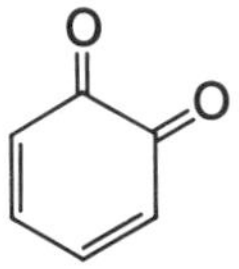

Either one or two carbonyl absorption bands may be observed for *para*-quinones, the range being 1690–1655 cm^{-1} (5.92–6.04 μm), even though only one might be expected from symmetry considerations.[68–72] On the other hand, *ortho*-quinones exhibit only one carbonyl band, which is in the same range although usually at about 1660 cm^{-1} (6.02 μm). The carbonyl absorption frequency of polycyclic quinones increases with the number of fused rings. Quinones with electronegative substituents absorb at the higher end of the frequency range given.

In the absence of hydroxyl and amino- groups, anthraquinones[73] absorb strongly in the region 1680–1650 cm^{-1} (5.95–6.06 μm) due to the carbonyl group. The presence of hydroxyl and amino- groups results in a lowering of this frequency. Charge-transfer complexes of benzoquinone and hydroquinone have been dealt with.[74]

Table 10.4. Quinone C=O stretching vibrations

Functional Groups	Region cm⁻¹	μm	Intensity	Comments
Quinones	1690–1655	5.92–6.04	vs	One or two bands
Polycyclic quinones	1655–1635	6.04–6.03	vs	
Anthraquinones (absence of OH and NH_2 groups)	1680–1650	5.95–6.06	vs	
1-Hydroxyl anthraquinones	1675–1645	6.01–6.08	vs	
	1640–1620	6.10–6.17	vs	
1,4- or 1,5-dihydroxyl anthraquinones	1645–1605	6.08–6.23	vs	
1,8-Dihydroxyl anthraquinones	1680–1660	5.95–6.02	vs	
	1625–1615	6.16–6.19	vs	
1,4,5-Trihydroxyl anthraquinones	1615–1590	6.19–6.29	vs	
1,4,5,8-Tetrahydroxyl anthraquinones	1590–1570	6.29–6.37	vs	
Tropones	1600–1575	6.25–6.35	vs	
Tropolones, 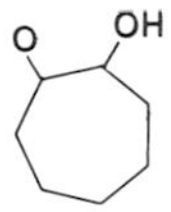	1620–1600	6.17–6.25	vs	Intramolecular bonding to CO group

Table 10.5. Quinone C—H out-of-plane deformation vibrations

Functional Groups	Region cm⁻¹	μm	Intensity	Comments
Monosubstituted	915–900	10.93–11.11	w–m	
p-benzoquinones	865–825	11.56–12.12	m–s	
2,3-Disubstituted *p*-benzoquinones	860–800	11.63–12.50	s	
2,5- and 2,6-disubstituted *p*-benzoquinones	920–895	10.87–11.17	s	

Aldehydes, —CHO

Aldehyde C=O stretching vibrations

The C=O stretching vibration is influenced in a similar manner to that observed for ketones (see earlier). In non-polar solvents, saturated aliphatic aldehydes absorb strongly in the region 1740–1720 cm^{-1} (5.75–5.82 μm),[75,76] aryl aldehydes at 1715–1695 cm^{-1} (5.83–5.90 μm),[77,78] and α,β-unsaturated aliphatic aldehydes at 1705–1685 cm^{-1} (5.87–5.93 μm), with additional unsaturation lowering the frequency only slightly (approximately 5–10 cm^{-1}). In the solid or liquid phase, the absorption frequencies are lowered by 10–20 cm^{-1} compared with those for dilute solution in non-polar solvents. A study has been made of the temperature dependence of the acetaldehyde C=O stretching vibration.[169]

Aldehydic C—H vibrations

Two bands are usually observed due to the stretching vibrations of the aldehydic C—H, both of which are of weak-to-medium intensity, one at about 2820 cm^{-1} (3.55 μm) and the other in the region 2745–2650 cm^{-1} (3.64–3.77 μm).

Benzaldehydes with bulky *ortho*- substituents such as nitro-, halogen, or methoxy groups absorb at 2900–2860 cm^{-1} (3.45–3.50 μm) and 2765–2745 cm^{-1} (3.62–3.65 μm). Otherwise, aryl aldehydes absorb at 2830–2810 cm^{-1} (3.53–3.56 μm) and 2745–2720 cm^{-1} (3.65–3.68 μm).

The presence of a sharp band at about 2720 cm^{-1} (3.68 μm) and a band due to the carbonyl stretching vibration in the region 1740–1685 cm^{-1} (5.75–5.95 μm) may usually be taken as indicating the presence of an aldehyde.

The presence of two bands in the region 2830–2650 cm^{-1} (3.53–3.77 μm) is probably due to an interaction between the C—H stretching vibration and the overtone of the C—H bending vibration near 1390 cm^{-1} (7.19 μm). This may involve Fermi resonance since aldehydes for which the latter band is shifted have only one band, this being in the region 2870–2830 cm^{-1} (3.48–3.53 μm).

A weak-to-medium intensity band due to the aldehydic C—H deformation vibration is found in the region 975–780 cm^{-1} (10.26–12.82 μm). However, because of its variable position and intensity, this band may be difficult to identify.

Other aldehyde bands

Aliphatic aldehydes absorb weakly in the region 1440–1325 cm^{-1} (6.94–7.55 μm) and aromatic aldehydes absorb weakly at 1415–1350 cm^{-1} (7.07–7.41 μm), 1320–1260 cm^{-1} (7.58–7.94 μm), and 1230–1160 cm^{-1} (8.13–8.62 μm), the last band being due to the C—C stretching vibration. These bands are not normally useful for assignment purposes.

In general, aromatic aldehydes have a strong absorption at 650–580 cm^{-1} (15.38–17.24 μm) due to in-plane deformation vibrations of the C—CHO group.[80]

Table 10.6. Aldehyde C=O stretching vibrations

Functional Groups	Region cm⁻¹	μm	Intensity	Comments
Saturated aliphatic aldehydes	1740–1720	5.75–5.81	vs	
α,β-Unsaturated aliphatic aldehydes	1705–1685	5.87–5.93	vs	
α,β-γ,δ-Conjugated aliphatic aldehydes	1690–1660	5.91–6.02	vs	Further conjugation has little effect
Aryl aldehydes	1715–1695	5.83–5.90	vs	
o-Hydroxy- and *o*-amino- aryl aldehydes	1665–1625	6.01–6.16	vs	Frequency lowered due to hydrogen bonding
α-Keto aldehydes in enol form, —C(OH)=C(—)—CHO	1670–1645	6.17–6.25	vs	Frequency lowered due to hydrogen bonding
α-Di- and trichloroaldehydes	1770–1740	5.65–5.75	vs	
$-CF_2CHO$	1785–1755	5.60–5.70	vs	

Table 10.7. Aldehydes: other bands

Functional Groups	Region cm⁻¹	μm	Intensity	Comments
Aldehydes, —CHO	2830–2810	3.53–3.56	w–m	C—H str
	2745–2650	3.64–3.77	w–m	C—H str, usually ~2720 cm⁻¹
	975–780	10.26–12.82	w	C—H def
Aliphatic aldehydes	1440–1325	6.94–7.55	m	In-plane C—H rocking
	695–665	14.39–15.04	m–s	C—C—CO in-plane def
	535–520	18.69–19.23	s	C—CO in-plane def
Aryl aldehydes	1415–1350	7.07–7.41	m	In-plane C—H rocking
	1320–1260	7.58–7.94	m	Due to aromatic ring
	1230–1160	8.13–8.62	m	Possibly ring C—CHO str
	650–580	15.38–17.24	s	=C—CHO in-plane def
α-Branched aliphatic aldehydes	665–635	15.04–15.75	s	C—C—CO in-plane def
	565–540	17.70–18.52	s	C—CO in-plane def

Carboxylic Acids, —COOH

Due to the presence of strong intermolecular hydrogen bonding, carboxylic acids normally exist as dimers. Their spectra exhibit a broad band due to the O—H stretching vibration and a strong band due to the C=O stretching vibration. The marked spectral changes which occur when a carboxylic acid is converted to its salt may be used to distinguish it from other C=O containing compounds.

Carboxylic acid O—H stretching vibrations

As a result of the presence of hydrogen bonding, carboxylic acids in the liquid and solid phases exhibit a broad band at 3300–2500 cm⁻¹ (3.03–4.00 μm), due to the O—H stretching vibration,[81, 82] which sometimes, in the lower half of the frequency range, has two or three weak bands superimposed on it. In the main, it is only chelated O—H groups, e.g. the OH group of the enol form of β-diketones, and carboxylic acids which absorb in the region 2700–2500 cm⁻¹ (3.70–4.00 μm), and these two structural groups may be distinguished by their C=O stretching vibrations. Although other groups absorb in the region 3300–2500 cm⁻¹ (3.04–4.00 μm), e.g. C—H, P—H, S—H, Si—H, their bands are all sharp. The O—H deformation band may also be useful for distinguishing between groups.

Carboxylic acid monomers have a weak, sharp band at 3550–3500 cm⁻¹ (2.82–2.86 μm). Usually, monomers only exist in the vapour phase, but of course some dimeric structure may also be present in this phase too.

Carboxylic acid C=O stretching vibrations

In general, the C=O stretching vibration for carboxylic acids gives rise to a band which is stronger than that for ketones or aldehydes. In the solid or liquid phases, the C=O group of saturated aliphatic carboxylic acids[83] absorbs very strongly in the region 1725–1700 cm⁻¹ (5.80–5.88 μm).

As mentioned above, most carboxylic acids exist as dimers. However, in very dilute solution in non-polar solvents, or in the vapour phase, when the acid may exist as a monomer, the C=O stretching vibration band is at about 1760 cm⁻¹ (5.68 μm).

The frequency of the C=O stretching vibration for saturated *n*-aliphatic acids usually decreases with increase in chain-length. Electronegative atoms or groups adjacent to carboxylic acid groups have the effect of increasing the C=O stretching vibration frequency, while hydrogen bonding tends to decrease it.[84, 85] For example, α-halo-carboxylic acids[7, 86] absorb strongly in the region 1740–1715 cm⁻¹ (5.75–5.83 μm) and intramolecularly hydrogen-bonded acids absorb at 1680–1650 cm⁻¹ (5.95–6.06 μm). Sometimes, α-halo-carboxylic acids exhibit two bands due to the C=O stretching vibration, this being the result of partially restricted rotation.

Aryl and α,β-unsaturated carboxylic acids absorb in the region 1715–1680 cm⁻¹ (5.83–5.95 μm). Further conjugation has little effect on the C=O stretching vibration. Aryl carboxylic acids with a hydroxyl group in the *ortho*- position absorb at about 50 cm⁻¹ lower (0.18 μm higher) and with an *ortho*-amino-group the frequency lowering is about 30 cm⁻¹ (0.09 μm). Aryl carboxylic acid monomers absorb at 1755–1735 cm⁻¹ (5.70–5.76 μm).

Some saturated dicarboxylic acids have a doublet structure for this C=O band in solid-phase spectra, even though both acid groups are chemically equivalent. This structure may be used to distinguish between optical isomers.

Association of the acid with a solvent such as pyridine, dioxane, etc., generally lowers the C=O stretching vibration frequency.

Other vibrations of carboxylic acids

C—H stretching vibration bands in the region 3100–2800 cm^{-1} (3.23–3.57 μm) sometimes have broad wings due to overlap with the bands due to the O—H stretching vibration. A band at about 1440–1395 cm^{-1} (6.95–7.17 μm), which may be overlooked because of its weak nature, is due to the combination of the C—O stretching and O—H deformation vibrations. A —CH_2CO— deformation vibration may further complicate matters since it gives rise to a medium-intensity band at 1410–1405 cm^{-1} (7.09–7.12 μm) which is characteristic of the group.

A medium-to-strong band at 1320–1210 cm^{-1} (7.58–8.28 μm) is observed but this band is not usually much help in identification as other compounds containing the carbonyl group have bands in this region. Carboxylic acid dimers absorb in the narrower range 1320–1280 cm^{-1} (7.58–7.81 μm) and also have a broad, usually asymmetric, band of medium-to-strong intensity in the region 955–915 cm^{-1} (10.47–10.93 μm) due to the out-of-plane deformation of the carboxylic acid OH···O group. This latter band is usually very weak or absent for hydroxy aliphatic acids, but is often more prominent and narrow for aromatic acids.

In the solid phase, the spectra of aliphatic long-chain carboxylic acids exhibit band patterns in the range 1345–1180 cm^{-1} (7.43–8.47 μm). The number of these almost equally-spaced weak bands is related to the length of the aliphatic chain.[87,88] For acids with an even number of carbon atoms, the number of bands observed equals half the number of carbon atoms. For acids with an odd number of carbon atoms, the number of bands is half (the number of carbon atoms plus one). Unfortunately, the band due to the C—O stretching vibration also occurs in this region so that these weak bands may appear as shoulders.

Normal-aliphatic monocarboxylic acids,[89] except those smaller than *n*-butyric acid, exhibit, in liquid-phase spectra, three strong bands that are not usually well-resolved in the region 675–590 cm^{-1} (14.81–16.95 μm) due to the in-plane vibration of the O—CO group. In addition, a strong band is found at 495–465 cm^{-1} (20.20–21.51 μm) which is attributed to the in-plane vibration of the C—C=O group. This may be coalesced with a sharp, strong band observed at about 500 cm^{-1} (20.00 μm). If branching occurs, it affects the position of these bands, as does the physical state of the sample. For example, the in-plane vibrations mentioned occur, in the solid phase, at 680–625 cm^{-1} (14.71–16.00 μm) and 550–525 cm^{-1} (18.18–19.05 μm). α-Branched aliphatic carboxylic acids have three strong bands in the region 665–610 cm^{-1} (15.04–16.39 μm) and a strong band in the region 555–520 cm^{-1} (18.02–19.32 μm). Other branched monocarboxylic acids have three medium-to-strong bands in the region 700–600 cm^{-1} (14.29–16.67 μm).

In the far infrared spectra of acetic acid derivatives,[90–92] a band due to the deformation of the OH···O group is observed at 185–100 cm^{-1} (57.14–100.00 μm) for monosubstituted compounds, at 125–95 cm^{-1} (80.00–105.26 μm) for disubstituted compounds, and at 105–80 cm^{-1} (95.24–125.00 μm) for tri-substituted compounds. A study of halogenated acids has been published.[93]

Aromatic acids and esters have a strong band at 570–545 cm^{-1} (17.54–18.35 μm) which is due to the rocking vibration of the CO_2 group. They also have a band of medium-to-strong intensity which is usually broad for acids and is observed at 370–270 cm^{-1} (27.03–37.04 μm). For *para*-substituted aromatic acids, the bending vibration of the CO_2 group results in a band at 620–610 cm^{-1} (16.13–16.39 μm).

The C=C stretching vibration band of α,β-unsaturated acids occurs at 1660–1630 cm^{-1} (6.02–6.14 μm), *trans* isomers absorbing 10–20 cm^{-1} higher than *cis* isomers.

Carboxylic acid salts

Carboxylic acid salts[94–97] have a very strong, characteristic band in the region 1610–1550 cm^{-1} (6.21–6.45 μm) due to the asymmetric stretching vibration of CO_2^-. The symmetric stretching vibration of this group gives rise to a band in the range 1420–1335 cm^{-1} (7.04–7.49 μm) and is of medium intensity, broad, and generally has two or three peaks. Unfortunately, water, which may be present in the sample, has an absorption at around 1640 cm^{-1} (6.10 μm) and may cause difficulties in identification, as might also the presence of primary or secondary amides due to their amide II band which also occurs in this region.

For acid salts with a strongly electron-withdrawing group such as CF_3, the asymmetric stretching vibration band may be found outside the normal range quoted and as high as 1690 cm^{-1} (5.92 μm). The symmetric vibration band for $CF_3COO^-Na^+$ occurs at about 1450 cm^{-1} (6.90 μm), for $CBr_3COO^-Na^+$ at about 1340 cm^{-1} (7.46 μm) and 1355 cm^{-1} (7.38 μm) (two bands), and for acetic acid salts at about 1415 cm^{-1} (7.07 μm).

The stretching vibration of the —CO_2^- group depends both on the metal ion and the organic portion of the salt.

Due to the rocking in-plane and out-of-plane deformation vibrations of the carboxylic ion, medium-to-strong bands are observed in the region 760–400 cm^{-1} (13.16–25.00 μm).

The salts of complexes of carboxylic acids and their derivatives are reviewed elsewhere.[98]

Table 10.8. Carboxylic acid C=O stretching vibrations

Functional Groups	Region cm^{-1}	Region μm	Intensity	Comments
Saturated aliphatic carboxylic acids (hydrogen-bonded or as dimer	1725–1700	5.80–5.88	vs	
Saturated aliphatic carboxylic acids (as monomer)	1800–1740	5.56–5.75	vs	In very dilute solution or as vapour
Aryl carboxylic acids (as dimer)	1700–1680	5.88–5.95	vs	
α,β-Unsaturated aliphatic carboxylic acids (as dimer)	1715–1690	5.83–5.92	vs	Band for triple bond compounds usually at 1690–1680 cm^{-1}
α-Halo-carboxylic acids (as dimer)	1740–1715	5.75–5.83	vs	(Band for $-CF_2COOH$ is at 1785–1750 cm^{-1})
Intramolecular hydrogen-bonded carboxylic acids	1680–1650	5.95–6.06	vs	Sharp—medium width band
Saturated dicarboxylic acids	1740–1700	5.75–5.88	vs	Sometimes broad
α-Unsaturated dicarboxylic acids	1700–1685	5.88–5.94	vs	Sometimes broad
Peroxy acids, —CO—OOH	1760–1730	5.68–5.77	vs	
γ-Ketocarboxylic acids	1750–1700	5.71–5.88	vs	Compounds exist in keto–lactol equilibrium, 2 or 1 band(s)
Thiol acids, —COSH	1700–1690	5.88–5.92	s	Also see ref. 99 (band due to C—S stretching vibration at 990–945 cm^{-1})

Table 10.9. Carboxylic acids: other vibrations

Functional Groups	Region cm^{-1}	Region μm	Intensity	Comments
—OH (associated carboxylic acid	3300–2500	3.00–4.00	m	br, —OH str, hydrogen bonding present, multiple structure
—OH (free carboxylic acid)	3550–3500	2.82–2.86	w–m	sh, as monomer
—OD (deuterated carboxylic acids)	690–650	14.49–15.38	v	O—D out-of-plane deformation, usually broad
Carboxylic acids, —COOH (dimer)	1440–1395	6.95–7.17	w	Combination band due to C—O str and O—H def
	1320–1210	7.58–8.26	m–s	C—O str, sometimes a doublet
	955–915	10.47–10.93	m	O—H...O out-of-plane def, usually broad
Carboxylic acids, —COOH (monomer)	1380–1280	7.25–7.81	m–s	O—H def
	1190–1075	8.40–9.39	s	C—O str
Long-chain aliphatic carboxylic acids	1345–1180	7.43–8.48	w	CH_2 def, number of bands determined by aliphatic chain length
Peracids, —CO—OOH	~3280	~3.05	m	O—H str
	~950	~10.53	m	O—H out-of-plane bending
Thiol acids, —CO—SH	2595–2560	3.81–3.91	w	S—H str
	840–825	11.90–12.12		S—H in-plane def
n-Aliphatic monocarboxylic acids	675–590	14.81–16.95	s	O—CO in-plane def, three bands usually at ~665, ~630, and ~600 cm^{-1}
	~500	~20.00	s	sh
	495–465	20.20–21.51	s	C—C=O in-plane def
α-Branched aliphatic monocarboxylic acids	665–610	15.04–16.39	s	O—CO in-plane def, three bands usually at ~655, ~635, and ~620 cm^{-1}
	555–520	18.18–19.23	s	C—CO in-plane def
β- and γ-branched aliphatic monocarboxylic acids	700–600	14.29–16.67	s	
	495–465	20.20–21.51	s	
Aromatic carboxylic acids	570–545	17.54–18.35	s	CO_2 out-of-plane rocking
	370–270	27.03–37.04	m–s	br, esters also absorb in this region but band usually narrower

Table 10.10. Carboxylic acid salts (solid-phase spectra)

Functional Groups	Region cm^{-1}	Region μm	Intensity	Comments
Carboxylic acid salts, $-CO_2^-$	1610–1550	6.21–6.45	s	Asymmetric CO_2^- stretching
	1420–1335	7.04–7.49	m	br, symmetric CO_2^- stretching, usually two or three peaks
$-CF_2CO_2^-$	1695–1615	5.90–6.19	s	
Thiol acid salts, $-CO-S^-$	~1525	~6.56	s	COS^- str
Monothiol carbonic acid salts, $R-O-CO-S^-$	~1580	~6.33	s	COS^- str, see ref. 100
α-Halo-carboxylic acid salts	1625–1580	6.15–6.33	s	Fluoro compounds at higher end of frequency range
$-CCl_2CO_2-$	1680–1640	5.95–6.10	s	

Carboxylic acid anhydrides, —CO—O—CO—

Due to the asymmetric and symmetric stretching vibrations of the two C=O groups, saturated aliphatic anhydrides[21, 31] absorb at 1840–1800 cm^{-1} (5.44–5.56 μm) and at 1780–1740 cm^{-1} (5.62–5.75 μm) respectively, both bands being sharp and strong. In most cases, these two bands are separated by about 60 cm^{-1} (0.18 μm). For acyclic anhydrides, the higher frequency band is usually the more intense.[8] The presence of conjugation results in a shift of about 20 cm^{-1} downward (0.05 μm upward) for both bands. α,β-Unsaturated acid anhydrides and aryl anhydrides absorb at 1830–1780 cm^{-1} (5.46–5.62 μm) and at 1755–1710 cm^{-1} (5.70–5.85 μm). All these frequencies are increased in strained-ring situations and also by electronegative atoms on the α-carbon atom.

Acid anhydrides also have a strong band in the range 1135–980 cm^{-1} (8.81–10.20 μm) due to the C—O—C stretching vibration which appears at 1300–1210 cm^{-1} (7.69–8.26 μm) for strained-ring compounds (five-membered ring anhydrides). Straight-chain alkyl anhydrides absorb in the narrow range 1050–1040 cm^{-1} (9.52–9.62 μm), the band usually being broad, an exception to this being acetic anhydride which absorbs at about 1135 cm^{-1} (8.81 μm). Acyclic anhydrides absorb at about 1050 cm^{-1} (9.52 μm), but branching at the α-carbon atom tends to decrease the frequency of this vibration. Cyclic anhydrides[14, 30] (five-membered ring) have a strong band at 955–895 cm^{-1} (10.47–11.17 μm) and often a weak band near 1060 cm^{-1} (9.44 μm) is observed also. Unconjugated cyclic anhydrides absorb strongly at 1130–1000 cm^{-1} (8.85–10.00 μm). All these bands are believed to involve the stretching vibration of the C—O—C group.

Table 10.11. Carboxylic acid anhydride C=O stretching vibrations

Functional Groups	Region cm^{-1}	Region μm	Intensity	Comments
Saturated aliphatic acid anhydrides	1840–1800	5.44–5.56	vs	Asymmetric stretching
	1780–1740	5.62–5.75	s	Symmetric stretching
Aryl and α,β-unsaturated acid anhydrides	1830–1780	5.46–5.62	vs	
	1755–1710	5.70–5.85	s	
Saturated five-membered ring acid anhydrides	1870–1820	5.35–5.50	s	Separation ~70 cm^{-1} except for aromatic compounds for which it is ~50 cm^{-1}
	1800–1775	5.56–5.63	vs	
α,β-Unsaturated five-membered ring acid anhydrides	1860–1850	5.38–5.41	s	
	1780–1760	5.62–5.68	vs	
Saturated six-membered ring acid anhydrides	1820–1780	5.49–5.62	s	Separation ~40 cm^{-1}
	1780–1740	5.62–5.75	vs	

Table 10.12. Carboxylic acid anhydrides: other bands

Functional Groups	Region cm^{-1}	Region μm	Intensity	Comments
Acyclic aliphatic and cyclic six-membered ring acid anhydrides	1135–980	8.81–10.20	s	C—O—C str, often a doublet at ~1050 cm^{-1}
Cyclic five-membered ring acid anhydrides	1300–1210	7.69–8.26	s	C—O—C str
	955–895	10.47–11.17	s	C—O—C str

Carboxylic acid halides, —CO—X

Due to the C=O stretching vibration, aliphatic acid chlorides[101] absorb strongly in the region 1815–1790 cm^{-1} (5.51–5.59 μm). Acid bromides and iodides absorb in the same region or at very slightly lower wavenumbers than acid chlorides, whereas the fluorides absorb at about 50 cm^{-1} higher (0.16 μm lower). Some α-methyl substituted acid halides exhibit a doublet.

Aryl[46, 102, 103] and α,β-unsaturated acid halides[11] (of Cl, Br, I) absorb in the range 1790–1735 cm^{-1} (5.59–5.76 μm) with fluorides absorbing at higher wavenumbers. In non-polar solvents, a double peak is often observed for aryl acid halides. The second band is probably an overtone band of the C—Cl (or C—Br) stretching vibration which occurs at about 850 cm^{-1} (11.76 μm). Fluorides exhibit a single band. The carbonyl stretching vibration frequency for α,β-unsaturated acid halides has been observed to decrease in the order

$$\text{fluoride} > \text{bromide} > \text{chloride}.$$

Compounds with one or more halogen atoms directly bonded to a carbonyl group absorb strongly, due to the carbonyl stretching vibration, in the region 1870–1790 cm^{-1} (5.35–5.59 μm), $F_2.CO$ absorbing outside this range at about 1930 cm^{-1} (5.18 μm).

Table 10.13. Carboxylic acid halide C=O stretching vibrations

Functional Groups	Region cm^{-1}	Region μm	Intensity	Comments
Saturated aliphatic acid chlorides	1815–1790	5.51–5.59	vs	Fluorides at higher wavenumbers (1845–1815 cm^{-1}), bromides and iodides at slightly lower ones
Aryl and α,β-unsaturated acid chlorides	1790–1765	5.59–5.66	vs	
	1750–1735	5.71–5.76	m	Involves overtone of band at 890–850 cm^{-1}
$-CF_2COF$	1900–1870	5.26–5.35	s	See ref. 104
$-CF_2COCl$	1820–1795	5.50–5.57	s	

Table 10.14. Carboxylic acid halides: other bands

Functional Groups	Region cm⁻¹	μm	Intensity	Comments
Saturated aliphatic acid chlorides	965–920	10.36–10.87	m	Probably C—C= str
	440–420	22.73–23.81	s	Cl—C=O in-plane def
Aryl acid chlorides	~1200	~8.33	m	Probably C—C str
	890–850	11.24–11.76	s	Probably C—Cl str

Diacyl peroxides, R—CO—O—O—CO—R, (acid peroxides) and peroxy acids, —CO—OOH

All acid peroxides[21] have a weak absorption band in the region 900–830 cm^{-1} (11.11–12.05 μm) due to the —O—O— stretching vibration. Acid peroxides also have strong bands due to their C=O stretching vibration. For saturated aliphatics, two bands are usually observed, one at 1820–1810 cm^{-1} (5.50–5.53 μm) and the other at 1800–1780 cm^{-1} (5.56–5.62 μm), this latter band being more intense. For aryl and α,β-unsaturated acid peroxides, these bands occur at 1805–1780 cm^{-1} (5.54–5.62 μm) and 1785–1755 cm^{-1} (5.60–5.70 μm). The nature and position of the substituent(s) in the aromatic portion of acid peroxides may significantly influence the position of these bands.

The C—O stretching vibrations are not very useful in the characterization of acid peroxides. They are found in the region 1300–1050 cm^{-1} (7.69–9.52 μm).

Table 10.15. Diacyl peroxide and peroxy acid C=O stretching vibrations

Functional Groups	Region cm⁻¹	μm	Intensity	Comments
Saturated aliphatic acid peroxides, —CO—O—O—CO—	1820–1810	5.50–5.53	s	Separation ~25 cm^{-1}, see ref. 21
	1800–1780	5.56–5.62	vs	
Aryl and α,β-unsaturated acid peroxides	1805–1780	5.54–5.62	s	
	1785–1755	5.60–5.70	vs	
Peroxy acids, —CO—OOH	1760–1730	5.68–5.77	vs	Intramolecular hydrogen bonding

Table 10.16. Diacyl peroxides and peroxy acids: other bands

Functional Groups	Region cm⁻¹	μm	Intensity	Comments
Peroxides, —O—O—	900–830	11.11–12.05	w	All peroxides, O—O str at ~865 cm^{-1} for peroxy acids
Peroxy acids, —CO—OOH	~3280	~3.05	m–s	Associated intramolecularly, due to O—H str
	1460–1430	6.85–7.00	m	O—H bending, near 1430 cm^{-1} for long-chain linear acids
	1300–1050	7.69–9.52	m–s	C—O str, often near 1175 cm^{-1}

Esters, —CO—O—, Carbonates, —O—CO—O—, and Chloroformates, —O—CO—Cl

All esters have two strong characteristic bands, one due to the C=O stretching vibration and the other due to the C—O stretching vibration. The frequency of the C=O stretching vibration for esters is influenced in a very similar way to that observed for ketones, except that the decrease in wavenumber for aliphatic esters due to the presence of α,β-unsaturation[105] is less, being approximately 10–20 cm^{-1}.

Ester C=O stretching vibrations

With the exception of formates,[113] which absorb in the region 1725–1720 cm^{-1} (5.80–5.81 μm), saturated aliphatic esters absorb at 1750–1725 cm^{-1} (5.71–5.80 μm). Electronegative groups or atoms directly bonded to the alcoholic oxygen atom of the ester group tend to increase the frequency of the C=O stretching vibration. Aryl and α,β-unsaturated esters (>C=C—CO—O—, β and α carbons marked)[112] absorb at 1730–1705 cm^{-1} (5.78–5.83 μm). Further conjugation has almost no effect on the C=O stretching vibration frequency. Strongly polar groups substituted on the benzene ring of aryl esters tend to increase the frequency of the C=O stretching vibration. Esters with electronegative α-substituents (>XC—CO—O—), e.g. α-halo-esters,[106,107] absorb at 1770–1745 cm^{-1} (5.65–5.73 μm), i.e. about 10–20 cm^{-1} higher than for the normal aliphatic ester. α,α′-Dihalo-esters[107] also absorb in the same region, but in general two closely-spaced bands are observed. Vinyl and phenyl esters (—CO—O—C=C< absorb at about 1770 cm^{-1} (5.65 μm), e.g. vinyl acetate absorbs at 1760 cm^{-1} (5.63 μm).

No change is observed in the position of the band due to the C=O stretching vibration when a carbonyl group is present in the α-position of an ester, —CO—CO—O—, e.g. α-keto-esters and α-diesters both absorb in the range 1760–1740 cm^{-1} (5.68–5.75 μm). In general, for esters of saturated dicarboxylic acids, the C=O band occurs in approximately the same range, 1760–1735 cm^{-1} (5.68–5.76 μm), as for monoesters and the same influences on the position of this band are observed. If the two ester groups are close together in the molecule then a doublet is observed, otherwise a single band is observed. Geminal diesters may absorb at slightly higher wavenumbers than those given above. A study of glycidic esters has been published.[123]

With β-keto-esters, —CO—CH—CO—O—, keto–enol tautomerism is possible:

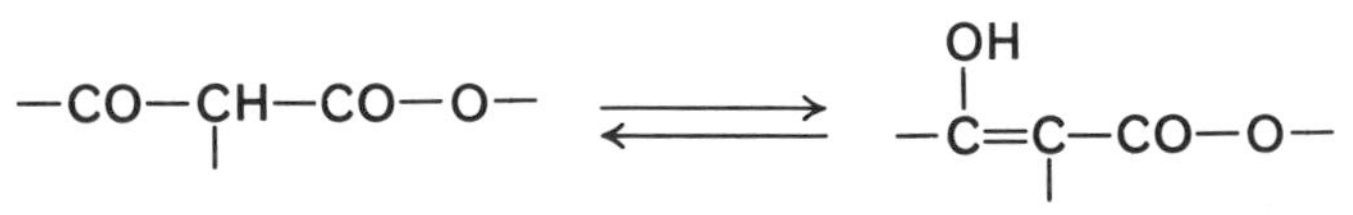

OH
—CO—CH—CO—O—
—C=C—CO—O—

In this case, a strong band at about 1650 cm^{-1} (6.07 μm) is observed due to the C=O stretching vibration of the hydrogen-bonded C=O group, i.e.

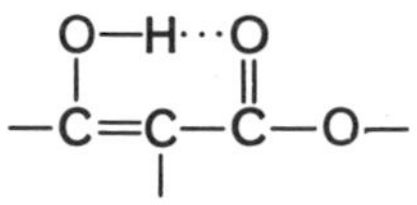

The C=O stretching vibration frequency is lowered due to the presence of the hydrogen bonding. There is also a band due to the C=C stretching vibration at about 1630 cm^{-1} (6.14 μm) and a sharp band due to the O—H stretching vibration at 3590–3420 cm^{-1} (2.79–2.92 μm). In addition, other bands due to the carbonyl stretching vibration, etc., may be observed, these being due to the keto- form of the ester. The relative intensities of these bands of the β-keto-esters depend on the relative amounts of each tautomer.

Due to intramolecular hydrogen bonding, *o*-hydroxyl (or *o*-amino-) benzoates absorb at 1690–1670 cm^{-1} (5.92–5.99 μm).

The effect of converting a methyl ester to a phenyl ester is normally to increase the wavenumber of the band due to the carbonyl stretching vibration by 10–20 cm^{-1} (a decrease of 0.03–0.07 μm).

Intensity correlations for the carbonyl band of esters have been studied extensively.[6, 10, 15, 109]

The carbonyl band for aliphatic chloroformates (—CO.Cl)[53, 105, 120, 121] is observed at higher wavenumbers than that for esters, at about 1780 cm^{-1} (5.62 μm), and for aryl chloroformates at about 1785 cm^{-1} (5.60 μm).

Most noncyclic carbonates[53, 108–110] absorb strongly in the region 1790–1740 cm^{-1} (5.59–5.75 μm) whilst five-membered-ring cyclic carbonates[27, 32] absorb at 1850–1790 cm^{-1} (5.41–5.59 μm).

The carbonyl band of thiol carbonyl esters —S—CO—[54, 111] occurs at lower frequencies than that of normal esters.

A weak band due to the overtone of the C=O stretching vibration of esters occurs at about 3450 cm^{-1} (2.90 μm) and may sometimes be used in confirming the presence of a C=O group.

Ester C—O—C stretching vibrations

The bands due to the ester C—O stretching vibration are strong, partly due to an interaction with the C—C vibration, and occur in the range 1300–1100 cm^{-1} (7.69–9.09 μm). Often a series of strong overlapping bands is observed. Caution is required when using these bands in making assignments since the C—O stretching vibrations of alcohols and acids, and possibly ketones also, occur in this region.

The band due to the C—O—C asymmetric stretching vibration for aliphatic esters occurs at 1275–1185 cm^{-1} (7.85–8.44 μm) and that due to the symmetric stretching vibration occurs at 1160–1050 cm^{-1} (8.62–8.70 μm). Both these bands are strong, the former band being usually more intense than that due to the C=O stretching vibration.

Esters of aromatic acids and α,β-unsaturated aliphatic acids have two strong absorption bands, one at 1310–1250 cm^{-1} (7.63–8.00 μm) and the other at 1200–1100 cm^{-1} (8.33–9.09 μm). For esters G—CO.OG′, where G′ is an aromatic or α,β-unsaturated group, a very strong absorption near 1210 cm^{-1} (8.26 μm) is observed. If, in addition, the other group G is aromatic in nature, then the band due to the asymmetric stretching vibration occurs at 1310–1250 cm^{-1} (7.64–8.00 μm) and that due to the symmetric stretching vibration at 1150–1080 cm^{-1} (8.70–9.26 μm). The C—O stretching vibration frequencies do not appear to vary as much as in alcohols, ethers, and acids. Some of the C—O asymmetric stretching vibration band positions are given in Table 10.17. Although it is not possible to distinguish between neighbouring esters in a homologous series, Table 10.17 is still useful in a more general sense.

Table 10.17. Some C—O asymmetric stretching vibration band positions

Ester	Approximate position cm^{-1}	μm	Ester	Approximate position cm^{-1}	μm
Formates	1190	8.40	Acetates	1245	8.03
Propionates	1190	8.40	*n*-Butyrates	1200	8.33
Isobutyrates	1200	8.33	Isovalerates	1195	8.33
Adipates	1175	8.51	Oleates	1170	8.54
Stearates	1175	8.51	Citrates	1180	8.46
Sebacates	1170	8.53	Laurates	1165	8.59
Benzoates	1280	7.81	Phthalates	1120	8.93
	(sym) 1120	8.91		(sym) 1070	9.35

The position of the band due to the C—O stretching vibration is dependent on the nature of both the acidic and the alcoholic components, although the latter is less important.

Methyl esters of long-chain aliphatic acids normally exhibit three bands, the strongest of which is at 1175 cm^{-1} (8.50 μm), the others being near 1250 cm^{-1} (8.00 μm) and 1205 cm^{-1} (8.30 μm).

Acetates of primary alcohols have a medium-intensity band at 1060–1035 cm^{-1} (9.39–9.64 μm) due to the asymmetric stretching of the O—CH_2—C group. For acetates of other than primary alcohols, this band is shifted to higher wavenumbers.

Other ester bands

Acetates have a medium-to-strong band near 1375 cm^{-1} (7.30 μm), due to the CH_3 symmetric deformation, and medium-to-weak bands near 1430 cm^{-1} (6.99 μm) and 2990 cm^{-1} (3.34 μm), due to the asymmetric deformation and stretching vibrations respectively of this group. For other saturated esters containing the —CH_2CO—O— group, the CH_2 deformation band occurs near 1420 cm^{-1} (7.04 μm).

Most aliphatic esters[66, 114] have bands in the regions 645–585 cm^{-1} (15.50–17.09 μm) and 350–300 cm^{-1} (28.57–33.33 μm).

All acetates absorb strongly at 665–635 cm^{-1} (15.04–15.75 μm) due to the bending of the O—C—O group and at 615–580 cm^{-1} (16.29–17.24 μm) due to the out-of-plane deformation vibration of the acetate group. A band at 325–305 cm^{-1} (30.77–32.79 μm) is also often observed. This last band decreases in intensity with increase in molecular weight.

Branched alkyl formates absorb at 520–485 cm^{-1} (19.23–20.62 μm) and 340–285 cm^{-1} (29.41–35.09 μm), whereas *n*-alkyl formates (ethyl to amyl) have three bands: near 620 cm^{-1} (16.13 μm), in the region 475–460 cm^{-1} (21.05–21.74 μm), and near 340 cm^{-1} (29.41 μm). This last band is always strong and the first, weak. The first two (higher-frequency) bands decrease in intensity as the molecular weight of the formate increases.

Methyl esters[88] have bands near 2960 cm^{-1} (3.38 μm) and 1440 cm^{-1} (6.94 μm) due to the CH_3 asymmetric stretching and deformation vibrations and weak bands near 1425 cm^{-1} (7.02 μm) and 1360 cm^{-1} (7.35 μm). In addition, with the exception of the formate and isobutyrate, methyl esters have a band of medium intensity at 450–430 cm^{-1} (22.22–23.26 μm). The characteristic absorptions of ethyl esters are given in Table 10.18.

Table 10.18. Characteristic absorptions of ethyl esters

	Approximate position cm^{-1}	μm	Intensity	Comments
Ethyl esters, —O—CH_2CH_3	2980	3.36	m	asym CH_3 str
	2950	3.39	w	CH_3 str
	2900	3.45	w	CH_3 str
	1475	6.78	m	O—CH_2
	1455	6.87	m	asym CH_3 def
	1400	7.14	m	O—CH_2 wagging
	1375	7.27	s	sym CH_3 def

For the —O—CH(R)—CH_3 group, a medium intensity band is observed near 1380 cm^{-1} (7.25 μm).

n-Propyl esters have a band near 1390 cm^{-1} (7.19 μm) and bands of variable intensities at 605–585 cm^{-1} (16.53–17.09 μm), near 495 cm^{-1} (20.20 μm), and at 350–340 cm^{-1} (28.57–29.41 μm). The band near 600 cm^{-1} is not present for the formate. Isopropyl esters have bands of variable intensity at 605–585 cm^{-1} (16.53–17.09 μm) and 505–480 cm^{-1} (19.80–20.83 μm), and strong bands near 435 cm^{-1} (22.99 μm) and at 425–410 cm^{-1} (23.53–24.39 μm), but isopropyl formate exhibits only the band near 435 cm^{-1}.

n-Butyl esters have medium-to-strong absorptions near 505 cm^{-1} (19.80 μm) and 435 cm^{-1} (22.99 μm) and a weak band at 350–335 cm^{-1} (28.57–29.85 μm). Isobutyl esters have a band of medium intensity near 505 cm^{-1} (19.80 μm), a strong band near 430 cm^{-1} (23.26 μm), and a band of variable intensity near 385 cm^{-1} (25.97 μm), the formate and isobutyrate not exhibiting the band near 505 cm^{-1}.

α,β-Unsaturated esters (e.g. acrylates, methacrylates, fumarates) have a band at 695–645 cm^{-1} (14.39–15.50 μm) due to the wagging vibration of the C=O group. These esters, of course, have a band due to the C=C stretching vibration and also bands due to the =C—H and =CH_2 groups, for instance, acrylates and methacrylates have a medium-to-strong band at 820–805 cm^{-1} (12.20–12.42 μm) and a strong band at 970–935 cm^{-1} (10.31–10.70 μm) due to the twisting and wagging respectively of the =CH_2 group. For acrylates, the C=C stretching vibration results in a doublet at 1640–1620 cm^{-1} (6.10–6.17 μm) due to the interaction with the overtone of the band near 810 cm^{-1} (12.35 μm). Benzoates with an unsubstituted ring have a strong band near 710 cm^{-1} (14.08 μm) and other bands, due to ring vibrations, of medium intensity near 1070 cm^{-1} (9.35 μm) and 1030 cm^{-1} (9.71 μm). Disubstituted aromatic esters often do not have the usual band pattern expected in the region 880–750 cm^{-1} (11.36–11.33 μm), which may be due to an interaction with the CO—O group. Because of their centre of symmetry, terephthalates do not have a band near 1600 cm^{-1} (6.25 μm).

Aromatic acids and esters absorb strongly at 570–545 cm^{-1} (17.54–18.35 μm) due to the rocking of the CO_2 group and also have a band of medium-to-strong intensity, which is usually broad for acids, at 370–270 cm^{-1} (27.03–37.04 μm). Aromatic esters have a band of variable intensity in the range 650–585 cm^{-1} (15.38–17.09 μm) which is due to a deformation vibration of the CO_2 group. A study of phthalides has also been published.[123]

Thiol formates have medium-intensity bands at 2835–2825 cm^{-1} (3.53–3.54 μm) and near 1340 cm^{-1} (7.46 μm) due to the stretching and deformation vibrations respectively of the CH group, and a weak band at 2680–2660 cm^{-1} (3.73–3.76 μm) which is an overtone of the CH deformation vibration.

Table 10.19. Ester, chloroformate, and carbonate C=O stretching vibrations

Functional Groups	Region cm⁻¹	μm	Intensity	Comments
Formates	1725–1720	5.80–5.81	vs	
Acetates	1750–1740	5.70–5.75	vs	
Saturated aliphatic esters	1750–1725	5.71–5.80	vs	Except formates
Aryl and α,β-unsaturated aliphatic esters (esters of aromatic acids, etc.), >C=C–CO–O– (β α)	1730–1705	5.78–5.87	vs	Usually at lower end of frequency range in cases of olefinic conjugation
Vinyl and phenyl esters, R–CO–O–C=C<	1800–1750	5.56–5.71	vs	Phenyl acetates at ~1775 cm^{-1}
α-Halo- and α-cyano-esters	1770–1740	5.65–5.75	vs	
α,α-difluoro esters, $-CF_2CO.O-$	1800–1775	5.56–5.63	vs	See ref. 106
α-Keto-esters and α-diesters, –CO.COOR	1760–1740	5.68–5.75	vs	
β-Keto-ester, enol form, –C=C–C–OR (with OH on C=C and O on C)	1655–1635	6.04–6.12	vs	Sometimes broad, usually ~1650 cm^{-1} (intramolecular hydrogen bonding)
o-Hydroxyl (or *o*-amino-) benzoates	1690–1670	5.92–5.99	vs	sh, intramolecular hydrogen bonding
Esters, CH_3COOX (X ≠ carbon atom)	1810–1710	5.53–5.85	vs	
Aliphatic chloroformates, R–O–CO–Cl	1780–1775	5.62–5.63	vs	Unsaturation tends to increase frequency, strong band near 690 cm^{-1} due to C–Cl str
Aryl chloroformates	1800–1780	5.56–5.62	vs	
Carbamoyl chlorides, NR_2COCl	1745–1735	5.73–5.76	vs	
Alkyl and aryl thiol chloroformates, –S–COCl	1775–1765	5.63–5.67	vs	
Peresters, –CO–O–O–	1785–1750	5.60–5.71	vs	
Dialkyl thiolesters, R–S–CO–R′	1700–1690	5.88–5.92	vs	
Alkyl aryl thiolesters:				
Ar–S–CO–R	1710–1690	5.85–5.92	vs	
R–S–CO–Ar	1670–1665	5.99–6.01	vs	*Ortho*- halogen-substituted compounds absorb at higher frequencies
Diaryl thiolesters	1700–1650	5.88–6.06	vs	*Ortho*- halogen-substituted compounds absorb at higher frequencies
HCO–S–R	~1675	~5.97	vs	
HCO–S–Ar	~1700	~5.88	vs	
Alkyl carbonates, –O–CO–O–	1760–1740	5.68–5.75	vs	
Alkyl aryl carbonates	1790–1755	5.59–5.70	vs	
Diaryl carbonates	1820–1775	5.50–5.63	vs	See ref. 170
Cyclic carbonates (five-membered ring)	1860–1750	5.38–5.71	vs	Halogen substitution of ring increases frequency
Cyclic carbohydrate carbonates	1845–1800	5.42–5.56	vs	
Dialkyl thiolcarbonates, R–S–CO–O–R	1720–1700	5.81–5.88	vs	See ref. 111 [cyclic compounds (five-membered ring) at 1760–1735 cm^{-1}]
Alkyl aryl thiolcarbonates:				
Ar–S–CO–O–R	1730–1715	5.78–5.83	vs	
R–S–CO–O–Ar	1740–1730	5.75–5.78	vs	

Table 10.19 (*cont.*)

Functional Groups	Region cm⁻¹	μm	Intensity	Comments
Dialkyl dithiolcarbonates, R–S–CO–S–R	1655–1640	6.04–6.10	vs	
Diaryl dithiolcarbonates, Ar–S–CO–S–Ar	1720–1715	5.81–5.83	vs	
R–O–CO–NH–R	1740–1730	5.75–5.78	s	See ref. 117
R–S–CO–NH–R	~1695	~5.90	vs	
G–S–CO–NH–Ar	1665–1650	6.01–6.06	vs	
$R-S-CO-NH_2$	~1700	~5.88	vs	
Silyl esters, $R.CO.SiR_3$	~1620	~6.17	vs	

Table 10.20. Ester, chloroformate, and carbonate C–O–C stretching vibrations

Functional Groups	Region cm⁻¹	μm	Intensity	Comments
R–CO.OR′	1275–1185	7.85–8.44	vs	asym str
	1160–1050	8.62–8.70	s	sym str
Formates, H.CO–OR	1195–1180	8.37–8.47	vs	Also a strong band near 1160 cm^{-1} (8.62 μm)
Acetates, CH_3COOR	1265–1230	7.91–8.13	vs	Often split
Propionates and higher	1200–1150	8.33–8.70	vs	Two bands in region 1275–1050 cm^{-1} due to asym and sym C–O–C str, band at higher wavenumbers (asym) usually the more intense
Esters of aromatic acids (e.g. benzoates, phthalates, etc.)	1310–1250	7.63–8.00	vs	asym C–O–C str
	1150–1100	8.70–9.09	s	sym C–O–C str
Unsubstituted benzoates, 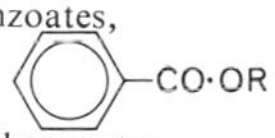	1280–1270	7.81–7.87	vs	Weak shoulder at ~1315 cm^{-1}
	~1110	~9.01	s	Weak shoulder at ~1175 cm^{-1}
Ortho-substituted benzoates, 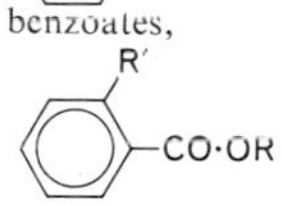	1265–1250	7.91–8.00	vs	Shoulder at ~1300 cm^{-1}
	1120–1070	8.93–9.35	s	
Meta-substituted benzoates, 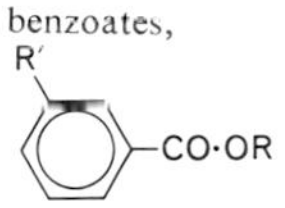	1295–1280	7.72–7.81	vs	Shoulder at ~1305 cm^{-1}
	1135–1105	8.81–9.05	s	
Para-substituted benzoates, 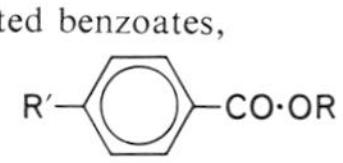	~1310	~7.63	s	Very strong doublet
	~1275	~7.84	vs	Very strong doublet
	~1180	~8.48	s	
	1120–1100	8.93–9.09	s	
α,β-Unsaturated aliphatic esters (e.g. acrylates, fumarates, etc.), >C=C–CO–OR	1310–1250	7.63–8.00	vs	asym C–O–C str
	1200–1130	8.33–8.85	s	sym C–O–C str
Acrylates, $CH_2=CH-CO-O-R$	1290–1280	7.75–7.81	vs	Shoulder at ~1300 cm^{-1}
	1200–1195	8.33–8.36	s	

Table 10.20 (*cont.*)

Functional Groups	Region cm^{-1}	μm	Intensity	Comments
Methacrylates, $CH_2{=}C(CH_3)CO.OR$	1305–1295	7.66–7.72	vs	Shoulder at ~1330 cm^{-1}
	1180–1165	8.48–8.58	s	
Crotonates, $CH_3CH{=}CH{-}CO.OR$	1290–1275	7.75–7.84	vs	Usually two shoulders
	1195–1180	8.36–8.48	s	
Cinnamates, 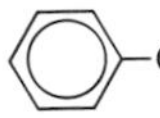 —CH=CH—CO·OR	1290–1250	7.75–8.00	vs	Usually two shoulders
	1185–1165	8.44–8.58	s	
R—CO.OR′ (R′ vinyl or aromatic)	~1210	~8.26	vs	asym str, vinyl C=C str at 1690–1650 cm^{-1} of greater intensity than usual
Ar—CO.OAr′	1310–1250	7.64–8.00	vs	asym str
	1150–1080	8.70–9.26	s	sym str
Methyl esters, $R{-}CO.OCH_3$	~1200	~8.33	s	$O{-}CH_3$ str
	1175–1155	8.51–8.66	s	O—C str
Aliphatic chloroformates	1200–1165	8.33–8.58	s	br, asym C—O—C str
	1150–1140	8.70–8.77	s	br, sym C—O—C str
Aromatic chloroformates	~1110	~9.01	s	br, usually difficult to identify
Dialkyl carbonates, RO.R′O.CO	1290–1260	7.75–7.94	s	
R—O—CO—O—Ar	1250–1210	8.00–8.26	s	
Diaryl carbonates	1220–1205	8.20–8.30	s	
Dialkyl thiolcarbonates, R—O—CO—S—R	1165–1140	8.58–8.77	s	
Alkyl aryl thiolcarbonates:				
R—O—CO—S—Ar	1140–1125	8.77–8.88	s	
Ar—O—CO—S—R	1105–1055	9.05–9.48	s	

Table 10.21. Esters, chloroformates, and carbonates: other bands

Functional Groups	Region cm^{-1}	μm	Intensity	Comments
Esters, —CO—O—	~3450	~2.90	w	C=O str overtone
n-Alkyl formates (ethyl to amyl)	~620	~16.13	w	
	475–460	21.05–21.74	v	Frequency increases with molecular weight increase
	~340	~29.41	s	
Branched alkyl formates	520–485	19.23–20.62	v	
	340–285	29.41–35.09	v	
Acetates	~2990	~3.34	m	
	~1430	~6.99	w–m	
	~1375	~7.27	m–s	
	1050–1020	9.52–9.80	m	
	990–980	10.10–10.20	w	
	845–835	11.83–11.93	w	
	665–635	15.04–15.75	w–s	Weak for tertiary (and sometimes secondary) acetates
	615–580	16.26–17.24	s	
	325–305	30.77–32.79	v	Absent for isopropyl and *sec*-butyl acetates

Table 10.21 (*cont.*)

Functional Groups	Region cm^{-1}	μm	Intensity	Comments
Propionates	1085–1080	9.21–9.26	m	Skeletal vib
	~1020	~9.80	m	
	~810	~12.35	w	
Butyrates	~1095	~9.13	m	
	1050–1040	9.52–9.62	m	
	930–865	10.75–11.56	w	
	850–830	11.76–12.05	w	
α,β-Unsaturated aliphatic esters	845–765	11.83–13.07	m	Mainly C—O—C def
	695–645	14.39–15.50	m	Mainly C—O—C def
Acrylates	1070–1065	9.35–9.40	m	Skeletal vib
	990–980	10.10–10.20	m	CH def
	970–960	10.30–10.40	s	$=CH_2$ wagging
	810–800	12.35–12.50	m–s	Skeletal vib
	675–660	14.81–15.15	m	Mainly C—O—C def
Methacrylates	~1010	~9.90	m	Skeletal vib
	~1000	~10.00	m	Skeletal vib
	950–935	10.53–10.70	s	$=CH_2$ wagging
	~815	~12.27	m–s	Skeletal vib
	660–645	15.15–15.50	m	Mainly C—O—C def
Crotonates	1105–1100	9.05–9.09	m	Skeletal vib
	970–960	10.31–10.42	s	CH=CH twisting
	920–900	10.87–11.11	m	Skeletal vib
	840–830	11.90–12.05	m	Skeletal vib
	695–675	14.39–14.81	m	Mainly C—O—C def
Methyl esters	~2960	~3.38	m	asym CH_3 str
	~1440	~6.94	m	CH_3 def
	1430–1420	6.99–7.04	w	
	~1360	~7.53	w	
	450–430	22.22–23.26	m–s	Not formate or isobutyrate
$-CO{-}O{-}CH_2-$	~1475	~6.78	m	$-OCH_2$ def
	~1400	~7.14	m	OCH_2 wagging
n-Propyl esters	605–585	16.53–17.09	v	
	~495	~20.20	v	
	350–340	28.57–29.41	v	
Isopropyl esters	605–585	16.53–17.09	v	Not formate
	505–480	19.80–20.83	v	Not formate
	~435	~22.09	s	
	425–410	23.53–24.39	s	Not formate
n-Butyl esters	~505	~19.08	m–s	
	~435	~22.99	m–s	
	350–335	28.57–29.85	w	
Isobutyl esters	~505	~19.08	m	Not formate
	~430	~23.26	s	
	~385	~25.97	v	
Aromatic esters	650–585	15.38–17.09	v	Rocking or in-plane def of CO_2 group
	370–270	27.03–37.04	m–s	Acids also absorb in this region
α-Hydroxy esters	1300–1260	7.69–7.94	s	br, O—H def
Acetylated pyranose sugars	670–625	14.93–16.00	s	See refs 115, 116
	610–600	16.39–16.67	v	
	405–365	24.69–27.40	v	
Aliphatic chloroformates	~850	~12.00	s	Most alkyl compounds
	695–690	14.39–14.49	s	C—Cl str
	485–470	20.62–21.28	s	C—Cl def

Table 10.21 (*cont.*)

Functional Groups	Region cm^{-1}	μm	Intensity	Comments
Thiocarbonyl compounds:				
R—CO—S—R	1140–1070	8.77–9.35	w	C—C str
	1030–930	9.71–10.75	s	C—S str
R—CO—S—Ar	1110–1060	9.01–9.43	w	C—C str
	1020–920	9.80–10.87	s	C—S str
Ar—CO—S—R	1210–1190	8.26–8.40	w	C—C str
	940–905	10.64–11.05	s	C—S str
H—CO—S—Ar and H—CO—S—R	780–730	12.82–13.70	s	C—S str
R—O—CO—NHR	1250–1210	8.00–8.26	s	C—N str, see ref. 117
R—S—CO—NHR	1230–1170	8.13–8.55	s	C—N str
Ar—S—CO—NHAr	1160–1150	8.62–8.70	s	C—N str

Lactones, >C—(C<)$_n$—CO (ring closed through —O—)

Lactones have bands due to the stretching of the C=O and C—O groups. The C=O stretching vibration for saturated γ-lactones[29,118] (five-membered ring) is at higher frequencies, 1790–1770 cm^{-1} (5.59–5.65 μm), than for aliphatic esters. Electronegative substituents on the γ-carbon atom tend to increase the frequency. The absorptions of δ-lactones[119] (six-membered ring) and larger lactones are similar to those of open-chain esters. α,β-Unsaturated γ-lactones have two bands due to the carbonyl stretching vibration, at 1790–1775 cm^{-1} (5.59–5.63 μm) and 1765–1740 cm^{-1} (5.67–5.75 μm), even though only one carbonyl group is present. This is probably due to Fermi resonance.[14] α,β–γ,δ-Unsaturated δ-lactones similarly have two carbonyl absorption bands which are at 1775–1740 cm^{-1} (5.63–5.75 μm) and 1740–1715 cm^{-1} (5.75–5.83 μm).

The band due to the C—O stretching vibration of lactones occurs in the region 1370–1160 cm^{-1} (7.29–8.62 μm), usually being at 1240–1220 cm^{-1} (8.07–8.12 μm) for δ-lactones.

Table 10.22. Lactone C=O and C—O stretching vibrations

Functional Groups	Region cm^{-1}	μm	Intensity	Comments
β-Lactones (four-membered ring)	1840–1815	5.44–5.51	s	C=O str, halogen substitution results in higher frequencies
γ-Lactones (saturated five-membered ring)	1790–1770	5.59–5.65	s	C=O str
α,β-Unsaturated γ-lactones (unsaturated five-membered ring)	1790–1775	5.59–5.63	s	C=O str } Doublet due to Fermi resonance
	1765–1740	5.67–5.75	s	C=O str }
β,γ-Unsaturated γ-lactones (unsaturated five-membered ring)	1815–1785	5.51–5.60	s	C=O str
δ-Lactones and larger	As for open-chain ester			
α,β-Unsaturated δ-lactones	1745–1730	5.73–5.78	s	C=O str
α,β-γ,δ-Unsaturated δ-lactones (unsaturated six-membered ring)	1775–1740	5.63–5.75	s	C=O str } Doublet due to Fermi resonance
	1740–1715	5.75–5.83	s	C=O str }
2-Benzofuranones, 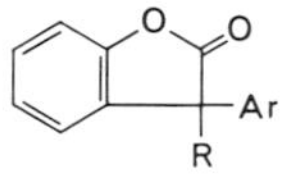	1820–1800	5.50–5.56	s	C=O str, see ref. 124
Phthalides, 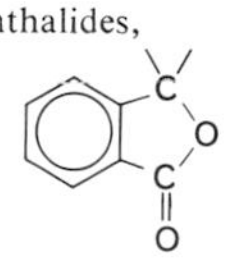	1775–1710	5.63–5.85	s	C=O str
	1290 1280	7.75–7.81	m–s	Characteristic phthalide ring vibrations
	1120–1100	8.93–9.01	m–s	
	1020–1010	9.80–9.90	w–m	
	515–490	19.42–20.41	w–m	
	490–470	20.41–21.28	w–m	
Lactones	1370–1160	7.30–8.62	s	C—O str

Amides, —CO—N<

All amides exhibit a band due to the C=O stretching vibration, with primary and secondary amides also having bands due to the N—H stretching and deformation vibrations. The positions of the carbonyl band and the N—H bands (if present) are dependent on the amount of hydrogen bonding occurring. The position of the carbonyl band depends also on the substituents on the nitrogen atom. Overtones of the bands due to the N—H stretching vibration for primary and secondary amides occur in the near infrared region.[125]

The absorption bands of even quite small molecules cannot strictly be considered as arising from a single vibration source. In other words, a given absorption band is never due solely to, say, the stretching vibration of the A—B group since in reality the whole molecule is involved. However, because of the complexity of the actual situation, the tendency is to simplify (in some cases, to oversimplify) mainly because it is useful to identify the major cause of any given band. In fact, all statements as to the vibration source of any band should always be interpreted as meaning that the stated type of vibration is the major, not the sole, contribution to that band. In the case of amides, in acknowledgement of the complexity of the situation, the bands observed are given names such as amide I, amide II, etc., rather than C=O stretching, etc.

Amide N—H stretching vibrations

For primary amides, two sharp bands of medium intensity are observed due to the asymmetric and symmetric stretching vibrations. In dilute, non-polar solvents, i.e. in the absence of hydrogen bonding, these bands occur at about 3500 cm^{-1} (2.86 μm) and 3400 cm^{-1} (2.94 μm).[126, 127] In the solid state and in the presence of hydrogen bonding, these bands are shifted by about 150 cm^{-1} (0.16 μm) to about 3350 cm^{-1} (2.99 μm) and 3200 cm^{-1} (3.13 μm). Both primary and secondary amides may exhibit a number of bands due to different hydrogen-bond states, e.g. dimers, trimers, etc. The bands are concentration- and solvent-dependent. Free (unassociated) secondary amides have a sharp, strong band at 3460–3300 cm^{-1} (2.89–3.03 μm).[128] This band may appear as a doublet due to the presence of *cis–trans* isomerism.[129, 144] In the solid or liquid phases, secondary amides generally exhibit a strong band at about 3270 cm^{-1} (3.06 μm) and a weak band at 3100–3070 cm^{-1} (3.23–3.26 μm).

The *cis*- and *trans*- forms of secondary amides may be distinguished by examination of their N—H vibration bands, as indicated in Table 10.23.

Table 10.23. The N—H vibration bands of secondary amides

Type of secondary amide	Region cm^{-1}	μm	Intensity	Comments
Hydrogen-bonded *trans*- form (solid or liquid phase)	3370–3270	2.97–3.06	m	N—H str
	3100–3070	3.23–3.26	w	Overtone of amide II band
	1570–1515	6.37–6.60	s	Amide II band
Hydrogen-bonded *cis*- form (solid or liquid phase) (may be as dimers)	3180–3140	3.15–3.19	m	N—H str
	~3080	~3.25	w	
	1450–1440	6.90–6.94	s	N—H def
Trans- form (in dilute solution)	3460–3420	2.89–2.92	m	N—H str
	1550–1510	6.45–6.62	s	Amide II band
Cis- form (in dilute solution)	3440–3300	2.91–3.03	m	N—H str, *cis*- form remains mainly associated even in very dilute solution whereas *trans*- form does not

Amide C=O stretching vibrations: amide I band

The amide band due to the C=O stretching vibration is often referred to as the amide I band.[130] Primary amides[6] have a very strong band due to the C=O stretching vibration at 1670–1650 cm^{-1} (5.99–6.06 μm) in the solid phase, the band appearing at 1690–1670 cm^{-1} (5.92–5.99 μm) for a dilute solution using a non-polar solvent. In the solid phase, secondary amides absorb strongly at 1680–1630 cm^{-1} (5.95–6.14 μm), and in dilute solution at 1700–1665 cm^{-1} (5.88–6.01 μm).[131–133, 145] The carbonyl absorption band of tertiary amides[134, 135] is independent of physical state, since hydrogen bonding to another amide molecule is not possible, and occurs in the region 1670–1630 cm^{-1} (5.99–6.14 μm). If the substituent on the nitrogen is an aromatic for either secondary or tertiary amides then the carbonyl absorption occurs at the higher end of the frequency ranges given,[136–138] whereas aliphatic secondary amides absorb at 1650–1630 cm^{-1} (6.06–6.14 μm). The carbonyl absorption band is obviously greatly influenced by solvents with which hydrogen bonds may be formed.

Primary α-halogenated amides[15] absorb at higher frequencies than the corresponding alkyl compound, up to about 1750 cm^{-1} (5.71 μm), and may, in fact, have two carbonyl bands due to the presence of rotational isomerism. The carbonyl band of *N*-halogen secondary amides also occurs at higher frequencies than that of the corresponding *N*-alkyl compound.[131c, 132]

In dilute solution in non-polar solvents, acetanilides and benzanilides absorb in the region 1710–1695 cm^{-1} (5.85–5.90 μm).[131c, 136, 137] *Ortho*-nitro-substituted anilides, in the solid phase, exhibit two carbonyl bands, one near 1700 cm^{-1} (5.88 μm) and the other at about 1670 cm^{-1} (6.00 μm). Compounds of the type $CH_3(Ar)NCOCH_3$ absorb in the region 1685–1650 cm^{-1} (5.93–6.06 μm).

The carbonyl stretching vibration frequency of *N*-acetyl and *N*-benzoyl groups in compounds where the nitrogen atom forms part of a heterocyclic ring increases as the resonance energy is increased, e.g. by increasing the number of nitrogen atoms in the ring.[139] For example, in the case of pyrroles, the carbonyl band occurs near 1730 cm^{-1} (5.78 μm) and in the case of tetrazoles, at about 1780 cm^{-1} (5.62 μm).

Oxamides,[146] thioamides,[146–149] amides of *n*-fatty acids,[150] polyamides,[142] phosphonamides,[131b] polyglycines,[141] and numerous other related compounds have been studied.

Amide N—H deformation and C—N stretching vibrations: amide II band

In the solid phase, primary amides have a weak-to-medium intensity band at 1650–1620 cm^{-1} (6.06–6.17 μm) which is generally too close to the strong carbonyl band to be resolved. In dilute solution, this band occurs at 1620–1590 cm^{-1} (6.17–6.31 μm). The position of this band is not greatly influenced by the nature of the primary amide, e.g. aliphatic or aromatic. This band is known as the amide II band and is due to a motion combining both the N—H bending and the C—N stretching vibrations of the group —CO—NH— in its *trans*- form. The amide II band appears to be mainly due to the N—H bending motion. Secondary amides in the solid phase have a characteristic, strong absorption at 1570–1515 cm^{-1} (6.37–6.60 μm) and in dilute solution, at 1550–1510 cm^{-1} (6.45–6.62 μm). In general, the amide II band of primary amides is more intense than that of secondary amides. In fact, it has been observed tha the amide II band is absent in *trans-N*-halogen secondary amides although it is present for *N*-iodo-amides in the solid phase.[131a, b, 132, 140, 143]

Other amide bands

Primary amides absorb at 1420–1400 cm^{-1} (7.04–7.14 μm) and secondary amides at 1305–1200 cm^{-1} (7.67–8.33 μm) and at about 700 cm^{-1} (14.3 μm). This last band may not be observed in that position in the spectra of dilute solutions.

In general, all amides have one or more bands of medium-to-strong intensity, which may be broad, in the region 695–550 cm^{-1} (14.39–18.18 μm) which are probably due to the bending motion of the O=C—N group.[141, 142] Primary aliphatic amides absorb at 635–570 cm^{-1} (15.75–17.54 μm), probably due to the out-of-plane bending of the C=O group, whereas α-branched primary amides absorb at 665–580 cm^{-1} (15.04–17.24 μm). Secondary aliphatic amides absorb at 610–590 cm^{-1} (16.39–16.95 μm) and in the case of α-branching, at 670–625 cm^{-1} (14.93–16.00 μm). With the exception of formamides, anilides, and diamides, amides have a medium-to-strong absorption at 520–430 cm^{-1} (19.23–23.26 μm) and, with the exception of *N*-methyl secondary amides, *N*-substituted anilides, lactams, and diamides (also acetamide and propionamide), a band which is usually observed at 390–305 cm^{-1} (25.64–32.79 μm). This last band is sensitive to conformational changes and has been observed as low as 215 cm^{-1} (46.51 μm). Formamide has a strong, broad absorption in the range 700–500 cm^{-1} (14.29–20.00 μm).

Hydroxamic acids, —CO—NHOH

Hydroxamic acids have a strong carbonyl absorption at about 1640 cm^{-1} (6.10 μm). In the solid phase, three medium-intensity bands are observed at 3300–2800 cm^{-1} (3.03–3.57 μm), a strong amide II band is observed near 1550 cm^{-1} (6.45 μm), a strong band at about 900 cm^{-1} (11.11 μm), and a band of variable intensity at 1440–1360 cm^{-1} (6.94–7.35 μm).

Table 10.24. Amide N—H stretching vibrations (and other bands in same region)

Functional Groups	Region cm^{-1}	Region μm	Intensity	Comments
(Free) primary amides, $-CO-NH_2$	3540–3480	2.83–2.88	m–s	asym N—H str
	3420–3380	2.92–2.96	m–s	sym N—H str
(Associated) primary amides	3360–3320	2.97–3.01	m–s	asym N—H str
	3220–3180	3.11–3.15	m–s	sym N—H str
(Free) secondary amides, —CO—NH—	3460–3420	2.89–2.93	m–s	Doublet if *cis–trans* isomerism present, N—H str
(Associated) secondary amides:				
trans- form	3370–3270	2.97–3.06	m	N—H str
cis- form	3180–3140	3.15–3.19	m	N—H str
trans- form	3100–3070	3.23–3.26	w	Overtone of amide II band near 1550 cm^{-1}
Hydroxamic acids (solid phase), —CO—NHOH	3300–2800	3.03–3.57	w–m	Three bands, N—H str and O—H str

Table 10.25. Amide C=O stretching vibrations: amide I bands

Functional Groups	Region cm^{-1}	Region μm	Intensity	Comments
Primary amides (solid phase)	1670–1650	5.99–6.06	s	Usually a doublet involving NH_2 def at ~1620 cm^{-1}
Primary amides (dilute solution)	1690–1670	5.92–5.99	s	
Secondary amides (solid phase)	1680–1630	5.95–6.14	s	
Secondary amides (dilute solution)	1700–1665	5.88–6.01	s	Strongly electron-accepting groups on nitrogen increase frequency
Acetanilides (dilute solution), $ArNH.CO.CH_3$	1710–1695	5.85–5.90	s	
Secondary amides of the type ArCO.NH—(dilute solution)	~1660	~6.02	s	
Tertiary amides (dilute solution or solid phase)	1670–1630	5.99–6.14	s	Strongly electron-accepting groups on nitrogen increase frequency
Amides containing —CO—NH—CO— (diacylamines)	1790–1720	5.59–5.81	s	Doublet, separation usually small, but larger for ring amides
	1720–1670	5.81–5.99	s	
Monosubstituted hydrazides, $-CONHNH_2$	1700–1640	5.88–6.10	s	Acid hydrazides, see ref. 138
Disubstituted hydrazides, —CONH.NHCO—	1745–1700	5.73–5.88	s	Doublet, usually marked difference between solid phase and solution spectra, amide II band for aliphatic compounds at 1500–1480 cm^{-1} (6.67–6.76 μm)
	1710–1680	5.85–5.95	s	
Alkyl hydroxamic acids (solid phase), R—CO.NH.OH	~1640	~6.10	s	
Amides of the type $Ar-SO_2.NHCOCH_3$ (solid phase)	1720–1685	5.81–5.93	s	
Aromatic isocyanates (dimers)	1785–1775	5.60–5.64	s	
Aliphatic isocyanurates (isocyanate trimers)	1700–1680	5.88–5.95	s	Shoulder at ~1755 cm^{-1}
Aromatic isocyanurates	1715–1710	5.83–5.85	s	Shoulder at ~1780 cm^{-1}
$-CF_2CONH_2$	1730–1700	5.78–5.88	s	
CF_3CONH-	1740–1695	5.75–5.90	s	
Carbamoyl chlorides, Cl.CO.N<	~1740	~5.75	s	See ref. 170
Polypeptides	~1650	~6.06	s	Mainly C=O str but coupled with C=N str also (due to group —CO—NH—)
$CHCl_2CONH-$	1715–1700	5.83–5.88	s	
CCl_3CONH-	~1730	~5.78	s	

Table 10.26. Amide N—H deformation and C—N stretching vibrations: amide II band

Functional Groups	Region cm^{-1}	Region μm	Intensity	Comments
Primary amides (solid phase)	1650–1620	6.06–6.17	w–m	(Exception is *o*-cyanobenzamide at 1667 cm^{-1})
Primary amides (dilute solution)	1620–1590	6.17–6.31	w–m	*n*-alkylamides ~1590 cm^{-1}
Secondary amides (*trans*- form) (solid phase)	1570–1515	6.37–6.60	s	
Secondary amides (*trans*- form) (dilute solution)	1550–1510	6.45–6.62	s	
Aliphatic disubstituted hydrazides —CONH.NHCO—	1500–1480	6.67–6.76	s	
$-CF_2CONH-$	1630–1610	6.14–6.21	m	

Table 10.27. Amides: other bands

Functional Groups	Region cm^{-1}	Region μm	Intensity	Comments
Primary amides	1420–1400	7.04–7.14	m	C—N str, known as amide III band
	~1150	~8.70	w	NH_2 in-plane rocking, not always seen
	750–600	13.33–16.67	m	br, NH_2 def
Secondary amides (*trans*- form)	1305–1200	7.67–8.33	w–m	amide III band, usually at ~1260 cm^{-1}
	770–620	13.00–16.13	m	br, out-of-plane N—H def, for hydrogen-bonded amides usually at ~700 cm^{-1}
Secondary amides (*cis*- form)	1450–1440	6.90–6.94	m	N—H bending
	1350–1310	7.41–7.63	w–m	C—N str (amide III band)
	~800	~12.50	m–s	br, N—H wagging
Monosubstituted hydrazides	1150–950	8.70–10.53	m–s	Two bands, NH_2 def
Primary aliphatic amides, $R-CH_2-CONH_2$	635–570	15.75–17.54	s	N—C=O def (Not formamides or anilides)
	480–450	20.83–22.22	m–s	C—C=O in-plane def
	360–320	27.78–31.25	s	C—CO—N def
Primary α-branched aliphatic amides, $>C-CO-NH_2$	665–580	15.04–17.24	s	
	520–495	19.23–20.20	m–s	
	320–305	31.25–32.79	s	
n-Aliphatic secondary amides and *N*-methyl aliphatic amides, $R_1-CH_2-CO-NHR_2$	610–590	16.39–16.95	m–s	
	480–430	20.83–23.26	s	
	380–330	26.32–30.30	m–s	Absent for *N*-methyl aliphatic amides
α-Branched aliphatic secondary amides, 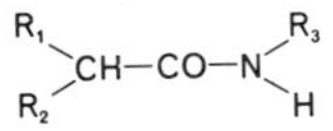	670–625	14.93–16.00	m	
	520–510	19.23–19.61	s	
	350–330	28.57–30.30	s	
Tertiary amides, 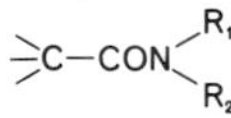	600–570	16.67–17.54	s	Absent for formamides
	480–440	20.83–22.73	m–s	Absent for formamides
	390–320	25.64–31.25	m	
Tertiary formamides, $H-CO-NR_1R_2$	700–645	14.29–15.50	s	Usually broad
	390–340	25.64–29.41	m–s	
N-Substituted anilides, 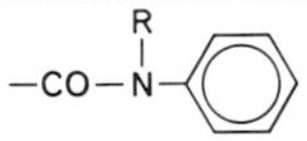	630–610	15.87–16.39	m	
	570–530	17.54–18.87	s	
	~445	~22.47	m	
	~405	~24.69	s	
Primary aromatic amides	645–590	15.50–16.95	s	N—C=O bending
Lactams, 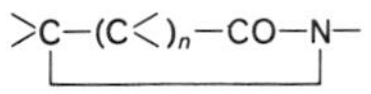	695–655	14.39–15.27	m–s	
	500–470	20.00–21.28	s	
Diamides, 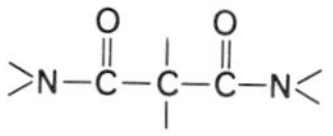	675–600	14.81–16.67	s	
	595–540	16.81–18.52	m–s	

Hydrazides, —CO—NH—NH_2 and —CO—NH—NH—CO—

Amides of the type —CO—NH—NH_2[138] have a number of medium-intensity bands in the region 3350–3180 cm^{-1} (2.99–3.15 μm) due to the NH and NH_2 stretching vibrations. The band due to the carbonyl stretching vibration, which is very strong, occurs at 1700–1640 cm^{-1} (5.88–6.10 μm). A medium-intensity band due to the deformation of the NH_2 group occurs at 1635–1600 cm^{-1} (6.12–6.25 μm). The amide II band, which is strong, occurs at 1545–1520 cm^{-1} (6.47–6.58 μm) and a weak-to-medium intensity band, due to the C—N stretching vibration, occurs at 1150–1050 cm^{-1} (8.70–9.52 μm).

For solid-phase spectra, aliphatic amides with the —CO—NH—NH—CO— group usually have only one very strong absorption due to the carbonyl groups, at 1625–1580 cm^{-1} (6.16–6.33 μm), whereas aromatic compounds usually have two strong bands, one at 1730–1670 cm^{-1} (5.78–5.99 μm) and the other at 1660–1635 cm^{-1} (6.02–6.12 μm). There are usually marked differences in the carbonyl band positions between the solution and solid-phase spectra of hydrazides.

Table 10.28. Hydrazides

Functional Groups	Region cm^{-1}	Region μm	Intensity	Comments
Amides with —CO—NH—NH_2 group	3350–3180	2.99–3.15	m	N—H str
	1700–1640	5.88–6.10	vs	C=O str
	1635–1600	6.12–6.25	m	NH_2 def
	1545–1520	6.47–6.58	s	Amide II band
	1150–1050	8.70–9.52	w–m	C—N str
Aliphatic and aryl amides with —CO—NH—NH—CO— group (in solution)	3330–3280	3.00–3.05	m	N—H str
	1745–1700	5.73–5.88	vs	C=O str
	1710–1680	5.85–5.95	vs	C=O str
	1535–1480	6.52–6.64	m	Amide II band, aliphatics at 1500–1480 cm^{-1}
Aliphatic amides with —CO—NH—NH—CO— group (solid phase)	3210–3100	3.12–3.23	m	N—H str
	3060–3020	3.27–3.31	m	N—H str
	1625–1580	6.15–6.33	vs	C=O str
	1505–1480	6.65–6.76	s	Amide II band
	1260–1200	7.94–8.33	m	C—N str
Aromatic amides with —CO—NH—NH—CO— group (solid phase)	3280–2980	3.05–3.36	m	N—H str
	1730–1670	5.78–5.99	vs	C=O str
	1660–1635	6.02–6.12	vs	C=O str
	1535–1525	6.52–6.56	s	Amide II band
	1285–1245	7.78–8.03	m	C—N str
Phthalhydrazides, 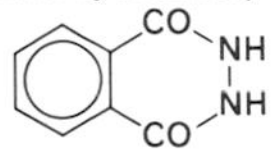	~3000	~3.33	m	Very br, N—H str
	1670–1635	5.99–6.12	vs	C=O str

Lactams —C—(C<)$_n$—CO (ring closed through —NH—) (cyclic amides)

The N—H and C=O stretching vibrations of lactams[34] give rise to bands in the same regions as those for secondary amides. Where ring strain occurs, as for β-lactams (four-membered ring) and γ-lactams (five-membered ring), the carbonyl stretching frequency is increased, the band regions being 1760–1730 cm^{-1} (5.68–5.78 μm) and 1720–1700 cm^{-1} (5.81–5.88 μm) respectively. The amide II band is not exhibited by lactams unless the ring consists of nine or more members, this band being associated with the group —CO—NH— in the *trans-* form. The N—H out-of-plane deformation band occurs at about 700 cm^{-1} (14.3 μm) and is generally broad as for secondary amides.

α,β-Unsaturation results in an increase in the carbonyl stretching vibration frequency by about 15 cm^{-1} (0.05 μm). Fused ring β- and γ-lactams have the frequency of their carbonyl stretching vibration increased by about 20–30 cm^{-1} (0.06–0.07 μm) compared with that of simple β- and γ-lactams. β-Lactams fused to unoxidized thiazolidine rings absorb at 1780–1770 cm^{-1} (5.62–5.65 μm).[149]

Table 10.29. Lactam C=O stretching vibrations: amide I band

Functional Groups	Region cm^{-1}	Region μm	Intensity	Comments
β-Lactams (four-membered ring) (dilute solution)	1760–1730	5.68–5.78	s	
γ-Lactams (five-membered ring) (dilute solution)	1720–1700	5.81–5.88	s	
δ-Lactams (six-membered ring) (dilute solution)	1690–1670	5.92–5.99	s	
β-Lactams (ring fused) (dilute solution)	1780–1770	5.62–5.65	s	
γ-Lactams (ring fused) (dilute solution)	1750–1700	5.71–5.88	s	

Table 10.30. Lactams: other bands

Functional Groups	Region cm^{-1}	Region μm	Intensity	Comments
Lactams	1315–1250	7.60–8.00	w	Amide III band, C—N str
	695–655	14.39–15.27	m–s	
	500–470	20.00–21.28	s	

Imides, —CO—NH—CO—

Imides[151] may exist in two forms: (a) the two carbonyl groups both *trans-* to the NH group, (b) one carbonyl group being *cis-* and the other *trans-* to the NH group. The *trans–trans* type has, in its solid-phase spectra, a medium-intensity absorption at 3280–3200 cm^{-1} (3.05–3.13 μm), due to the N—H stretching vibration, and strong bands at 1740–1730 cm^{-1} (5.75–5.78 μm) (the carbonyl band), at 1510–1500 cm^{-1} (6.62–6.67 μm) (the amide II band), at 1235–1165 cm^{-1} (8.10–8.58 μm) (the amide III band), and at 740–730 cm^{-1}

(13.51–13.70 μm) (due to the N—H wagging vibration). The spectra of the *cis–trans* forms differ from the above in that the carbonyl band occurs near 1700 cm^{-1} (5.88 μm) with weaker bands near 1630 cm^{-1} (6.14 μm) and 1650 cm^{-1} (6.06 μm). The band due to the N—H stretching vibration occurs at 3250 cm^{-1} (3.08 μm) with weak bands on either side and the band due to the N—H wagging vibration occurs at 835–815 cm^{-1} (11.98–12.27 μm).

The carbonyl band of cyclic imides is shifted to higher frequencies if the ring is strained. Cyclic imides do not have an amide II band near 1510 cm^{-1} (6.62 μm).

In general, acylic imides exhibit two amide I bands and weak amide IV bands have also been observed near 610 cm^{-1} (16.39 μm) and 560 cm^{-1} (17.86 μm).

Table 10.31. Imides

Functional Groups	Region cm^{-1}	μm	Intensity	Comments
Imides (solid phase)	3280–3200	3.05–3.13	m	N—H str
	1740–1670	5.75–5.99	vs	C=O str, amide I band
	1510–1500	6.62–6.67	vs	br, amide II band
	1235–1165	8.10–8.58	s	amide III band
	740–730	13.51–13.70	m–s	br, N—H wagging (N—D wagging ~540 cm^{-1})
Cyclic imides (five-membered ring), 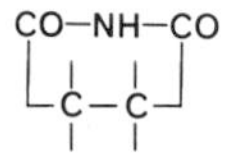	~1770	~5.65	s	Unsaturation results in an increase in wavenumber of about 15 cm^{-1}
	~1700	~5.88	s	
Cyclic imides (six-membered ring), 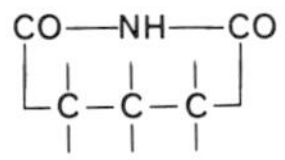	~1710	~5.85	s	
	~1700	~5.88	s	Usually of greater intensity than the other band
Maleimides, 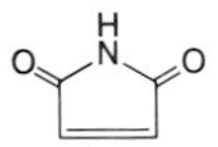	1805–1745	5.54–5.73	s	C=O str } doublet, see refs 152, 153
	1730–1685	5.78–5.93	s	C=O str }
	1650–1630	6.06–6.13	s	C=C str
	1550–1450	6.45–6.90	s	
	1365–1340	7.33–7.46	s	C—N str
	1080–1040	9.26–9.62	m	
	780–730	12.82–13.70	s	
Phthalimides, 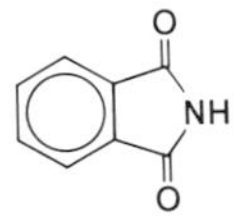	1790–1735	5.59–5.76	s	See ref. 154
	1745–1670	5.73–5.99	s	

Ureas, >N—CO—N< (carbamides)

The band due to the stretching vibration of the carbonyl group of ureas[29, 155–157] occurs at 1705–1635 cm^{-1} (5.82–6.12 μm). The presence of ring strain tends to increase the frequency of this vibration. Strongly electron-accepting groups on the nitrogen also raise this frequency (amides behave in a similar manner).

In dilute carbon tetrachloride solution, *N*-monoalkyl ureas have three bands due to the N—H stretching vibrations, the bands due to the NH_2 asymmetric and symmetric vibrations are at about 3515 cm^{-1} (2.85 μm) and 3415 cm^{-1} (2.93 μm) respectively and that due to the N—H stretching vibration, which varies according to the alkyl substituent, occurs at 3465–3440 cm^{-1} (2.89–2.91 μm). The carbonyl stretching vibration gives rise to a band near 1705 cm^{-1} (5.86 μm).

In dilute solution (non-polar solvent), sym-*N*,*N*′-dialkylureas have, essentially, a single band due to the N—H stretching vibration in the region 3465–3435 cm^{-1} (2.89–2.91 μm) and a strong band due to the C=O stretching vibration at about 1695 cm^{-1} (5.90 μm). Also in dilute solution (non-polar solvent), unsym-*N*,*N*′-dialkylureas may exhibit one or two bands due to the N—H stretching vibration.

The amide II band of ureas has been found at 1555–1515 cm^{-1} (6.43–6.60 μm). For associated (hydrogen-bonded) ureas, the band due to the N—H stretching vibration occurs at 3400–3360 cm^{-1} (2.94–2.98 μm) and that due to the C=O stretching vibration at about 1635 cm^{-1} (6.11 μm). For the monomer, this last band is found at about 1690 cm^{-1} (5.92 μm).

Ureas have a strong, characteristic band at 1490–1465 cm^{-1} (6.71–6.82 μm) due to the asymmetric stretching vibration of the N—C—N group, the band due to the symmetric vibration being of medium intensity and occurring at about 1010 cm^{-1} (9.90 μm).

Table 10.32. Urea C=O stretching vibrations: amide I band

Functional Groups	Region cm^{-1}	μm	Intensity	Comments
Ureas (solid phase)	1680–1635	6.33–6.12	s	br
Ureas (in solution)	1705–1660	5.86–6.02	s	
Cyclic ureas (five-membered ring) (in solution)	1735–1685	5.76–5.93	s	Ketone groups in ring increase frequency
Diaryl ureas, ArNH—CO—NHAr (solid phase)	~1640	~6.10	s	
N-Chloro diaryl ureas, ArNCl—CO—NClAr (solid phase)	1735–1710	5.76–5.85	s	

Table 10.33. Ureas: other bands

Functional Groups	Region cm⁻¹	μm	Intensity	Comments
Ureas	~1600	~6.25	m	Amide II band
	1490–1465	6.71–6.82	s	asym N—C—N str
	~1155	~8.65	m	
	~1010	~9.90	m	sym N—C—N str
	~715	~13.99	m–s	
	~530	~18.87	m	NH_2 def

Urethanes, >N—CO—O— (carbamates)

The band due to the carbonyl stretching vibration (amide I band) of urethanes[158–163] occurs in the region 1740–1680 cm^{-1} (5.75–5.95 μm). Primary urethanes have a number of absorptions in the region 3450–3200 cm^{-1} (2.90–3.13 μm) due to the N—H stretching vibration. Secondary urethanes absorb near 3300 cm^{-1} (3.03 μm) if hydrogen-bonding occurs and at 3450–3390 cm^{-1} (2.92–2.95 μm) if it is absent. For alkyl primary urethanes in chloroform solution, the amide I band is observed at 1730–1720 cm^{-1} (5.78–5.81 μm), for secondary urethanes (*N*-monosubstituted) at 1720–1705 cm^{-1} (5.81–5.87 μm), and for tertiary urethanes (*N,N*-disubstituted) at 1690–1680 cm^{-1} (5.92–5.95 μm). These ranges may be slightly lower in frequency than for other solvents. In the solid phase, primary urethanes may have very broad C=O bands and absorb as low as 1690 cm^{-1} (5.92 μm), otherwise the same general absorption pattern is observed.

Primary urethanes have a medium-to-strong band near 1620 cm^{-1} (6.17 μm) due to the deformation vibrations of the NH_2 group. Associated secondary urethanes absorb strongly at 1540–1530 cm^{-1} (6.49–6.54 μm) due to the CHN group vibration (similar to that of secondary amides) and in dilute solution this band is found at 1530–1510 cm^{-1} (6.54–6.62 μm).

Table 10.34. Urethane N—H stretching vibrations

Functional Groups	Region cm⁻¹	μm	Intensity	Comments
Primary urethanes, H_2N—CO—O—	3450–3200	2.90–3.13	m	
(Associated) secondary urethanes, —HN—CO—O—	3340–3250	2.99–3.08	m	Hydrogen-bonded
(Unassociated) secondary urethanes	3450–3390	2.92–2.95	m	
N-Aryl urethanes (associated), Ar—NH—CO.OR	3460–3295	2.89–3.03	m	
N-Aryl urethanes (unassociated)	3460–3410	2.89–2.93	m	

Table 10.35. Urethane C=O stretching vibrations: amide I band

Functional Groups	Region cm⁻¹	μm	Intensity	Comments
Alkyl urethanes, >NCO—O—	1740–1680	5.75–5.95	s	
N-Aryl urethanes, Ar—NH—CO.OR (associated) (solid phase)	1735–1705	5.75–5.87	vs	One or two peaks, strong hydrogen bonding may result in band as low as 1690 cm^{-1}
N-Aryl urethanes (unassociated)	1760–1730	5.68–5.78	vs	See ref. 163
Alkyl thiocarbamates, —NH—CO—S—	~1695	~5.90	s	See ref. 53
N-Aryl thiocarbamates	1700–1660	5.88–6.02	s	
Alkyl carbamoyl chlorides, NR_2.COCl	1745–1735	5.73–5.75	vs	
Cyclic urethanes (five-membered ring) 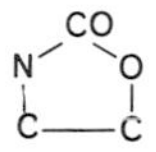	1785–1745	5.60–5.73	vs	Ring carbonyl groups increase frequency (N-acetyloxazolidones have two bands: at ~1795 cm^{-1} and ~1710 cm^{-1}

Table 10.36. Urethane combination N—H deformation and C—N stretching vibrations (amide II band) and other bands

Functional Groups	Region cm⁻¹	μm	Intensity	Comments
Primary urethanes	~1620	~6.17	m–s	NH_2 def
Secondary urethanes (dilute solution)	1530–1510	6.54–6.62	s	Absorption due to CHN group
Secondary urethanes (associated or in solid phase)	1540–1530	6.49–6.54	s	
Urethanes	1265–1200	7.90–8.33		Amide IV band (coupled C—N and C—O stretching vibrations)
N-Aryl urethanes (unassociated)	1550–1500	6.45–6.67	s	In-plane N—H bending
	1285–1235	7.78–8.10	m	Probably Ar—N str, does not alter significantly on phase change
	1225–1195	8.14–8.36	vs	Amide II band, stronger than C=O band in solution spectra, in solid-phase band occurs at 1260–1220 cm^{-1} (7.94–8.20 μm)
	1090–1040	9.17–9.62	m–s	C—O str
	570–500	17.54–20.00	w–m	Out-of-plane N—H def, in solid phase band occurs at 680–625 cm^{-1} (14.71–16.00 μm), *ortho*-halogen substituted compounds absorb in range 570–550 cm^{-1} (17.54–18.18 μm)

Amino Acids

Amino acids[164–168] are amine derivatives of carboxylic acids and may, in fact, contain a number of amino and carboxylic acid groups. In the simplest case, with one acid and one amino group, the amino group may occupy α or β or γ, etc., positions. Amino acids, polypeptides, and proteins are related compounds and their infrared spectra reflect this to a certain extent.[165]

Amino acids may be found in three forms:

(a) as a free acid,

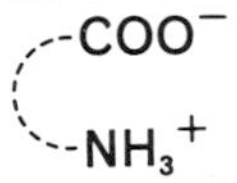

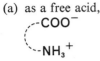

where the dotted line represents any carbon backbone structure.

(b) as the salt, e.g. sodium, of the acid,

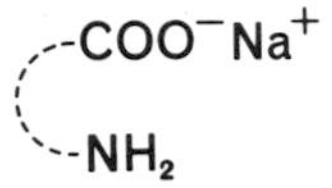

(c) as the amine hydrohalide,

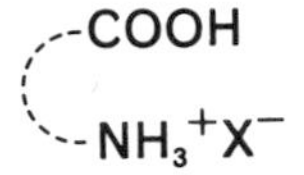

Free amino acid —NH_3^+ vibrations[167, 168]

Free amino acids have —NH_3^+ stretching and deformation vibrations. In the solid phase, a broad absorption of medium intensity is observed in the region 3200–3000 cm^{-1} (3.13–3.33 μm) due to the asymmetric —NH_3^+ stretching vibration. Weak bands due to the symmetric stretching vibration of the NH_3^+ group are observed near 2600 cm^{-1} (3.85 μm) and 2100 cm^{-1} (4.76 μm). A fairly strong —NH_3^+ deformation band is observed at 1550–1485 cm^{-1} (6.46–6.74 μm) and a weaker band, which is not resolved for most amino acids, at 1660–1590 cm^{-1} (6.03–6.29 μm).

Free amino acid carboxyl bands[164–168]

Free amino acids also have carboxylate ion CO_2^- stretching vibrations, a strong band occurring in the region 1600–1560 cm^{-1} (6.25–6.41 μm). Dicarboxylic acids have a strong band due to the C=O stretching vibration of the carboxyl group at 1755–1700 cm^{-1} (5.70–5.88 μm) and another strong band at 1230–1215 cm^{-1} (8.13–8.23 μm) due to the stretching vibration of the C—O bond. A band of medium intensity and of uncertain origin is usually observed near 1320 cm^{-1} (7.68 μm). A strong band at 560–500 cm^{-1} (17.86–20.00 μm) which is due to the CO_2^- or C—C—N group deformation vibrations is observed for amino acids, except for cyclic amino acids. Skeletal deformation bands occur in the region 500–285 cm^{-1} (20.00–35.09 μm).

Amino acid hydrohalides

In addition to the —NH_3^+ stretching and deformation absorption bands, which are as given above, a series of weak, fairly broad bands are observed in the region 3000–2500 cm^{-1} (3.33–4.00 μm). Also, the band due to the C—O stretching vibration is the same as for free amino acids. The band due to the C=O stretching vibration of the carboxyl group is observed in the range 1755–1700 cm^{-1} (5.70–5.88 μm).

Amino acid salts

For amino acid salts, two bands of medium intensity are observed at 3400–3200 cm^{-1} (2.94–3.13 μm) due to the asymmetric and symmetric stretching vibrations of the —NH_2 group. A strong band at 1600–1560 cm^{-1} (6.25–6.42 μm) due to the carboxylate ion is observed.

Table 10.37. Amino acid —NH_3^+ and N—H vibrations

Functional Groups	Region cm^{-1}	Region μm	Intensity	Comments
Free amino acids $(NH_3^+)\cdots COO^-$	3200–3000	3.13–3.33	m	asym —NH_3^+ str
and amino acid hydrohalides	2760–2530	3.62–3.95	w–m	br } —NH_3^+ sym str
$X^-(NH_3^+)\cdots COOH$	2140–2050	4.67–4.88	w–m	} —NH_3^+ sym str
(X = halogen)	1660–1590	6.03–6.29	w	asym —NH_3^+ def
	1550–1485	6.46–6.74	w–m	sym NH_3^+ def
	1295–1090	7.72–9.18	w	NH_3^+ rocking
Deuterated amino acids	1190–1150	8.40–8.70	w	ND_3^+ def
	~800	~12.50	w	ND_3^+ rocking
Amino acid salts $(NH_2)\cdots COO^-M^+$ (M = metal atom, e.g. Na)	3400–3200	2.94–3.13	m	Two bands, —NH_2 str

Table 10.38. Amino acid carboxyl group vibrations

Functional Groups	Region cm^{-1}	Region μm	Intensity	Comments
Free amino acids and amino acid salts	1600–1560	6.25–6.41	s	asym CO_2^- str
	1425–1390	7.02–7.19	w	sym CO_2^- str
Amino acid hydrohalides and dicarboxylic amino acids	1755–1700	5.70–5.88	s	C=O str of —COOH, α-amino acids absorb at 1755–1730 cm^{-1}, other amino acids absorb at 1730–1700 cm^{-1}

Table 10.39. Amino acids: other bands

Functional Groups	Region cm⁻¹	μm	Intensity	Comments
Free amino acids	1340–1315	7.46–7.61	m	CH def
	560–500	17.86–20.00	s	
Amino acid hydrohalides	3000–2500	3.33–4.00	w	Series of broad bands
	~1300	~7.69	m	
	1230–1215	8.13–8.23	s	C—O str

Amido Acids, >N—CO—···COOH

In the solid phase, α-amido acids have a medium-intensity absorption at 3390–3260 cm^{-1} (2.95–3.07 μm) due to the stretching vibration of the N—H bond and strong bands at 1750–1695 cm^{-1} (5.71–5.90 μm), 1620–1600 cm^{-1} (6.14–6.25 μm), and 1565–1505 cm^{-1} (6.39–6.64 μm). The first two of these three bands are due to the carbonyl stretching vibration, the first being due to the acid and the other to the amide group. The third band is an amide II band. The band near 1610 cm^{-1} (6.21 μm) is characteristic of α-amido acids, other amido acids having this absorption at 1650–1620 cm^{-1} (6.06–6.14 μm).

Table 10.40. Amido acids

Functional Groups	Region cm⁻¹	μm	Intensity	Comments
Amido acids	3390–3260	2.95–3.07	m	N—H str
	2640–2360	3.79–4.24	w	Not always present
	1945–1835	5.14–5.45	w	Not always present
	1750–1695	5.71–5.90	s	Acid C=O str
	1565–1505	6.39–6.64	s	Amide II band
	1230–1215	8.13–8.23	s	C—O str
α-Amido acids	1620–1600	6.14–6.25	s	Amide C=O str
Other amido acids	1650–1620	6.06–6.14	s	Amide C=O str

References

1. P. Combelas *et al.*, *Ann. Chim. France*, 1970, **5**, 315
2. L. J. Bellamy and R. J. Pace, *Spectrochim. Acta*, 1963, **19**, 1831
3. W. A. Seth-Paul, *Spectrochim. Acta*, 1974, **30A**, 1817
4. V. G. Boitsov and Y. Y. Gotlib, *Optics Spectrosc.*, 1961, **11**, 372
5. C. N. R. Rao and R. Venkataraghavan, *Canad. J. Chem.*, 1961, **39**, 1757
6. M. St. Flett, *Spectrochim. Acta*, 1962, **18**, 1537
7. J. Bellantano and J. R. Barcello, *Spectrochim. Acta*, 1960, **16**, 1333
8. C. Fayat and A. Foucaud, *Compt. Rend.*, 1966, **263B**, 860
9. T. L. Brown, *Spectrochim. Acta*, 1962, **18**, 1617
10. J. L. Mateos *et al.*, *J. Org. Chem.*, 1961, **26**, 2494
11. H. N. Al-Jallo and M. G. Jalhoom, *Spectrochim. Acta*, 1975, **31A**, 265
12. J. Dabrowski, *Spectrochim. Acta*, 1963, **19**, 475
13. K. Noack, *Spectrochim. Acta*, 1962, **18**, 1625
14. R. N. Jones *et al.*, *Canad. J. Chem.*, 1959, **37**, 2007
15. T. L. Brown *et al.*, *J. Phys. Chem.*, 1959, **63**, 1324
16. R. L. Erskine and E. S. Waight, *J. Chem. Soc.*, **1960**, 3425
17. R. Mecke and K. Noack, *Chem. Ber.*, 1960, **93**, 210
18. T. Gramstad and W. J. Fuglevik, *Spectrochim. Acta*, 1965, **21**, 343
19. J. A. Pullin and R. L. Werner, *Spectrochim. Acta*, 1965, **21**, 1257
20. A. R. Katritzky and R. A. Jones, *Spectrochim. Acta*, 1961, **17**, 64
21. L. J. Bellamy *et al.*, *Zeit. Elektrochem.*, 1960, **64**, 563
22. L. J. Bellamy and P. E. Rogash, *Spectrochim. Acta*, 1960, **16**, 30
23. L. H. Bellamy and R. L. Williams, *Trans. Faraday Soc.*, 1959, **55**, 14
24. K. B. Whetsel and R. E. Kagarise, *Spectrochim. Acta*, 1962, **18**, 315, 329 and 341
25. J. Dabrowski and K. Kamiensa-Trela, *Spectrochim. Acta*, 1966, **22**, 211
26. H. P. Figeys and J. Nasielski, *Spectrochim. Acta*, 1967, **23A**, 465
27. R. A. Pethrick and A. D. Wilson, *Spectrochim. Acta*, 1974, **30A**, 1073
28. P. Bassignana *et al.*, *Spectrochim. Acta*, 1965, **21**, 677
29. H. K. Hall and R. Zbinden, *J. Amer. Chem. Soc.*, 1958, **80**, 6428
30. W. G. Dauben and W. W. Epstein, *J. Org. Chem.*, 1959, **24**, 1595
31. F. Marquarat, *J. Chem. Soc. B*, **1966**, 1242
32. L. Hough and J. E. Priddle, *J. Chem. Soc.*, **1961**, 3178
33. B. F. Kucherov *et al.*, *Izvest. Acad. Nauk USSR Otdel. Khim. Nauk*, **1958**, 186
34. H. Zahn and J. Kunde, *Chem. Bar.*, 1961, **94**, 2470
35. G. L. Caldow and W. H. Thompson, *Proc. Roy. Soc.*, 1960, **254A**, 1
36. A. D. Buckingham, *Trans. Faraday Soc.*, 1960, **56**, 753
37. L. J. Bellamy, In *Spectroscopy*, *Rept Conf. Organ. Hydrocarbon Res. Group, Inst. Petroleum*, **1962**, p. 205
38. H. H. Freedman, *J. Amer. Chem. Soc.*, 1960, **82**, 2454
39. D. Peltier *et al.*, *Compt Rend.*, 1959, **248**, 1148
40. C. J. W. Brooks, *J. Chem. Soc.*, **1961**, 106
41. K. Shimzu *et al.*, *Spectrochim. Acta*, 1966, **22**, 1528
42. P. J. Kruger, *Canad. J. Chem.*, 1973, **51**, 1363
43. W. A. Seth-Paul and A. van Duyse, *Spectrochim. Acta*, 1972, **28A**, 211
44. F. A. Long and R. Bakule, *J. Amer. Chem. Soc.*, 1963, **85**, 2313
45. R. Bauke and F. A. Long, *J. Amer. Chem. Soc.*, 1963, **85**, 2309
46. C. N. R. Rao and R. Venkataraghavan, *Spectrochim. Acta*, 1962, **18**, 273
47. R. Cataliotti and R. N. Jones, *Spectrochim. Acta*, 1971, **27A**, 2011
48. C. I. Angell *et al.*, *Spectrochim. Acta*, 1959, **15**, 926
49. G. Allen *et al.*, *J. Chem. Soc.*, **1960**, 1909
50. R. N. Jones *et al.*, *Canad. J. Chem.*, 1959, **37**, 2007
51. R. N. Jones and B. S. Gallagher, *J. Amer. Chem. Soc.*, 1959, **81**, 5242
52. L. J. Bellamy and R. L. Williams, *Trans. Faraday Soc.*, 1959, **55**, 14
53. R. A. Nyquist and W. J. Potts, *Spectrochim. Acta*, 1961, **17**, 679
54. R. A. Nyquist and W. J. Potts, *Spectrochim. Acta*, 1959, **15**, 514
55. B. Subrahmanyam *et al.*, *Current Sci.*, 1964, **33**, 304
56. N. L. Allinger *et al.*, *J. Amer. Chem. Soc.*, 1960, **82**, 5876
57. J. Cantacuzene, *J. Chim. Phys.*, 1962, **59**, 186
58. E. M. Marek *et al.*, *Zh. Prinkl. Spektrosk.*, 1973, **19**, 130
59. C. E. Griffin, *Spectrochim. Acta*, 1960, **16**, 1464
60. R. N. Jones and E. Spinner, *Canad. J. Chem.*, 1958, **36**, 1020
61. R. D. Campbell and H. M. Gilow, *J. Amer. Chem. Soc.*, 1962, **84**, 1440
62. R. A. Abromovitch, *Canad. J. Chem.*, 1959, **37**, 361 and 1146
63. R. Mecke and E. Funck, *Zeit. Elektrochem.*, 1956, **60**, 1124

64. A. N. Hambly and B. V. O'Grady, *Austral. J. Chem.*, 1963, **16,** 459
65. M. A. Gianturco and R. G. Pitcher, *App. Spectrosc.*, 1965, **19,** 109
66. J. K. Katon and F. F. Bentley, *Spectrochim. Acta*, 1963, **19,** 639
67. E. I. Matrosov and M. I. Kabachnik, *Spectrochim. Acta*, 1972, **28A,** 191
68. R. L. Edwards, *J. App. Chem.*, 1960, **10,** 246
69. J. Y. Savoie and P. Brassard, *Canad. J. Chem.*, 1966, **44,** 2867
70. J. F. Baghi, *J. Amer. Chem. Soc.*, 1962, **84,** 177
71. T. L. Brown, *Spectrochim. Acta*, 1962, **18,** 1065
72. E. D. Becker *et al.*, *Spectrochim. Acta*, 1963, **19,** 1871
73. H. Bloom *et al.*, *J. Chem. Soc.*, **1959,** 178
74. M. A. Slifkin, *Spectrochim. Acta*, 1973, **29A,** 835
75. A. Marco *et al.*, *Compt. Rend.*, 1972, **274B,** 400
76. E. Sanicki and T. R. Hauser, *Anal. Chem.*, 1959, **31,** 523
77. M. R. Padhye and B. G. Vilader, *J. Sci. Instr.Res.*, 1960, **19B,** 45.
78. W. F. Forbes, *Canad. J. Chem.*, 1962, **40,** 1891
79. E. L. Saier *et al.*, *Anal. Chem.*, 1962, **34,** 824
80. J. V. Pustinger *et al.*, *App. Spectrosc.*, 1964, **18,** 36
81. R. Blinc *et al.*, *Zeit. Elektrochem.*, 1960, **64,** 567
82. M. Oki and M. Hirota, *Bull. Chem. Soc. Japan*, 1963, **36,** 290
83. M. Josien *et al.*, *Compt. Rend.*, 1960, **250,** 4146
84. C. J. W. Brooks *et al.*, *J. Chem. Soc.*, **1961,** 661
85. A. J. Collins and K. J. Morgan, *J. Chem. Soc.*, **1963,** 3437
86. J. R. Barcello and C. Otero, *Spectrochim. Acta*, 1962, **18,** 1231
87. H. Susi, *Anal. Chem.*, 1959, **31,** 910
88. R. N. Jones, *Canad. J. Chem.*, 1962, **40,** 301
89. R. J. Jakobsen *et al.*, *App. Spectrosc.*, 1968, **22,** 641
90. J. W. Brasch *et al.*, *App. Spectrosc. Rev.*, 1968, **1** 187
91. R. J. Jakobsen *et al.*, *App. Spectrosc.*, 1968, **22,** 641
92. F. F. Bentley *et al.*, *Spectrochim. Acta*, 1964, **20,** 685
93. R. J. Jakobsen and J. E. Katon, *Spectrochim. Acta*, 1973, **29A,** 1953
94. E. Spinner, *J. Chem. Soc.*, **1964,** 4217
95. M. K. Hargreaves and E. A. Stevenson, *Spectrochim. Acta*, 1965, **21,** 1681
96. F. Vratny *et al.*, *Anal. Chem.*, 1961, **33,** 1455
97. J. H. S. Green *et al.*, *Spectrochim. Acta*, 1961, **17,** 486
98. L. L. Shevchenko, *Russ. Chem. Rev.*, 1963, **32,** 201
99. A. S. N. Murthy *et al.*, *Trans. Faraday Soc.*, 1962, **58,** 855
100. F. G. Pearson and R. B. Stasiak, *App. Spectrosc.*, 1958, **12,** 116
101. L. J. Bellamy and R. L. Williams, *J. Chem. Soc.*, **1958,** 3465
102. S. Pinchas *et al.*, *J. Chem. Soc.*, **1961,** 2382
103. H. N. Al-Jallo and M. G. Jalhoon, *Spectrochim. Acta*, 1972, **28A,** 1655
104. K. R. Loos and R. G. Lord, *Spectrochim. Acta*, 1965, **21,** 119
105. A. R. Katritzky *et al.*, *Spectrochim. Acta*, 1960, **16,** 964
106. J. Radell and L. A. Harrah, *J. Chem. Phys.*, 1962, **36,** 1571
107. R. N. Jones and E. Spinner, *Canad. J. Chem.*, 1958, **36,** 1020
108. B. J. Hales *et al.*, *J. Chem. Soc.*, **1957,** 618
109. H. W. Thompson and D. A. Jameson, *Spectrochim. Acta*, 1958, **13,** 236
110. J. R. Durig *et al.*, *J. Mol. Struct.*, 1970, **5,** 67
111. A. W. Baker and G. H. Harris, *J. Amer. Chem. Soc.*, 1960, **82,** 1923
112. F. Dalton *et al.*, *J. Chem. Soc.*, **1960,** 3681
113. J. K. Wilmshurst, *J. Mol. Spectrosc.*, 1957, **1,** 201
114. J. L. Lucier and F. F. Bentley, *Spectrochim. Acta*, 1964, **20,** 1
115. R. S. Tipson and H. S. Isbell, *J. Res. NBS*, 1960, **64A,** 405
116. R. S. Tipson and H. S. Isbell, *J. Res. NBS*, 1961, **65A,** 249
117. J. P. Freeman, *J. Amer. Chem. Soc.*, 1958, **80,** 5954
118. B. F. Kucherov *et al.*, *Izvest. Acad. Nauk USSR Otdel. Khim. Nauk*, **1958,** 186
119. F. Korte *et al.*, *Angew. Chem.*, 1959, **71,** 523
120. R. A. Nyquist, *Spectrochim. Acta*, 1972, **28A,** 285
121. H. A. Ory, *Spectrochim. Acta*, 1960, **16,** 1488
122. E. P. Blanchard and G. Buchi, *J. Amer. Chem. Soc.*, 1963, **85,** 955
123. R. J. Jakobsen and R. E. Wyant, *App. Spectrosc.*, 1960, **14,** 61
124. W. H. Washburn, *App. Spectrosc.*, 164, **18,** 61
125. S. E. Krikorian and M. Mahpour, *Spectrochim. Acta*, 1973, **29A,** 1233
126. A. R. Katritzky and R. A. Jones, *J. Chem. Soc.*, **1959,** 2067
127. A. R. Katritzky and R. A. Jones, *J. Chem. Soc.*, **1960,** 679
128. R. L. Jones, *J. Mol. Spectrosc.*, 1963, **11,** 411
129. R. A. Russell and H. W. Thompson, *Spectrochim. Acta*, 1956, **8,** 138
130. J. Jakes and S. Krimm, *Spectrochim. Acta*, 1971, **27A,** 35
131. R. A. Nyquist, *Spectrochim. Acta*, 1963, **19,** 509, 713 and 1595
132. R. D. Mclachlan and R. A. Nyquist, *Spectrochim. Acta*, 1964, **20,** 1397
133. M. Bear *et al.*, *J. Chem. Phys.*, 1958, **29,** 1097
134. C. D. Schmulbach and R. S. Drago, *J. Phys. Chem.*, 1960, **64,** 1956
135. A. J. Speziale and R. C. Freeman, *J. Amer. Chem. Soc.*, 1960, **82,** 903
136. H. H. Freedman, *J. Amer. Chem. Soc.*, 1960, **82,** 2454
137. E. J. Forbes *et al.*, *J. Chem. Soc.*, **1963,** 835
138. M. Mashima, *Bull. Chem. Soc. Japan*, 1962, **35,** 332 and 1862
139. N. Ogata, *Bull. Chem. Soc. Japan*, 1961, **34,** 245 and 249
140. J. E. Devia and J. C. Carter, *Spectrochim. Acta*, 1973, **29A,** 613
141. T. Miyazawa, *Bull. Chem. Soc. Japan*, 1961, **34,** 691
142. C. G. Cannon, *Spectrochim. Acta*, 1960, **16,** 302
143. W. J. Klein and A. R. Plesman, *Spectrochim. Acta*, 1972, **28A,** 673
144. T. L. Brown, *J. Phys. Chem.*, 1959, **63,** 1324
145. Y. Kuroda *et al.*, *Spectrochim. Acta*, 1973, **29A,** 411
146. B. Milligan *et al.*, *J. Chem. Soc.*, **1961,** 1919
147. H. O. Desseyn and M. A. Herman, *Spectrochim. Acta*, 1967, **23A,** 2457
148. H. O. Desseyn, *Spectrochim. Acta*, 1974, **30A,** 503
149. P. J. F. Griffiths and G. D. Morgan, *Spectrochim. Acta*, 1972, **28A,** 1899
150. K. Machida *et al.*, *Spectrochim. Acta*, 1972, **28A,** 235
151. T. Uno and K. Machida, *Bull. Chem. Soc. Japan*, 1961, **34,** 545 and 551
152. R. H. Wiley and S. C. Slaymaker, *J. Amer. Chem. Soc.*, 1958, **80,** 1385
153. D. E. Ames and T. F. Gray, *J. Chem. Soc.*, **1955,** 631
154. N. A. Borisevitch and N. N. Khoratovitch, *Optics Spectrosc.*, 1961, **10,** 309
155. E. Spinner, *Spectrochim. Acta*, 1959, **15,** 95
156. Y. Mido, *Spectrochim. Acta*, 1973, **29A,** 1 and 431
157. D. F. Kutepov and S. S. Dubov, *J. Gen. Chem. Moscow.*, 1960, **30** (92), 3416
158. S. Pinchas and D. Ben Ishai, *J. Amer. Chem. Soc.*, 1957, **79,** 4099
159. M. Sato, *J. Org. Chem.*, 1961, **26,** 770
160. A. R. Katritzky and R. A. Jones, *J. Chem. Soc.*, **1960,** 676
161. J. C. Carter and J. E. Devia, *Spectrochim. Acta*, 1973, **29A,** 623
162. H. S. Randhawa *et al.*, *Spectrochim. Acta*, 1974, **30A,** 1915
163. R. A. Nyquist, *Spectrochim. Acta*, 1973, **29A,** 1635
164. R. J. Koegel *et al.*, *Ann. NY Acad. Sci.*, 1957, **69,** 94
165. G. B. B. M. Sutherland, *Adv. Protein Chem.*, 1952, **7,** 291
166. R. J. Koegel *et al.*, *J. Amer. Chem. Soc.*, 1955, **77,** 5708
167. M. Tsuboi *et al.*, *Spectrochim. Acta*, 1963, **19,** 271
168. E. Steger *et al.*, *Spectrochim. Acta*, 1963, **19,** 293
169. J. Weinman *et al.*, *Chim. Phys.-Chim. Biol.*, 1976, **73,** 331
170. G. Guiheneut and C. Laurence, *Spectrochim. Acta*, 1978, **34A,** 15

CHAPTER 11

Aromatic Compounds

For simplicity and convenience, the modes of vibration of aromatic compounds are considered as separate C—H or ring C=C vibrations. However, as with any 'complex' molecule, vibrational interactions occur and these labels really only indicate the predominant vibration.

The strongest absorptions for aromatic compounds[1,2] occur in the region 900–650 cm^{-1} (11.11–15.27 μm) and are due to the C—H vibrations out of the plane of the aromatic ring. Most mononuclear and polynuclear aromatic compounds have three or four peaks in the region 3080–3010 cm^{-1} (3.25–3.32 μm), these being due to the stretching vibrations of the ring C—H bonds. Ring carbon–carbon stretching vibrations occur in the region 1625–1430 cm^{-1} (6.16–6.99 μm). A number of weak combination and overtone bands occur in the region 2000–1650 cm^{-1} (5.00–6.06 μm). These bands are highly characteristic (Figure 11.1) and can be very useful in the evaluation of the number of substituents on the aromatic ring. Unfortunately, part of this region may be overlapped by strong absorptions due to carbonyl or alkene groups.

An in-plane vibrational mode of the aromatic ring may interact with the stretching vibration of an (aromatic carbon)—X bond. This type of stretching vibration is known as an X-sensitive mode and is dependent on the mass of the substituent X.

The X-sensitive bands normally occur in the regions 1300–1050 cm^{-1} (7.69–9.52 μm), 850–620 cm^{-1} (11.76–16.13 μm), and 580–200 cm^{-1} (17.24–50.00 μm). These vibrational modes are due to (a) the ring carbons 1, 3, and 5 moving radially in phase while the substituent on carbon 1 moves radially out of phase, (b) the in-plane bending of a quadrant of the ring in which the C—X bond-length increases as the distance between the carbons 1 and 4 decreases, and (c) the distance between carbons 1 and 4 and the C—X bond-length both increasing simultaneously.

For compounds of the type aryl–metal, a band at 1120–1050 cm^{-1} (8.93–9.52 μm) is observed whose position is dependent on the nature of the metal. References specific to aromatic compounds are: benzyl compounds,[22,23] bridged aromatic compounds,[24] dibenzene oxacyclanes,[30] and phenyl derivatives of Bi, Sb, Si, Ge, Sn, Pb, and P.[25]

Aromatic C—H Stretching Vibrations

As already mentioned, these bands occur in the region 3080–3010 cm^{-1} (3.25–3.32 μm)[3] and are of strong-to-medium intensity. A band with up to five peaks may be observed in this region. As might be expected, monosubstituted benzenes usually exhibit more peaks than di- or trisubstituted benzenes. Alkene C—H stretching vibrations also cause bands in this region, as do both O—H and N—H stretching vibrations, although the latter bands are much broader than those due to the aromatic C—H stretching vibration.

Aromatic In-plane C—H Deformation Vibrations

A number of C—H in-plane deformation bands (up to six) occur in the region 1290–1000 cm^{-1} (7.75–10.00 μm), the bands usually being sharp but of weak-to-medium intensity. However, these bands are not normally of importance for interpretation purposes although they can be used. In fact, a number of interactions are possible, thus necessitating great care in interpretation of bands in this region. Polar ring substituents may result in an increase in the intensity of these bands. Additional difficulties may also arise due to the presence of other bands in this region, e.g. due to C—C, C—O stretching vibrations.

Aromatic Out-of-plane C—H Deformation Vibrations and Ring Out-of-plane Vibrations in the Region 900–675 cm^{-1}

The frequencies of the C—H out-of-plane deformation vibrations are mainly determined by the number of adjacent hydrogen atoms on the ring and not very much affected by the nature of the substituent(s),[4,5] although strongly electron-attracting substituent groups, such as nitro-, can result in an increase of about 30 cm^{-1} in the frequency of the vibration. These bands give an important means for determining the type of aromatic substitution. Although normally strong, they are often not the only strong bands in the region since, for example, the carbon–halogen bond vibration may also give rise to absorptions in this region. As always, any interpretation should, if possible, be supported by the presence of more than one band.

The C—H out-of-plane deformation bands are as follows:

(a) Monosubstituted benzenes[4] have two strong absorptions, one at 770–730 cm^{-1} (12.99–13.70 μm) and the other at 710–690 cm^{-1} (14.08–14.49 μm). The first of these bands is usually not as intense as the second.
(b) *Ortho*-disubstituted benzenes[6] have a strong absorption at 770–735 cm^{-1} (12.99–13.61 μm).
(c) *Meta*-disubstituted benzenes have two medium-intensity absorptions, one at 960–900 cm^{-1} (10.42–11.11 μm), the other at 880–830 cm^{-1} (11.36–12.05 μm), a band of weak-to-medium intensity at 820–765 cm^{-1} (12.20–13.07 μm), and a strong band at 710–680 cm^{-1} (14.08–14.71 μm).
(d) *Para*-disubstituted benzenes[6] absorb strongly at 860–800 cm^{-1} (11.63–12.50 μm).
(e) 1,2,3-Trisubstituted benzenes[7,8] absorb strongly at 800–760 cm^{-1} (12.50–13.10 μm) and at 720–685 cm^{-1} (13.88–14.60 μm), the first band often not being as intense as the second.
(f) 1,2,4-Trisubstituted benzenes[7] have a medium absorption at 900–885 cm^{-1} (11.11–11.30 μm) and a strong band at 780–760 cm^{-1} (12.82–13.16 μm).
(g) 1,3,5-Trisubstituted benzenes[7] have a strong absorption at 865–810 cm^{-1} (11.56–12.35 μm) and a band of lesser intensity at 730–680 cm^{-1} (13.70–14.71 μm).
(h) 1,2,3,4-Tetrasubstituted benzenes absorb strongly at 860–800 cm^{-1} (11.63–12.50 μm).
(i) 1,2,3,5-Tetrasubstituted benzenes[9, 34] absorb strongly at 850–840 cm^{-1} (11.76–11.90 μm).
(j) 1,2,4,5-Tetrasubstituted benzenes absorb strongly at 870–860 cm^{-1} (11.49–11.63 μm).
(k) Pentasubstituted benzenes have a band of medium-to-strong intensity at 900–860 cm^{-1} (11.11–11.63 μm).

Mono-, 1,3-di-, and 1,3,5-trisubstituted benzenes have a strong band in the region 730–680 cm^{-1} (13.70–14.71 μm). In the same region, 1,2- and 1,4-disubstituted benzenes absorb weakly or not at all, depending on whether the two substituent groups are different or not. When the substituents are identical, symmetry results in this vibration being infrared inactive. Trisubstituted 1,2,3- and 1,2,4- benzenes also absorb in this range.

It is both useful and convenient to summarize the C—H out-of-plane vibrations in terms of the number of adjacent hydrogen atoms:

1. *Six adjacent hydrogen atoms* (e.g. benzene), band at 671 cm^{-1}.
2. *Five adjacent hydrogen atoms* (e.g. monosubstituted aromatics), band at 770–735 cm^{-1}.
3. *Four adjacent hydrogen atoms* (e.g. *ortho*-substituted aromatics), band at 770–735 cm^{-1}.
4. *Three adjacent hydrogen atoms* (*meta*- and 1,2,3-trisubstituted aromatics), band at 820–760 cm^{-1}.
5. *Two adjacent hydrogen atoms* (e.g. *para*- and 1,2,3,4-tetrasubstituted aromatics), band at 860–800 cm^{-1}.
6. *An isolated hydrogen atom* (e.g. *meta*-, 1,2,3,5-tetra-, 1,2,4,5-tetra-, and pentasubstituted aromatics), band at 900–830 cm^{-1}.

An additional band is observed at 745–690 cm^{-1} (13.42–14.49 μm) in the spectra of monosubstituted, 1,3-disubstituted, and 1,2,3-trisubstituted compounds.

A coupling between adjacent hydrogen atoms is also observed for naphthalenes, phenanthrenes, pyridines, quinolines (the nitrogen atom being treated as a substituted carbon atom of a benzene ring), and other aromatic compounds.

Nitro-substituted benzenes have a band, in addition to that expected, near 700 cm^{-1} which is believed to involve an NO_2 out-of-plane bending vibration.[32]

Aromatic C=C Stretching Vibrations

The ring carbon–carbon stretching vibrations occur in the region 1625–1430 cm^{-1} (6.16–6.99 μm).[10–14] For aromatic six-membered rings, e.g. benzenes and pyridines, there are two or three bands in this region due to skeletal vibrations, the strongest usually being at about 1500 cm^{-1} (6.67 μm). In the case where the ring is conjugated further, a band at about 1580 cm^{-1} (6.33 μm) is also observed.

In general, the bands are of variable intensity and are observed at 1625–1590 cm^{-1} (6.15–6.29 μm), 1590–1575 cm^{-1} (6.29–6.35 μm), 1525–1470 cm^{-1} (6.56–6.80 μm), and 1465–1430 cm^{-1} (6.83–6.99 μm). For substituted benzenes with identical atoms or groups on all *para*- pairs of ring carbon atoms, the vibrations causing the band at 1625–1590 cm^{-1} (6.15–6.29 μm) (and also the band at 730–680 cm^{-1} (13.70–14.71 μm)—see above) are infrared-inactive due to symmetry considerations, the compound having a centre of symmetry at the ring centre. If the groups on a para pair of carbon atoms are different then there is no centre of symmetry and the vibration(s) are infrared-active.

When a substituent is C=O, C=C, C=N, or NO_2 and is directly conjugated to the ring, or is a heavy element such as Cl, Br, I, S, P, or Si, a doublet is observed at 1625–1575 cm^{-1} (6.15–6.35 μm). Substituents resulting in conjugation, such as C=C and C=O, increase the intensity of this doublet.

For monosubstituted benzenes with strong electron acceptor or donor groups, the bands at 1625–1590 cm^{-1} (6.15–6.29 μm) and 1590–1575 cm^{-1} (6.29–6.35 μm) are of medium intensity, the second band being the weaker, but for weakly-interacting groups these bands are both weak.

For *meta*-disubstituted benzenes, the intensity of the band at about 1600 cm^{-1} (6.25 μm) is directly dependent on the sum of the electronic effects of the substituents whereas for *para*-disubstituted benzenes it is dependent on the difference of the electronic effects of the substituents. For example, due to the large dipole changes possible for *para*-disubstituted compounds in which one group is *ortho–para*-directing and the other is *meta*-directing, the band at 1625–1590 cm^{-1} (6.15–6.29 μm) is quite intense. In general, mono-, *meta*-di-, and 1,3,5-trisubstituted benzenes have strong bands at 1625–1590 cm^{-1} (6.15–6.29 μm) and at 730–680 cm^{-1} (13.70–14.71 μm).

A fairly weak band is observed in the region 1465–1430 cm^{-1} (6.83–6.99 μm) for aromatic compounds, except *para*-disubstituted benzenes for which the range is 1420–1400 cm^{-1} (7.04–7.14 μm). A band in the range 1510–1470 cm^{-1} (6.62–6.80 μm) is observed for monosubstituted, *ortho*- and *meta*-disubstituted, and 1,2,3-trisubstituted benzenes, whereas for *para*-disubstituted and 1,2,4-trisubstituted compounds this band occurs at 1525–1480 cm^{-1} (6.56–6.76 μm). (The differences noted for *para*- compounds are useful in isomer studies.) This last band (at ~1500 cm^{-1}) is relatively strong for electron donor groups but is otherwise weak or absent, e.g. for the carbonyl group it is very weak. The bands at 1500–1400 cm^{-1} (6.67–7.14 μm) cannot be misinterpreted as due to olefinic C=C stretching vibrations since the latter lie outside this range. However, the band near 1450 cm^{-1} (6.90 μm) may be obscured by the band due to the aliphatic C—H deformation vibration.

Overtone and Combination Bands

Overtone and combination bands due to the C—H out-of-plane deformation vibrations occur in the region 2000–1660 cm^{-1} (5.00–6.02 μm).[15] The absorption patterns observed are characteristic of different benzene ring substitutions (see Figure 11.1, which gives a guide as to what may be observed for a given compound). These bands are weak and it may, therefore, be necessary in some cases to use cells of longer path-length or to use a more concentrated sample. Interference in this region from olefinic C=C and carbonyl C=O absorptions may also occur.

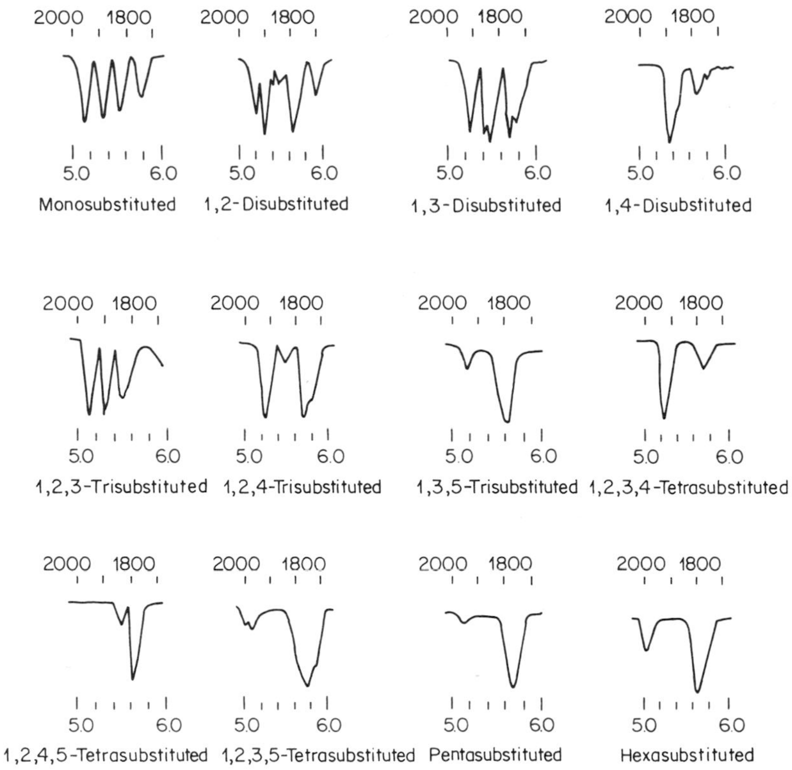

2000 1800
5.0 6.0
Monosubstituted
1,2-Disubstituted
1,3-Disubstituted
1,4-Disubstituted
1,2,3-Trisubstituted
1,2,4-Trisubstituted
1,3,5-Trisubstituted
1,2,3,4-Tetrasubstituted
1,2,4,5-Tetrasubstituted
1,2,3,5-Tetrasubstituted
Pentasubstituted
Hexasubstituted

Figure 11.1 C—H out-of-plane bending overtone and combination band patterns in the region 2000–1660 cm^{-1} (5.0–6.0 μm).

Chart 5 The absorption ranges and corresponding intensities of bands in the region 1000–300 cm⁻¹ due to substituted benzenes.

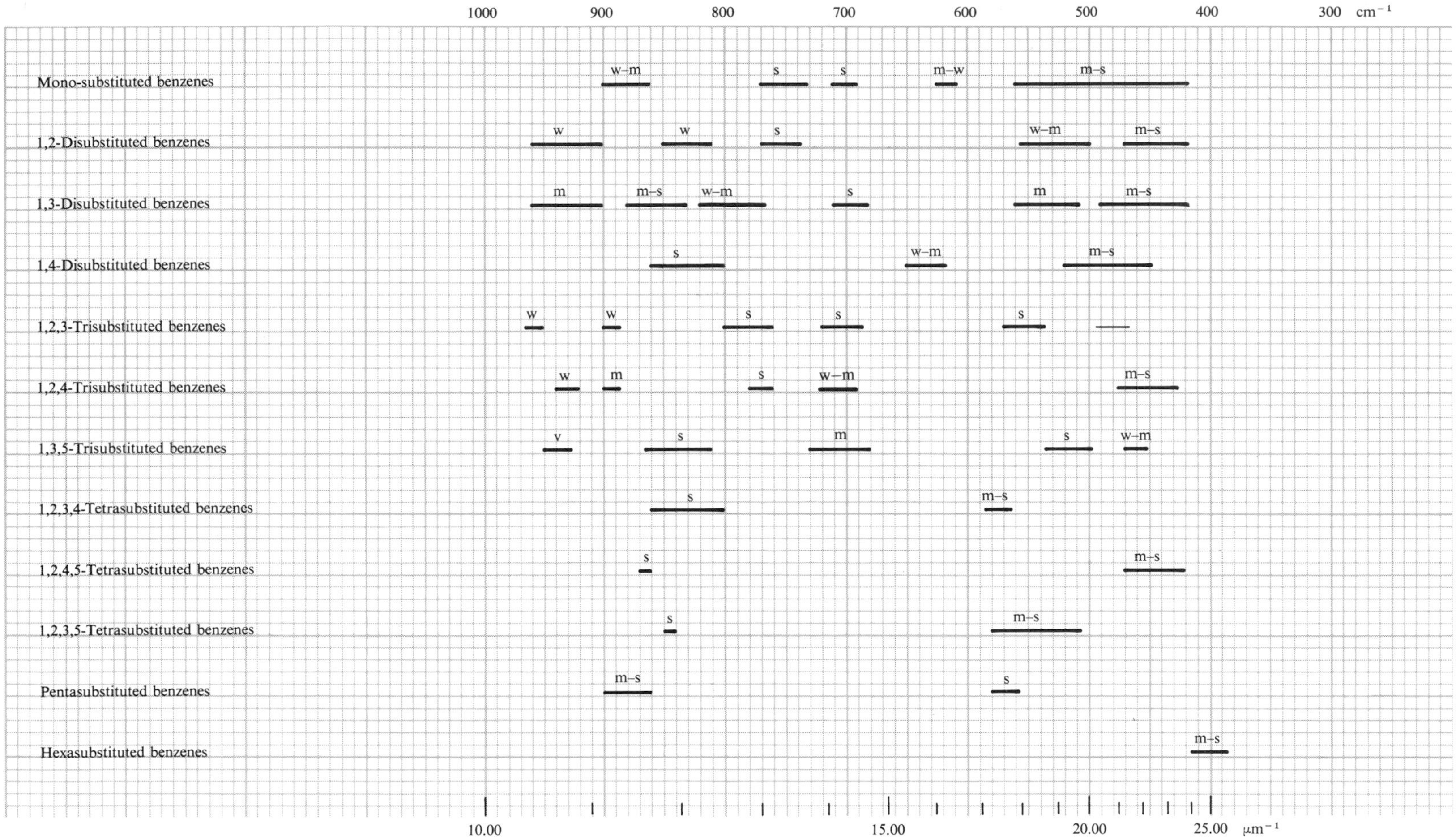
1000
900
800
700
600
500
400
300 cm⁻¹
Mono-substituted benzenes
w–m
s
s
m–w
m–s
1,2-Disubstituted benzenes
w
w
s
w–m
m–s
1,3-Disubstituted benzenes
m
m–s
w–m
s
m
m–s
1,4-Disubstituted benzenes
s
w–m
m–s
1,2,3-Trisubstituted benzenes
w
w
s
s
s
1,2,4-Trisubstituted benzenes
w
m
s
w–m
m–s
1,3,5-Trisubstituted benzenes
v
s
m
s
w–m
1,2,3,4-Tetrasubstituted benzenes
s
m–s
1,2,4,5-Tetrasubstituted benzenes
s
m–s
1,2,3,5-Tetrasubstituted benzenes
s
m–s
Pentasubstituted benzenes
m–s
s
Hexasubstituted benzenes
m–s
10.00
15.00
20.00
25.00 μm⁻¹

Aromatic Ring Deformation Vibrations below 700 cm^{-1}

Some bands in this region are quite sensitive to changes in the nature and position of substituents,[16–23] although other bands (due to certain vibrations of aromatic rings) depend mainly on the distribution and number of substituents rather than on their chemical nature or mass, so that these latter vibrations, together with the out-of-plane vibrations of the ring hydrogen atoms, are extremely useful in determining the positions of substituents.

Two bands usually observed are those due to the in-plane and out-of-plane ring deformation vibrations. The in-plane deformation vibration is at higher frequencies than the out-of-plane vibration and is generally weak for *mono*- and *para*-substituted benzenes, often also being masked by other stronger absorptions which may occur due to the substituent group.

For *mono*-substituted aromatics, the band due to the out-of-plane ring deformation vibration occurs as follows for the stated substituents:

(a) >C=C<, —C≡C—, or —C≡N: near 550 cm^{-1} (18.18 μm);
(b) an electron donor such as —OH or —NH_2: near 500 cm^{-1} (20.00 μm);
(c) a halogen or alkyl group: in the range 500–440 cm^{-1} (20.00–22.73 μm);
(d) an electron acceptor such as NO_2 or COOH: below 450 cm^{-1} (above 22.22 μm).

For *meta*-disubstituted compounds, this band occurs in the region 460–415 cm^{-1} (21.74–24.10 μm) except when the substituents are electron-accepting groups in which case the range is 490–460 cm^{-1} (20.41–21.74 μm). The band for *para*-disubstituted benzenes with electron-donating substituents occurs at 520–490 cm^{-1} (19.23–20.41 μm), exceptions being cyano- compounds which absorb at about 545 cm^{-1} (18.35 μm). Phthalides have bands at 520–490 cm^{-1} (19.23–20.41 μm) and 490–470 cm^{-1} (20.41–21.28 μm).

Alkyl-substituted diphenyl compounds exhibit three bands of medium-to-strong intensity, due to ring deformation vibrations, at 620–605 cm^{-1} (16.13–16.53 μm), 490–455 cm^{-1} (20.41–21.98 μm), and 410–400 cm^{-1} (24.39–25.00 μm). A number of 1,2-dialkyl-substituted diphenyls have a band at 560–545 cm^{-1} (17.86–18.35 μm) and 1,3-dialkyl-substituted diphenyls have a band near 530 cm^{-1} (18.87 μm).

Table 11.1. Aromatic =C—H and ring C=C stretching vibrations

Functional Groups	Region cm^{-1}	Region μm	Intensity	Comments
=C—H	3080–3010	3.25–3.32	m	A number of peaks, decreasing in number with increase in substitution
—C=C—	1625–1590	6.16–6.29	v	Usually ~1600 cm^{-1}
	1590–1575	6.29–6.35	v	Strongest band if conjugated, usually 1580 cm^{-1}
	1525–1470	6.56–6.80	v	Usually ~1470 cm^{-1} for electron acceptors and ~1510 cm^{-1} for electron donors
	1465–1430	6.38–6.99	v	

Table 11.2. Aromatic =C—H in-plane deformation vibrations

Functional Groups	Region cm^{-1}	Region μm	Intensity	Comments
Monosubstituted benzenes	1250–1230	8.00–8.13	w	
	~1175	~8.51	w–m	
	~1155	~8.66	w	
	~1075	~9.30	m	
	1030–1015	9.71–9.85	w–m	
1,2-Disubstituted benzenes	1290–1250	7.75–8.00	w	
	~1160	~8.62	w–m	
	1150–1110	8.70–9.01	w–m	
	1050–1020	9.52–9.80	m	
1,3-Disubstituted benzenes	1300–1240	7.69–8.06	w	
	~1160	~8.62	w–m	
	~1095	~9.13	w	
	~1075	~9.30	v	
1,4-Disubstituted benzenes	1270–1250	7.87–8.00	w–m	
	~1175	~8.51	v	
	1125–1110	8.89–9.01	v	
	~1015	~9.86	w–m	
	~985	~10.15	w	
1,2,3-Trisubstituted benzenes	~1160	~8.62	w	
	~1075	~9.30	m	
	~1020	~9.80	m	
1,2,4-Trisubstituted benzenes	~1210	~8.26		
	~1150	~8.70	m	
	~1130	~8.85	m	
1,3,5-Trisubstituted benzenes	~1265	~7.91	m	
	~1170	~8.55	m	
	1040–995	9.62–10.05	v	
1,2,4,5-Tetrasubstituted benzenes	~1270	~7.87	w	
	~1195	~8.37	w	

Table 11.3. Aromatic =C—H out-of-plane deformation vibrations and other bands in region 900–675 cm^{-1}

Functional Groups	Region cm^{-1}	Region μm	Intensity	Comments
Monosubstituted benzenes	900–860	11.11–11.63	w–m	Out-of-plane def (5H)
	770–730	12.99–13.70	s	Out-of-plane def (5H)
	710–690	14.08–14.49	s	Ring out-of-plane def
1,2-Disubstituted benzenes	960–900	10.42–11.05	w	Out-of-plane def (4H)
	850–810	11.76–12.35	w	Out-of-plane def (4H)
	770–735	12.99–13.61	s	Out-of-plane def (4H)
1,3-Disubstituted benzenes	960–900	10.42–11.11	m	Out-of-plane def (1H)
	880–830	11.36–12.05	m–s	Out-of-plane def (3H)
	820–765	12.20–13.07	w–m	Out-of-plane def (3H)
	710–680	14.08–14.71	s	Ring out-of-plane def
1,4-Disubstituted benzenes	860–800	11.63–12.50	s	Out-of-plane def (2H)
	~695	~14.39	w–m	
1,2,3-Trisubstituted benzenes	965–950	10.36–10.53	w	Out-of-plane def (3H)
	900–885	11.11–11.30	w	Out-of-plane def (3H)
	800–760	12.50–13.10	s	Out-of-plane def (3H)
	720–685	13.89–14.60	s	Out-of-plane def (3H)
1,2,4-Trisubstituted benzenes	940–920	10.64–10.87	w	Out-of-plane def (1H)
	900–885	11.36–11.30	m	Out-of-plane def (2H)
	780–760	12.82–13.16	s	Out-of-plane def (2H)
	~705	~14.18	w–m	
1,3,5-Trisubstituted benzenes	950–925	10.53–10.81	v	Out-of-plane def (1H)
	865–810	11.56–12.35	s	Out-of-plane def (1H)
	730–680	13.70–14.81	m–s	Ring out-of-plane def
1,2,3,4-Tetrasubstituted benzenes	860–800	11.63–12.50	s	Out-of-plane def (2H)
1,2,4,5-Tetrasubstituted benzenes	870–860	11.49–11.63	s	Out-of-plane def (1H)
	~805	~12.42	w–m	
1,2,3,5-Tetrasubstituted benzenes	850–840	11.76–11.90	s	Out-of-plane def (1H), see ref. 34
Pentasubstituted benzenes	900–860	11.11–11.63	m–s	Out-of-plane def (1H)

Table 11.4. Aromatic ring deformation vibrations

Functional Groups	Region cm^{-1}	Region μm	Intensity	Comments
Monosubstituted benzenes	625–605	16.00–16.53	m–w	In-plane ring def
	560–415	17.86–24.10	m–s	Out-of-plane ring def
1,2-Disubstituted benzenes	555–495	18.02–20.20	w–m	In-plane ring def
	470–415	21.28–24.10	m–s	Out-of-plane ring def
1,3-Disubstituted benzenes	560–505	17.86–19.80	m	In-plane ring def
	490–415	20.41–24.10	m–s	Out-of-plane ring def
1,4-Disubstituted benzenes	650–615	15.38–16.26	w–m	In-plane ring def
	520–445	19.23–22.47	m–s	Out-of-plane ring def (except for CN substituted benzenes)
1,2,3-Trisubstituted benzenes	570–535	17.54–18.69	s	Out-of-plane ring def
	~485	~20.62		
1,2,4-Trisubstituted benzenes	475–425	21.05–23.53	m–s	Out-of-plane ring def
1,3,5-Trisubstituted benzenes	535–495	18.69–20.20	s	Out-of-plane ring def
	470–450	21.28–22.22	w–m	
1,2,3,4-Tetrasubstituted benzenes	585–565	17.09–17.70	m–s	
1,2,3,5-Tetrasubstituted benzenes	580–505	17.24–19.80	m–s	Out-of-plane ring def
1,2,4,5-Tetrasubstituted benzenes	470–420	21.28–23.81	m–s	
Pentasubstituted benzenes	580–555	17.24–18.02	s	
Hexasubstituted benzenes	415–385	24.10–25.97	m–s	
Alkyl-substituted diphenyls	620–605	16.13–16.53	m–s	
	490–455	20.41–21.98	m–s	
	410–400	24.39–25.00	m–s	

Polynuclear Aromatic Compounds

Polynuclear, aromatic, condensed-ring compounds absorb in the same general regions as benzene derivatives[26–31] and therefore the previous section should be noted carefully. (A study of pyrenes has been published.[31])

Naphthalenes, 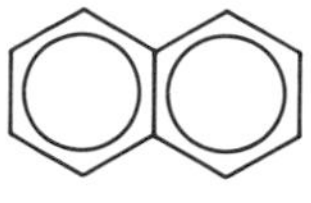

Naphthalenes[26–28] have a band of medium intensity in the region 1620–1580 cm^{-1} (6.17–6.33 µm) and a band near 1515 cm^{-1} (6.56 µm) and 1395 cm^{-1} (7.17 µm). As a result of C—H out-of-plane deformation vibrations, 1-substituted naphthalenes absorb at 810–785 cm^{-1} (12.35–12.74 µm) due to the presence of three adjacent hydrogen atoms on a ring, and at 780–760 cm^{-1} (12.82–13.16 µm) due to four adjacent hydrogen atoms. 2-Substituted naphthalenes absorb at 760–735 cm^{-1} (13.16–13.61 µm) due to four adjacent hydrogen atoms, at 835–805 cm^{-1} (11.98–12.42 µm) due to two adjacent hydrogen atoms, and at 895–855 cm^{-1} (11.17–11.70 µm) due to a single atom. Table 11.3 correlating the C—H out-of-plane bending vibrations to the number of adjacent hydrogen atoms on the aromatic ring is, of course, applicable here. There are also a number of bands in the region 1400–1000 cm^{-1} (7.14–10.00 µm).

Mono- and dialkyl substituted naphthalenes have a strong band at 645–615 cm^{-1} (15.50–16.26 µm) and a band of variable intensity at 490–465 cm^{-1} (20.41–21.51 µm). Both naphthalenes and anthracenes have a band at about 475 cm^{-1} (21.05 µm) due to the out-of-plane ring vibrations.

As a result of the C—H out-of-plane vibrations of adjacent aromatic hydrogen atoms, tetrahydronaphthalenes (tetralins), and polynuclear aromatic compounds in general, have absorption bands as follows:

four adjacent aromatic hydrogen atoms, 770–740 cm^{-1} (12.99–13.51 µm);
three adjacent aromatic hydrogen atoms { 815–785 cm^{-1} (12.27–12.74 µm), 760–730 cm^{-1} (13.10–13.70 µm);
two adjacent aromatic hydrogen atoms, 850–810 cm^{-1} (11.76–12.35 µm);
one isolated aromatic hydrogen atom, 900–830 cm^{-1} (11.11–12.05 µm).

Anthracenes, 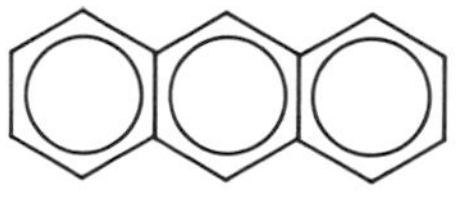 , *and phenanthrenes,* 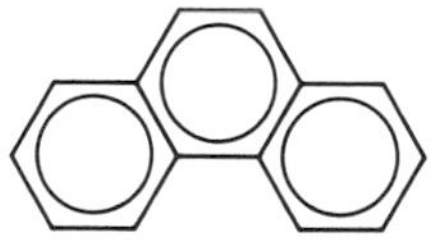

Anthracenes[26,29] absorb near 1630 cm^{-1} (6.14 µm) and near 1550 cm^{-1} (6.45 µm), whilst phenanthrenes have two bands near 1600 cm^{-1} (6.25 µm) and one band near 1500 cm^{-1} (6.67 µm). Anthracenes also have one or two strong bands in the region 900–650 cm^{-1} (11.11–15.38 µm). The higher frequency band near 900 cm^{-1} (11.11 µm) is associated with the 9,10 hydrogen atoms and disappears if these are substituted. Both anthracenes and naphthalenes have a band at about 475 cm^{-1} (21.05 µm) due to out-of-plane ring vibrations. Spectra of anthracene and acridene derivatives are available elsewhere,[33] although the complete normal infrared range is not covered.

Table 11.5. Polynuclear aromatic compounds

Functional Groups	Region cm^{-1}	Region µm	Intensity	Comments
Naphthalenes	1620–1580	6.17–6.33	m	C=C str, often a doublet
	1550–1505	6.45–6.65	m	C=C str
	1400–1390	7.14–7.19	m	
1-Monosubstituted naphthalenes	810–785	12.35–12.74	s	C—H out-of-plane def (3H)
	780–760	12.82–13.16	s	C—H out-of-plane def (4H)
2-Monosubstituted naphthalenes	895–855	11.17–11.70	s	Out-of-plane def (1H)
	835–805	11.98–12.42	s	Out-of-plane def (2H)
	760–735	13.16–13.61	m-s	Out-of-plane def (4H)
Mono- and dialkyl-substituted naphthalenes	645–615	15.50–16.25	m-s	
	490–465	20.41–21.51	v	Out-of-plane ring def
Phenanthrenes	~1600	~6.25	m	Two bands
	~1500	~6.67	m	
Anthracenes	1640–1620	6.10–6.17	m	
	~1550	~6.45	m	Not present with 9,10 substitution
	900–650	11.11–15.38	s	One or two bands
Mono- and dimethyl 1,2-benzanthracenes	~680	~14.71		Possibly skeletal

Table 11.6. Substituted naphthalenes: characteristic C—H out-of-plane vibrations

Hydrogen atom positions on one ring†	Region cm^{-1}	Region µm	Intensity	Comments
1, 2, 3, 4	800–760	12.50–13.10	s	Four adjacent hydrogen atoms
	770–725	12.99–13.79	s–vs	Four adjacent hydrogen atoms
1, 2, 3	820–775	12.20–12.90	s	Three adjacent hydrogen atoms
	775–730	12.90–13.70	s	Three adjacent hydrogen atoms
1, 2, 4	925–885	10.81–11.30	m	
	900–835	11.11–11.98	m–s	Isolated hydrogen atom
	850–805	11.76–12.42	vs	Two adjacent hydrogen atoms
1, 2	835–800	11.98–12.50	s	Two adjacent hydrogen atoms
	755–720	13.25–13.89	m–s	
2, 3	835–810	11.98–12.35	s–vs	Two adjacent hydrogen atoms
1, 3	905–865	11.05–11.56	m–s	Isolated hydrogen atom
	875–840	11.43–11.90	s	Isolated hydrogen atom
1, 4	890–870	11.24–11.49	s	Isolated hydrogen atom
1 or 2	900–855	11.11–11.70	m	Isolated hydrogen atom

† The numbers refer to the substitution pattern of hydrogen atoms so chosen as to give the lowest possible numbering.

References

1. G. Varsányi, *Vibrational Spectra of Benzene Derivatives*. Academic Press, New York, 1969
2. T. F. Ardyukova *et al.*, *Atlas of Spectra of Aromatic and Heterocyclic Compounds*. Nauka Sib. Otd., Novosibirsk, 1973
3. S. E. Wiberly *et al.*, *Anal. Chem.*, 1960, **32,** 217
4. S. Higuchi *et al.*, *Spectrochim. Acta*, 1974, **30A,** 463
5. D. H. Whiffen, *Spectrochim. Acta*, 1955, **7,** 253
6. A. Stojikjkovic and D. H. Whiffen, *Spectrochim. Acta*, 1958, **12,** 47 and 57
7. J. H. S. Green *et al.*, *Spectrochim. Acta*, 1971, **27A,** 793 and 807
8. V. P. Fedorov *et al.*, *Vorp. Mol. Spektrosk.*, **1971,** 41
9. G. Varsányi and P. Soyar, *Acta Chim. Budapest*, 1973, **76,** 243
10. A. R. Katritzky, *J. Chem. Soc.*, **1958,** 4162
11. A. R. Katritzky and P. Simmons, *J. Chem. Soc.*, **1959,** 2058
12. A. R. Katritzky and J. M. Lagowski, *J. Chem. Soc.*, **1958,** 4155
13. A. R. Katritzky and R. A. Jones, *J. Chem. Soc.*, **1959,** 3670
14. A. R. Katritzky, *J. Chem. Soc.*, **1959,** 2051
15. C. W. Young *et al.*, *Anal. Chem.*, 1951, **23,** 709
16. R. J. Jakobsen, *Wright–Patterson Air Force Base Tech. Report*, 1962, Documentary Report No. ASD–TDR–62–895 Oct.
18. W. S. Wilcox *et al.*, *WADD Tech. Report*, **1960,** 60–333
19. R. J. Jakobsen, *WADD Tech. Report*, **1960,** 60–204
20. R. J. Jakobsen and F. F. Bentley, *App. Spectrosc.*, 1964, **18,** 88
21. A. Mansingh, *J. Chem. Phys.*, 1970, **52,** 5896
22. L. Verdonck *et al.*, *Spectrochim. Acta*, 1973, **29A,** 813
23. L. Verdonck and G. P. van der Kelen, *Spectrochim. Acta*, 1972, **28A,** 51 and 55
24. B. H. Smith, *Bridged Aromatic Compounds*. Academic Press, New York, 1964, pp. 385–391
25. L. A. Harrah *et al.*, *Spectrochim. Acta*, 1962, **18,** 21
26. S. E. Wiberley and R. D. Gonzalez, *App. Spectrosc.*, 1961, **15,** 174
27. J. G. Hawkins *et al.*, *Spectrochim. Acta*, 1957, **10,** 105
28. B. W. Cox *et al.*, *Spectrochim. Acta*, 1965, **21,** 1633
29. S. Califano, *J. Chem. Phys.*, 1962, **36,** 903
30. G. Karogounis and J. Agathokli, *Prakt. Acad. Athenon, Greece*, 1970, **44,** 338
31. R. Mecke and W. E. Klee, *Zeit. Elektrochem.*, 1961, **65,** 327
32. J. H. S. Green and D. J. Harrison, *Spectrochim. Acta*, 1970, **26A,** 1925
33. V. A. Koptyug (ed.), *Atlas of Spectra of Aromatic and Heteroaromatic Compounds*, No. 7. Nauka Sib. Otd., Novosibirsk, 1974
34. G. Varsanyi *et al.*, *Acta Chim. Acad. Sci. Hungary*, 1977, **93,** 315

CHAPTER 12
Six-membered Ring Heterocyclic Compounds

Pyridine Derivatives, 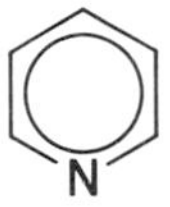
N

The spectra of pyridine compounds[1–10] have many of the features of the spectra of homonuclear aromatic compounds, such as bands due to the aromatic C—H stretching vibration, overtones in the region 2080–1750 cm^{-1} (4.81–5.88 μm) etc., with the nitrogen atom behaving in a similar fashion to that observed for a substituted carbon atom. Therefore, the contents of the previous chapter should be noted.

Aromatic C—H stretching vibrations

The aromatic C—H stretching vibration of nitrogen heterocyclic aromatic compounds gives rise to a band at 3100–3010 cm^{-1} (3.23–3.32 μm). This band is in the same region as that expected for benzene derivatives and is also similar in that the band is of medium-to-strong intensity and consists of a number of peaks.

Overtone and combination bands

As with benzene derivatives, weak overtone and combination bands are observed in the region 2080–1750 cm^{-1} (4.81–5.88 μm), these being characteristic of the position of the substitution (see Figure 12.1). These patterns are intended to serve as a guide as to what might be observed.

C=C and C=N stretching vibrations

Interactions between ring C=C and C=N stretching vibrations result in two strong-to-medium intensity absorptions about 100 cm^{-1} (0.4 μm) apart. These absorptions occur at 1615–1575 cm^{-1} (6.19–6.35 μm) and 1520–1465 cm^{-1} (6.58–6.83 μm), the higher-frequency band often having another medium-intensity band on its low-frequency side which is found at 1590–1555 cm^{-1} (6.29–6.43 μm). A strong band is usually observed in the region 1000–985 cm^{-1} (10.00–10.15 μm), but this band may be very weak or undetectable for 3-substituted pyridines.

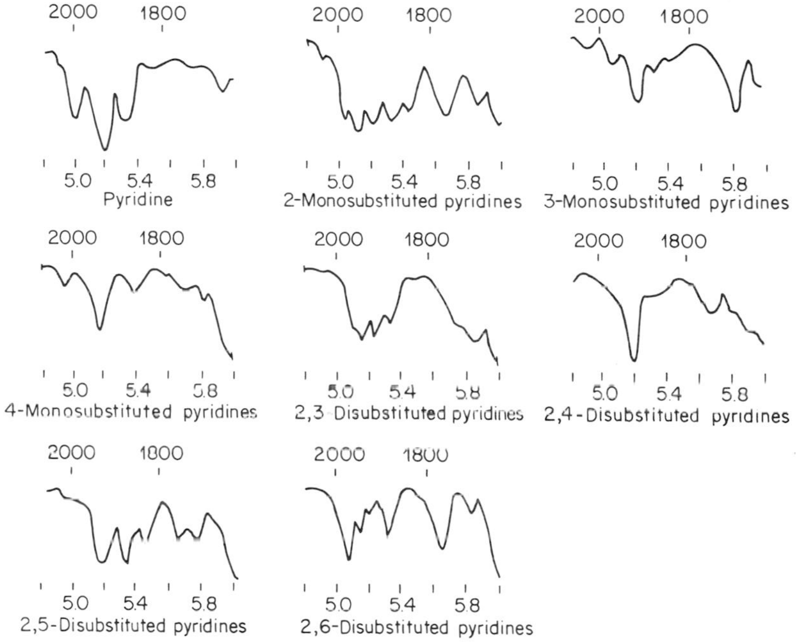
2000 1800
5.0 5.4 5.8
Pyridine
2-Monosubstituted pyridines
3-Monosubstituted pyridines
4-Monosubstituted pyridines
2,3-Disubstituted pyridines
2,4-Disubstituted pyridines
2,5-Disubstituted pyridines
2,6-Disubstituted pyridines

Figure 12.1 Overtone and combination bands of substituted pyridines

Ring C—H deformation vibrations

Bands of variable intensity are observed in the regions 1300–1180 cm^{-1} (7.69–8.48 μm) and 1100–1000 cm^{-1} (9.09–10.00 μm) due to in-plane deformation vibrations. Strong bands are observed in the region 850–690 cm^{-1} (11.76–14.49 μm) which are characteristic of the position of the substitution, these bands being due to C—H out-of-plane deformation vibrations. See Table 11.3 for the correlation of C—H out-of-plane vibrations with the number of adjacent hydrogen atoms on the aromatic ring.

Other bands

Monosubstituted pyridines,[7, 24] with the exception of 4-substituted pyridines,[24] have a medium-to-strong band at 635–600 cm^{-1} (15.75–16.67 μm) and a strong band at 420–385 cm^{-1} (23.81–25.97 μm). 4-Monosubstituted pyridines appear to have bands below 650 cm^{-1} (15.38 μm) similar to those for the corresponding monosubstituted benzenes. (Studies of di- and trisubstituted pyridines have been published.[4, 8])

Pyridine N-oxides

Pyridine *N*-oxides have similar absorptions to those of pyridines.

A particular feature of pyridine *N*-oxides[11] is a strong band in the region 1310–1220 cm^{-1} (7.64–8.20 μm) due to the N—O stretching vibration.

Other comments

Pyridine may form charge transfer complexes.[12] Studies on picolines,[13] bipyridines,[14] and pyrazine *N*-oxides[15] have also been published.

Table 12.1. Pyridine ring and C—H stretching vibrations

Functional Groups	Region cm^{-1}	Region μm	Intensity	Comments
Pyridines	3100–3010	3.23–3.32	m–s	C=H str, number of peaks
	1615–1575	6.19–6.35	m–s	C=C and C=N in plane vibrations (ring stretching vibrations) general ranges (1615–985)
	1590–1555	6.29–6.43	m	
	1520–1465	6.58–6.83	m–s	
	1450–1410	6.90–7.09	v	
	1000–985	10.00–10.15	w–m	
2-Monosubstituted pyridines	1615–1575	6.19–6.35	m	Ring str
	1575–1570	6.35–6.37	m	Ring str
	~1470	~6.80	m	Ring str
	1440–1425	6.94–7.02	v	Ring str
3-Monosubstituted pyridines	1600–1590	6.25–6.29	m	Ring str
	~1575	~6.35	m	Ring str
	1485–1465	6.73–6.83	m	Ring str
	~1420	~7.04	v	Ring str
4-Monosubstituted pyridines	1610–1595	6.21–6.27	m	Ring str
	1570–1550	6.27–6.45	m	Ring str
	1520–1480	6.58–6.76	m	Ring str
	1420–1410	7.04–7.09	v	Ring str
Polysubstituted pyridines	1610–1595	7.09–6.27	m	Ring str
	1590–1565	6.29–6.39	m	Ring str
	1555–1490	6.43–6.71	m	Ring str

Table 12.2. Pyridine C—H deformation vibrations

Functional Groups	Region cm^{-1}	Region μm	Intensity	Comments
2-Monosubstituted pyridines	1300–1265	7.69–7.90	w	
	~1150	~8.70	w	Also *N*-oxides
	1115–1090	8.97–9.17	w	Also *N*-oxides
	1055–1040	9.48–9.61	w	Also *N*-oxides
	780–740	12.82–13.51	s	Four adjacent hydrogen atoms
3-Monosubstituted pyridines	1200–1180	8.33–8.48	v	
	~1125	~8.89	w	
	~1105	~9.05	w	
	1045–1030	9.57–9.71	w	
	920–890	10.87–11.24	w	
	820–770	12.20–12.29	m–s	Also *N*-oxides, three adjacent hydrogen atoms
	730–690	13.70–14.49	m–s	Ring bend
4-Monosubstituted pyridines	1230–1210	8.13–8.26	v	
	~1070	~9.35	s	
	850–790	11.76–12.66	s	Also *N*-oxides, two adjacent hydrogen atoms
2,3-Disubstituted pyridines	815–785	12.27–12.74	m	
	740–690	13.51–14.49	m–s	
2,5-Disubstituted pyridines	825–810	12.12–12.35	m–s	
	735–725	13.60–13.75	m–s	
2,6-Disubstituted pyridines	815–770	12.27–12.99	m–s	See ref. 10
	750–720	13.33–13.89	m–s	
3,4-Disubstituted pyridines	860–840	11.63–11.90	m	
Trisubstituted pyridines	~725	~13.79	s	See refs 4, 8

Table 12.3. Pyridines: other bands

Functional Groups	Region cm^{-1}	Region μm	Intensity	Comments
2- and 3-Monosubstituted pyridines	635–600	15.75–16.67	m–s	In-plane bending of ring
	420–385	23.81–25.97	s	Out-of-plane bending of ring
Pyridinium salts (free) 	3340–3210	2.99–3.12	v	N^+—H str, a number of bands
Pyridinium salts (hydrogen-bonded)	3300–2370	3.03–4.22	v	N^+—H str, a number of bands

Table 12.4. Pyridine *N*-oxide C—H and ring stretching vibrations

Functional Groups	Region cm⁻¹	μm	Intensity	Comments
Pyridine *N*-oxides	3100–3010	3.23–3.32	m	C—H str
	1645–1600	6.08–6.25	v	C=C and C=N in-plane vibs general ranges
	~1570	~6.37	v	
	1540–1475	6.49–6.78	v	
	1450–1425	6.90–7.02	v	
	1310–1220	7.64–8.20	s	$N^+—O^-$ str
	880–845	11.36–11.83	m–s	
2-Monosubstituted pyridine *N*-oxides	1640–1600	6.10–6.25	v	
	1580–1555	6.33–6.43	v	
	1540–1480	6.49–6.76	v	
	1445–1425	6.92–7.02	v	
3-Monosubstituted pyridine *N*-oxides	~1605	~6.23	v	See ref. 11
	1565–1560	6.39–6.41	v	
	1490–1475	6.71–6.78	v	
	~1435	~6.97	v	
	~1015	~9.86	s	Ring vib
4-Monosubstituted pyridine *N*-oxides	1645–1610	6.08–6.21	v	
	1490–1475	6.71–6.78	v	
	1450–1435	6.90–6.97	v	

Table 12.5. Pyridine *N*-oxide C—H deformation vibrations

Functional Groups	Region cm⁻¹	μm	Intensity	Comments
2-Monosubstituted pyridine *N*-oxides	~1150	~8.70	w	Also pyridines
	1115–1090	8.97–9.17	w	Also pyridines
	1055–1040	9.48–9.61	w	Also pyridines
	990–960	10.10–10.42	m	
	790–750	12.66–13.33	m–s	Four adjacent hydrogen atoms
3-Monosubstituted pyridine *N*-oxides	~1160	~8.62	v	
	1120–1080	8.93–9.26	w–m	
	980–930	10.20–10.75	s	
	820–770	12.20–12.29	m–s	Also pyridines, three adjacent hydrogen atoms
	680–660	14.71–15.15	m	Ring bending vib
4-Monosubstituted pyridine *N*-oxides	~1170	~8.55	s	
	1110–1095	9.01–9.13	w	
	~1035	~9.66	m	
	850–820	11.76–12.20	s	Also pyridines, two adjacent hydrogen atoms
3,4-Disubstituted pyridine *N*-oxides	890–860	11.24–11.63	s	
	825–810	12.12–12.35	s	

Table 12.6. 2-Pyridols, 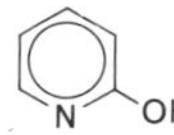 and 4-pyridols, 

Functional Groups	Region cm⁻¹	μm	Intensity	Comments
2-Pyridols	1670–1655	5.99–6.04	vs	C=O str
	1630–1590	6.14–6.29	vs	
	1570–1535	6.37–6.52	s	
	1500–1470	6.67–6.80	m	
	1445–1415	6.92–7.06	m–s	
4-Pyridols	1660–1620	6.02–6.17	vs	
	1580–1550	6.33–6.45	vs	C=O str
	1515–1485	6.60–6.74	w–m	
	1470–1400	6.80–7.14	m–s	
2-Pyridthiones, 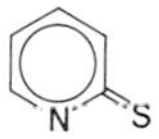	1145–1100	8.73–9.09	m–s	C=S str
4-Pyridthiones	1120–1105	8.93–9.05	vs	C=S str

Quinolines, 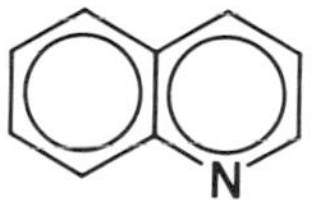, and isoquinolines, 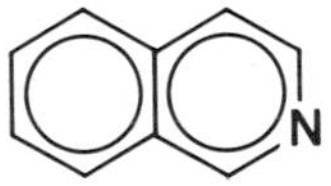

Quinolines and isoquinolines[1,16] have three bands near 1600 cm⁻¹ (6.25 μm) and five in the range 1500–1300 cm⁻¹ (6.67–7.69 μm). Disubstituted methyl quinolines have four bands in the region 1600–1500 cm⁻¹ (6.25–6.67 μm). The aromatic C—H out-of-plane deformation vibrations are similar to those observed for naphthalenes. (Reviews of the infrared spectra of acridines have been published.[17,18])

Pyrimidines, 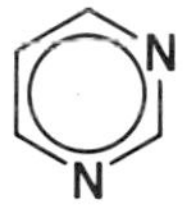

Pyrimidines[19] absorb strongly at 1600–1500 cm⁻¹ (6.25–6.67 μm) due to the C=C and C=N ring stretching vibrations. Absorptions are also observed at 1640–1620 cm⁻¹ (6.10–6.17 μm), 1580–1520 cm⁻¹ (6.33–6.58 μm), 1000–960 cm⁻¹ (10.00–10.42 μm), and 825–775 cm⁻¹ (12.12–12.90 μm). 2-Monosubstituted pyrimidines have three medium-to-strong absorption bands at 650–630 cm⁻¹ (15.38–15.87 μm), 580–475 cm⁻¹ (17.24–21.05 μm), and 515–440 cm⁻¹ (19.42–22.73 μm). 4-Monosubstituted pyrimidines have a band of variable intensity at 685–660 cm⁻¹ (14.60–15.15 μm) which is usually at 680 cm⁻¹ (14.71 μm), a medium-to-strong band at 555–500 cm⁻¹ (18.02–20.00 μm), and a strong band at 500–430 cm⁻¹ (20.00–23.26 μm). Due to tautomerism, pyrimidines substituted with hydroxyl groups are generally in the keto form and therefore have a band due to the carbonyl group.

Quinazolines, 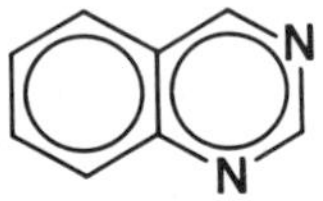

Due to aromatic ring vibrations, quinazolines[20] absorb strongly at 1635–1610 cm^{-1} (6.13–6.21 μm), 1580–1565 cm^{-1} (6.33–6.39 μm), and 1520–1475 cm^{-1} (6.58–6.78 μm), with six bands of variable intensity usually being observed at 1500–1300 cm^{-1} (6.67–7.69 μm). In the region 1000–700 cm^{-1} (10.00–14.29 μm), bands of variable intensity are observed due to the C—H out-of-plane deformation vibrations. These bands are useful for assignment purposes since different types of monosubstitution may be recognized. Bands of variable intensity, usually weak, due to C—H in-plane deformation vibrations, are observed at 1290–1010 cm^{-1} (7.75–9.90 μm), six bands often being observed.

Table 12.7. Acridines

Functional Groups	Region		Inten-	Comments
	cm^{-1}	μm	sity	
Acridines	3100–3010	3.23–3.32	m–s	See refs 17, 18
	1630–1360	6.13–7.35	m–s	ring vib, 7–9 bands
	~1000	~10.00	w–m	ring vib, 2 bands
9-Monosubstituted acridines	~1630	~6.13	m–s	
	1610–1595	6.21–6.27	m–s	
	~1545	~6.47	m–s	
	~1520	~6.58	m–s	
	~1460	~6.85	m–s	
	~1435	~6.97	m–s	
	~1400	~7.14	m–s	

Table 12.8. Pyrimidines

Functional Groups	Region		Inten-	Comments
	cm^{-1}	μm	sity	
Pyrimidines	3100–3010	3.23–3.32	m	=C—H str
	1580–1520	6.33–6.58	m–s	C=C, C=N str
	1480–1400	6.76–7.15	v	
	1410–1375	7.09–7.28	v	
	1000–960	10.00–10.42	m	C=C, C=N str
	825–775	12.12–12.90	m	C=C, C=N str
2-Pyrimidines	650–630	15.38–15.87	m–s	
	580–475	17.24–21.05	m–s	
	515–440	19.42–22.73	m–s	
4-Pyrimidines	685–660	14.60–15.15	v	Usually at ~680 cm^{-1}
	555–500	18.02–20.00	m–s	
	500–430	20.00–23.26	s	
Pyrimidine *N*-oxides	1300–1240	7.69–8.07	s–vs	N—O str, often ~1280 cm^{-1}

Table 12.9. Quinazoline aromatic ring stretching vibrations

Functional Groups	Region		Inten-	Comments
	cm^{-1}	μm	sity	
Quinazolines	1635–1565	6.13–6.39	s	Two or three bands
2-Monosubstituted	1630–1620	6.14–6.17	s	
quinazolines	1600–1580	6.25–6.33	m–s	
	1585–1570	6.31–6.37	s	
	1495–1480	6.69–6.76	s	
	1475–1445	6.78–6.92	m–s	
	1415–1395	7.07–7.17	m–s	
	1390–1355	7.19–7.38	s	
	1335–1325	7.49–7.55	w–m	
4-Monosubstituted	1620–1615	6.17–6.19	m–s	
quinazolines	1575–1565	6.35–6.39	s	
	1505–1485	6.65–6.73	s	
	1470–1455	6.80–6.87	w	
	1410–1365	7.09–7.33	m–s	
	1360–1340	7.35–7.46	m–s	
5-Monosubstituted	1630–1615	6.13–6.19	m–s	
quinazolines	1585–1575	6.31–6.35	s	
	1580–1560	6.33–6.41	s	
	1490–1480	6.71–6.76	s	
	1470–1445	6.80–6.92	w	
	1420–1415	7.04–7.07	v	
	1400–1395	7.14–7.17	w–m	
	1385–1360	7.22–7.35	s	
	1315–1305	7.61–7.67	w–m	
6-Monosubstituted	1630–1620	6.14–6.17	m–s	
quinazolines	1605–1595	6.23–6.27	v	
	1580–1565	6.33–6.39	s	
	1505–1490	6.65–6.71	s	
	1475–1430	6.78–7.00	w–m	
	1425–1405	7.02–7.12	v	
	1390–1380	7.19–7.25	s	
	1375–1360	7.27–7.35	s	
	1325–1310	7.55–7.63	v	
7-Monosubstituted	1630–1615	6.14–6.19	m–s	
quinazolines	1595–1575	6.27–6.35	m–s	
	1575–1545	6.35–6.47	m	
	1495–1475	6.69–6.78	v	
	1475–1445	6.78–6.92	w	
	1425–1410	7.02–7.07	v	
	1390–1380	7.19–7.25	s	
	1375–1360	7.27–7.35	s	
	1325–1305	7.55–7.66	w–m	
8-Monosubstituted	1635–1615	6.12–6.19	m–s	
quinazolines	1585–1580	6.31–6.33	s	
	1575–1560	6.35–6.41	m–s	
	1490–1475	6.71–6.78	m–s	
	1470–1460	6.80–6.85	w	
	1450–1445	6.90–6.92	w	
	1410–1390	7.09–7.19	v	
	1390–1380	7.19–7.25	s	
	1310–1300	7.63–7.69	w–m	

Purines, 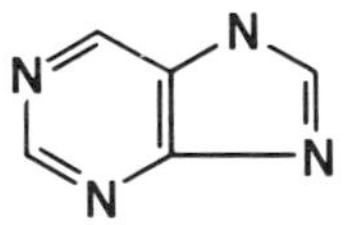

Purines[21] are not, in general, easily distinguished from pyrimidines.

All purines have a characteristic, strong band at about 640 cm^{-1} (15.63 μm). 2-Monosubstituted purines have two bands of medium-to-strong intensity at 650–610 cm^{-1} (15.38–16.39 μm) and 630–585 cm^{-1} (15.87–17.09 μm) and one of variable intensity at 495–375 cm^{-1} (20.20–26.67 μm). 6-Monosubstituted purines have a strong band at 650–625 cm^{-1} (15.38–16.00 μm) and a band of variable intensity at 690–645 cm^{-1} (14.49–15.50 μm), with the exception of 6-cyanopurine for which this last band is not observed. 8-Monosubstituted purines have a strong band at 630–610 cm^{-1} (15.87–16.39 μm) and one of variable intensity at 660–640 cm^{-1} (15.15–15.63 μm). Most 2,6-disubstituted purines have a strong band at 650–630 cm^{-1} (15.38–15.87 μm) and a band of variable intensity at 575–535 cm^{-1} (17.39–18.69 μm). For some 2,6- compounds, this last band is not observed, e.g. some methyl aminopurines.

Table 12.10. Purines

Functional Groups	Region cm^{-1}	Region μm	Intensity	Comments
2-Monosubstituted purines, 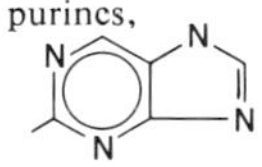	650–610	15.38–16.39	m–s	
	630–585	15.87–17.09	m–s	C—H out-of-plane bending
	495–375	20.20–26.67	v	Out-of-plane pyrimidine ring def
6-Monosubstituted purines, 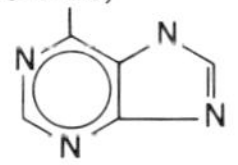	690–645	14.49–15.50	v	
	650–625	15.38–16.00	s	
8-Monosubstituted purines, 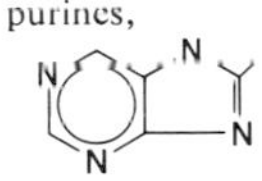	660–640	15.15–15.63	v	
	630–610	15.87–16.39	s	
2,6-Disubstituted purines	650–630	15.38–15.87	s	
	575–535	17.39–18.69	v	Not observed for di(methylamino) purine or 6-amino-2-methylamino purine

Table 12.11. Pyrazines, 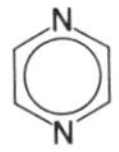, and pyrazine N-oxides

Functional Groups	Region cm^{-1}	Region μm	Intensity	Comments
Pyrazines and pyrazine *N*-oxides	1600–1575	6.25–6.35	v	See ref. 15
	1550–1520	6.45–6.58	w–m	
	1500–1465	6.67–6.83	m–s	
	1420–1370	7.04–7.30	m–s	
Pyrazine *N*-oxides	1350–1260	7.41–7.94	s	N—O str, pyrazine mono-*N*-oxide ~1320 cm^{-1}

Phenazines,[22] 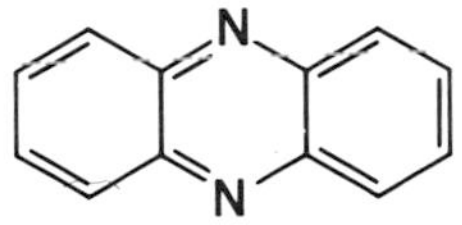

Bands due to the stretching vibrations of N—H and C—H are observed at 3500–3150 cm^{-1} (2.86–3.18 μm) and 3070–3050 cm^{-1} (3.26–3.28 μm) respectively.

A number of strong bands due to the C—H out-of-plane deformation vibrations are observed in the region 900–680 cm^{-1} (11.11–13.79 μm). As with aromatic hydrocarbons, the position of these bands correlates with the number of adjacent hydrogen atoms on the rings:

one hydrogen atom	900–850 cm^{-1} (11.11–11.76 μm)	s;
two adjacent hydrogen atoms	860–800 cm^{-1} (11.63–12.50 μm)	s;
three adjacent hydrogen atoms	810–750 cm^{-1} (12.35–13.33 μm)	s;
	725–680 cm^{-1} (13.79–14.71 μm)	m;
four adjacent hydrogen atoms	770–735 cm^{-1} (12.99–13.61 μm)	s.

Sym-triazines, 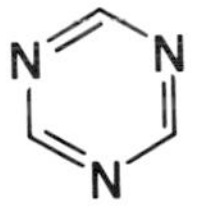

Alkyl and aryl sym-triazines[23] have at least one strong band at 1580–1520 cm^{-1} (6.33–6.58 μm) which may be a doublet, at least one band at 1450–1350 cm^{-1} (6.90–7.41 μm), and at least one weak band at 860–775 cm^{-1} (11.63–12.90 μm). This last band is due to an out-of-plane deformation of the ring, the others being due to in-plane stretching vibrations. 'Hydroxyl' substituted triazines have a strong band at 1775–1675 cm^{-1} (5.63–5.97 μm) due to the C=O stretching vibrations of the keto form and a sharp, medium-intensity band at 795–750 cm^{-1} (12.58–13.25 μm) due to the iso form (see melamines). The normal triazine ring out-of-plane bending vibration band is found at 825–795 cm^{-1} (12.12–12.58 μm).

Cyanuric acid may exist in tautomeric forms:

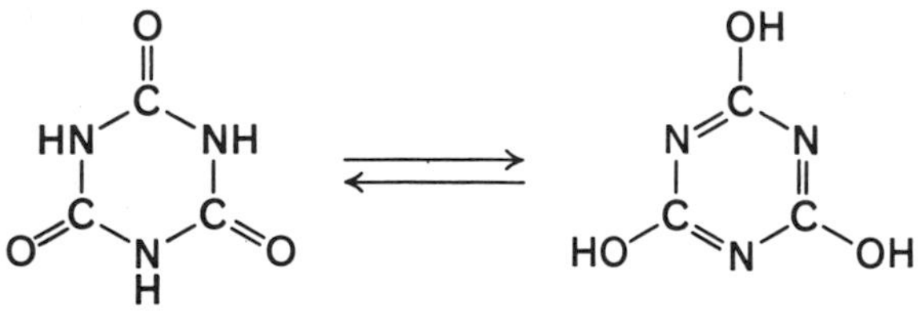

Since the trisodium salt of cyanuric acid is in the enol form, it has the band normally observed for triazines near 820 cm^{-1} (12.20 μm) (see melamine), as do trialkyl cyanurates.

Ammelide (6-amino-sym-triazine-2,4-diol) and ammeline (4,6-di-amino-sym-triazine-2-ol) have a broad absorption near 2650 cm^{-1} (3.77 μm) resulting from the ring NH group which is intramolecularly bonded to the C=O group, e.g. one form of ammeline is

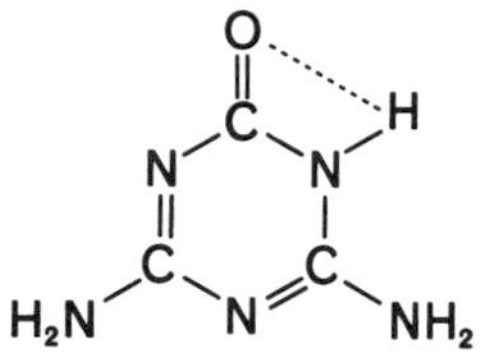

Table 12.12. Sym-triazines

Functional Groups	Region cm⁻¹	Region μm	Intensity	Comments
Sym-triazines	3100–3000	3.23–3.33	m	C—H str
	1580–1520	6.33–6.58	vs	Ring str, at least one band
	1450–1350	6.90–7.41	v	Ring str, at least one band
	860–775	11.63–12.90	w	Out-of-plane ring bending, at least one band
Amino-substituted triazines	1680–1640	5.95–6.10	m–s	NH_2 def
Trialkyl cyanurates	1600–1540	6.25–6.49	s	Ring str
	1380–1320	7.25–7.58	v	
	1160–1110	8.62–9.01	m	OCH_2 str
	820–805	12.20–12.42	m	Triazine out-of-plane bending
Ammelines and ammelides	~2650	~3.77	w–m	br, ring NH···O=C vib
Thioammelines	2900–2800	3.45–3.57	w–m	br, ring NH···S=C vib
	~1200	~8.33	s	C=S str
	~775	~12.90	m	Iso form ring out-of-plane bending

Melamines,

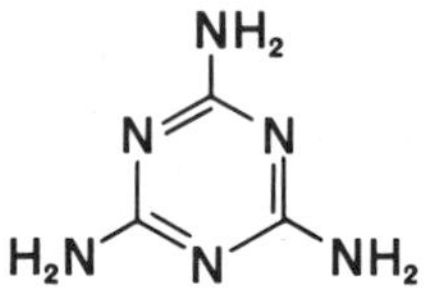

Melamine may exist in tautomeric forms, e.g.

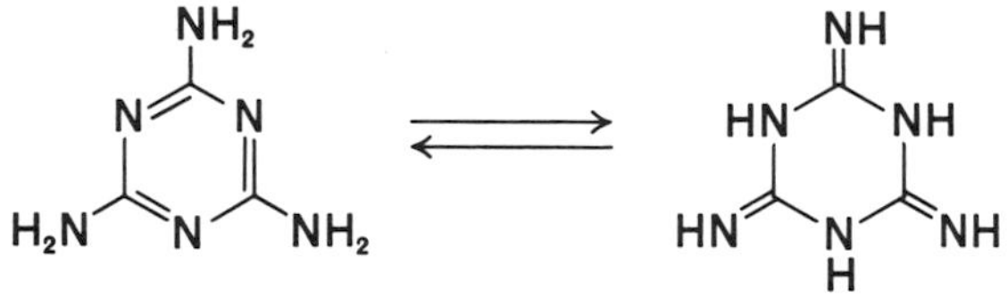

Melamines have an absorption of variable intensity at 3500–3100 cm^{-1} (2.86–3.23 μm) due to the NH_2 stretching vibrations and a band of medium intensity at 1680–1640 cm^{-1} (5.95–6.10 μm) due to NH_2 deformations. A strong band in the region 1600–1500 cm^{-1} (6.25–6.67 μm), usually at 1550 cm^{-1} (6.45 μm), and a number of absorptions at 1450–1350 cm^{-1} (6.90–7.41 μm) are also observed. A sharp, medium-intensity band is usually found at 825–800 cm^{-1} (12.12–12.50 μm) although this band may be at 795–750 cm^{-1} (12.58–13.25 μm) when the triazine ring is in the iso form in which at least one double bond is external to the ring. The ring *N*-alkyl iso-melamines and hydrohalide melamine salts also absorb in this region.

Table 12.13. Melamines

Functional Groups	Region cm⁻¹	Region μm	Intensity	Comments
Melamines	3500–3100	2.86–3.23	v	NH_2 str
	1680–1640	5.95–6.10	m	NH_2 def
	1600–1500	6.25–6.67	s	Ring str
	1450–1350	6.90–7.41	v	sh, number of bands
	825–800	12.12–12.50	m	} Only one of these two is present
	795–750	12.58–13.25	m	}

Table 12.14. Sym-tetrazines, 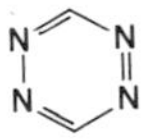

Functional Groups	Region cm^{-1}	μm	Intensity	Comments
Sym-tetrazines	1600–1500	6.25–6.67	m–	Ring str, absent for molecules with centre of symmetry
	1495–1320	6.69–7.58	m	Ring str
	970–880	10.31–11.36	m	

Table 12.15. α-Pyrones, 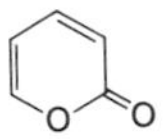, and γ-pyrones, 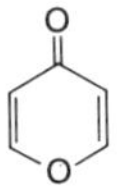

Functional Groups	Region cm^{-1}	μm	Intensity	Comments
α-Pyrones	~1730	~5.78	s	C=O str
	1650–1635	6.06–6.12	m	C=C str
	~1565	~6.39	s	C=C str
γ-Pyrones	1680–1600	5.92–6.25	vs	Combination of C=O and C=C stretching vibrations
	1635–1525	6.12–6.56	vs	
	1465–1445	6.83–6.92	m–s	
	1420–1400	7.04–7.14	m	
γ-Pyrthiones	~1100	~9.09	s	C=S str

Table 12.16. Pyrylium compounds, 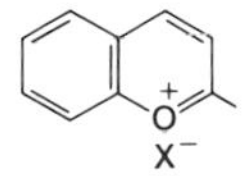

Functional Groups	Region cm^{-1}	μm	Intensity	Comments
Pyrylium derivatives	3100–3010	3.23–3.32	w–m	=C—H str, a number of bands
	1650–1615	6.06–6.19	vs	Ring in-plane vib
	1560–1520	6.41–6.58	vs	Ring in-plane vib
	1520–1465	6.58–6.83	m	Ring in-plane vib
	1450–1400	6.90–7.14	v	Ring in-plane vib
	1000–970	10.00–10.31	v	Ring in-plane vib
Unsubstituted pyrylium salts	~960	~10.42	s	C—H out-of-plane vib
	775	~12.92	m	
2,6-Disubstituted pyrylium compounds	~935	~10.70	m	C—H out-of-plane vib
	~800	~12.50	s	
2,4,6-Trisubstituted pyrylium compounds	960–900	10.42–11.11	v	C—H out-of-plane vib, two bands
	890–870	11.24–11.49	m	C—H out-of-plane vib
2,3,4,6-Tetrasubstituted pyrylium compounds	~920	~10.87	w	C—H out-of-plane vib
	900–880	11.11–11.36	w	C—H out-of-plane vib
2,3,5,6-Tetrasubstituted pyrylium compounds	~705	~14.18	m	

References

1. A. F. Ardyukova *et al.*, *Atlas of Spectra of Aromatic and Heterocyclic Compounds.* Nauka Sib. Otd., Novosibirsk, 1973
2. G. L. Cook and F. M. Church, *J. Phys. Chem.*, 1957, **61,** 458
3. A. R. Katritzky, *Quart. Rev.*, 1959, **13,** 353
4. J. H. S. Green and D. J. Harrison, *Spectrochim. Acta*, 1973, **29A,** 293
5. D. Hement *et al.*, *J. Amer. Chem. Soc.*, 1959, **81,** 3933
6. J. K. Wilmshurst and H. J. Bernstein, *Canad. J. Chem.*, 1957, **35,** 1183
7. R. Isaac *et al.*, *App. Spectrosc.*, 1963, **17,** 90
8. J. H. S. Green *et al.*, *Spectrochim. Acta*, 1973, **29A,** 1177
9. J. H. S. Green *et al.*, *Spectrochim. Acta*, 1963, **19,** 549
10. R. Tripathi, *Indian J. Pure Appl. Phys.*, 1973, **11,** 277
11. A. R. Katritzky *et al.*, *J. Chem. Soc.*, **1959,** 3680
12. J. Yarwood, *Spectrochim. Acta*, 1970, **26A,** 2077
13. G. Varsányi *et al.*, *Acta Chim. Hungary*, 1965, **43,** 205
14. J. S. Strukl and J. L. Walter, *Spectrochim. Acta*, 1971, **27A,** 209
15. H. Shindo, *Chem. Pharm. Bull. Japan*, 1960, **8,** 33
16. A. Leifer *et al.*, *Spectrochim. Acta*, 1964, **20,** 909
17. R. Acheson, *Chem. Heterocyclic Compds*, 1973, **9,** 665
18. J. Reisch *et al.*, *Pharmazie*, 1972, **27,** 208
19. A. F. Ardyukova *et al.*, *Atlas of Spectra of Aromatic and Heterocyclic Compounds*, No. 4, *Infrared Spectra of Pyrimidine Series*, Nauka Sib. Otd., Novosibirsk, 1974
20. A. R. Katritzky *et al.*, *Spectrochim. Acta*, 1964, **20,** 593
21. J. H. Lister, *Chem. Heterocyclic Compds*, 1971, **7,** 496
22. C. Stammer and A. Taurins, *Spectrochim. Acta*, 1963, **19,** 1625
23. W. A. Heckle *et al.*, *Spectrochim. Acta*, 1961, **17,** 600
24. O. P. Shkurko and I. K. Korobeinicheva, *Zh. Prinkl. Spektrosk.*, 1975, **23,** 860

CHAPTER 13

Five-membered Ring Heterocyclic Compounds

Heteroaromatic compounds of the type

X

generally have three bands due to C=C in-plane vibrations at about 1580 cm^{-1} (6.33 μm), 1490 cm^{-1} (6.71 μm), and 1400 cm^{-1} (7.14 μm). In addition, those with a CH=CH group have a strong band in the region 800–700 cm^{-1} (12.50–14.29 μm) due to an out-of-plane deformation vibration.

Pyrroles, 
N
, and indoles, 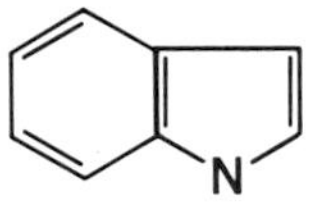
N

In dilute solution, the band due to the N—H stretching vibration occurs at 3500–3400 cm^{-1} (2.86–2.94 μm).[1] In the presence of hydrogen bonding, a broad absorption occurs at 3400–3000 cm^{-1} (2.94–3.33 μm). The bands due to the C=C and C=N stretching vibrations occur in the region 1580–1390 cm^{-1} (6.33–7.19 μm).[1]

Pyrroles have one or two bands in the region 1580–1545 cm^{-1} (6.33–6.47 μm) depending on whether or not there is substitution on the nitrogen atom. A very strong band is observed at 1430–1390 cm^{-1} (6.99–7.19 μm) and a weak band near 1470 cm^{-1} (6.80 μm). 1,2-Disubstituted pyrroles have a medium-intensity band at 1500–1475 cm^{-1} (6.67–6.78 μm) and a weak band at 1530 cm^{-1} (6.54 μm). This latter band is also observed for 1,2,5- and 1,3,4-trisubstituted pyrroles.

Indoles[2,3] absorb at 3480–3020 cm^{-1} (2.87–3.31 μm) and near 1460 cm^{-1} (6.85 μm), 1420 cm^{-1} (7.04 μm), and 1350 cm^{-1} (7.41 μm).

Pyrrolines, 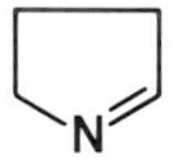
N

Pyrrolines[4] have a medium-to-strong band at 1660–1560 cm^{-1} (6.02–6.41 μm) due to the C=N stretching vibration. The other forms, such as

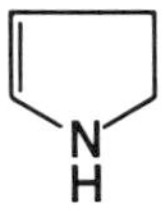
N
H

are normally unstable and therefore it is usual for no band due to the N—H stretching vibration to be observed.

Table 13.1. Pyrroles (and similar five-membered ring compounds): N—H, C—H, and ring stretching vibrations

Functional Groups	Region cm⁻¹	Region μm	Intensity	Comments
Pyrroles	3500–3400	2.86–2.94	v	N—H str, free pyrroles
	3400–3000	2.94–3.33	s	br, N—H str, hydrogen-bonded pyrroles
	3100–3010	3.23–3.32	m	=C—H str, multiple peaks
	1580–1545	6.33–6.47	w–m	Two bands for 1-substituted pyrroles, C=C and C=N in-plane vibs
	~1470	~6.80	w–m	C=C and C=N in-plane vibs
	1430–1390	6.99–7.19	vs	C=C and C=N in-plane vibs
	~480	~20.83	m–s	Ring def, not greatly affected by substitution
1,2-Disubstituted pyrroles	~1530	~6.54	w	C=C in-plane vib
	1500–1475	6.67–6.78	m–s	C=C in-plane vib
1,2,5- and 1,3,4-trisubstituted pyrroles	~1530	~6.54	w	C=C in-plane vib
Indoles, 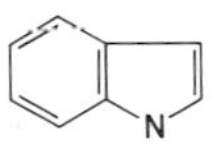	1625–1615	6.15–6.19	m	Ring vib
	1600–1575	6.25–6.35	m	Ring vib
	1565–1540	6.39–6.49	v	Sometimes absent, ring vib
	1520–1470	6.58–6.80	m	Ring vib
Pyrrolines, 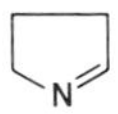	1660–1560	6.02–6.41		C=N str, see ref. 4
Oxazoles, 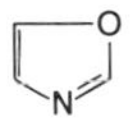	1585–1555	6.31–6.43	m	C=N str
Thiazoles, 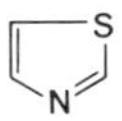	1550–1505	6.45–6.64	m	C=N str, see refs 6, 7
Imidazoles, 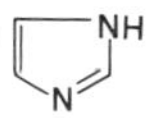	1560–1520	6.41–6.58	m	C=N str, see ref. 5
Benzimidazoles, 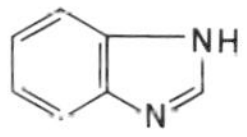	1560–1520	6.41–6.58	m	C=N str
Oxazolines, 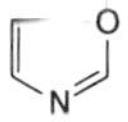	1695–1645	5.90–6.08	s	C=N str
1,2,4-Oxadiazoles, 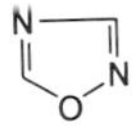	1590–1560	6.29–6.41	s	Ring vib
	1470–1430	6.80–6.99	s	Ring vib
	1390–1360	7.19–7.35	s	Ring vib
1,2,5-Oxadiazoles, 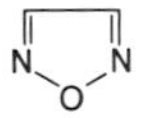	1625–1560	6.15–6.41	s	
	1400–1370	7.14–7.30	s	

Table 13.2. Substituted pyrroles: N—H and C—H deformation vibrations

Functional Groups	Region cm⁻¹	Region μm	Intensity	Comments
1-Substituted pyrroles	~1070	~9.35	s	Four adjacent hydrogen atoms
	1035–1015	9.66–9.85	m	Four adjacent hydrogen atoms
	~925	~10.81	m	Four adjacent hydrogen atoms
	~725	~13.79	vs	Four adjacent hydrogen atoms
2-Substituted pyrroles	~1115	~8.97	w–m	Three adjacent hydrogen atoms
	1105–1070	9.05–8.55	m–s	Three adjacent hydrogen atoms
	~1030	~9.71	m–s	Three adjacent hydrogen atoms
	~925	~10.81	w	Three adjacent hydrgen atoms
	~880	~11.36	w–m	Three adjacent hydrogen atoms
1,2-Disubstituted pyrroles	~1090	~9.17	m	Three adjacent hydrogen atoms
	1065–1050	9.39–9.52	v	Three adjacent hydrogen atoms
1,2,5-Trisubstituted pyrroles	~1035	~9.66	m	Two adjacent hydrogen atoms
	980–965	10.20–10.36	w	Two adjacent hydrogen atoms
	~755	~13.25	vs	Two adjacent hydrogen atoms
1,3,4-Trisubstituted pyrroles	~1055	~9.48	s	One hydrogen atom
	~930	~10.75	m	One hydrogen atom
	~770	~12.99	vs	One hydrogen atom

Furans, 

Furan derivatives[8,13,14] have medium-to-strong bands at 1610–1560 cm^{-1} (6.21–6.41 μm), 1520–1470 cm^{-1} (6.58–6.80 μm), and 1400–1390 cm^{-1} (7.14–7.19 μm) which are due to the C=C ring stretching vibrations. Furans with electronegative substituents usually have strong bands in these regions.

For 2-substituted furans,[13] the out-of-plane deformation vibrations of the C—H group give bands at 935–915 cm^{-1} (10.70–10.93 μm), 885–880 cm^{-1} (11.29–11.34 μm), and 835–780 cm^{-1} (11.98–12.82 μm).

All furans have a strong absorption near 595 cm^{-1} (16.81 μm) which is probably due to a ring deformation vibration.

Tetrahydrofurans[11,12] have a strong band at 1100–1075 cm^{-1} (9.09–9.30 μm) due to the C—O stretching vibration and another band near 915 cm^{-1} (10.93 μm).

Table 13.3. Furans

Functional Groups	Region cm⁻¹	Region μm	Intensity	Comments
Furan derivatives	3100–3000	3.23–3.33	m	=C—H str
	1610–1560	6.21–6.41	m–s	C=C str
	1520–1470	6.58–6.80	m–s	C=C str
	1400–1390	7.14–7.19	m–s	C=C str
	1025–1000	9.76–10.00	m–s	
	595–515	16.81–19.42	s	Ring def
2-Monosubstituted furans	1240–1200	8.07–8.33	v	C—H def, see ref. 14
	1175–1145	8.51–8.73	m–s	C—H def
	1085–1070	9.22–9.35	m	C—H def
	935–915	10.70–10.93	w–m	Out-of-plane C—H def
	885–880	11.29–11.34	w–m	Out-of-plane C—H def
	835–780	11.98–12.82	w–m	Out-of-plane C—H def
	595–515	16.81–19.42	s	Ring def
3-Monosubstituted furans	1170–1150	8.55–8.70	s	C—H def, see ref. 14
	1080–1050	9.26–9.52	m–s	C—H def
	1025–1000	9.76–10.00	vs	C—H def
	~920	~10.87	v	C—H def
	~875	~11.43	s	C—H def
	790–720	12.66–13.89	s	Usually two bands
2,5-Disubstituted furans	1255–1225	7.97–8.17	w–m	See ref. 13
	1165–1140	8.58–8.77	w–m	
	~1020	~9.90	m	Ring def
	990–960	10.10–10.42	m	
	930–915	10.75–10.93	w–m	C—H out-of-plane def
	835–780	11.98–12.82	w–m	C—H out-of-plane def
Polysubstituted furans	~1560	~6.41	m–s	C=C str
	~1510	~6.62	m–s	C=C str
Oxazoles, 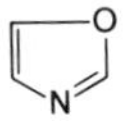	1585–1555	6.31–6.43	m	C=N str
Iso-oxazoles, 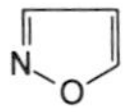	~1600	~6.25	m–s	
	~1460	~6.85	m–s	
	~1380	~7.25	m–s	
1,2,4-Oxadiazoles, 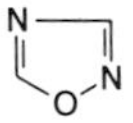	1590–1560	6.29–6.41	m–s	Ring str, see ref. 9
	1470–1430	6.80–6.99	m–s	Ring str
	1390–1360	7.19–7.35	m–s	Ring str
	1070–1050	9.35–9.52	m	
	915–885	10.93–11.30	m–s	
	750–710	13.33–14.08	m–s	
1,2,5-Oxadiazoles (furazanes), 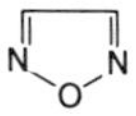	1625–1560	6.15–6.41	m–s	Ring str
	~1570	~6.37	m–s	
	~1425	~6.78	m–s	
	1395–1370	7.17–7.30	m–s	Ring str
1,2,5-Oxadiazole oxides, 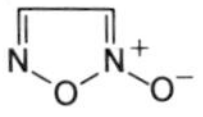	1635–1600	6.12–6.25	m–s	Ring str, see ref. 10
	1530–1515	6.54–6.60	m–s	Ring str
	1475–1410	6.78–7.09	m–s	Ring str
1,3-Dioxolanes 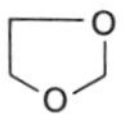	1170–1145	8.55–8.73	s	Ring vib
	1100–1050	9.09–9.52	s	Ring vib
	~940	~10.64	m–s	Ring vib, may be absent

Table 13.3 (*cont.*)

Functional Groups	Region cm⁻¹	Region μm	Intensity	Comments
Oxalolidines, 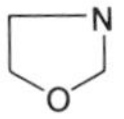	1190–1050	8.40–9.52	m	Ring def, at least three bands
Tetrahydrofurans, 	2980–2700	3.36–3.70	s	Several bands, see refs 11, 12
	1500–1450	6.67–6.90	v	CH_2 def
	1375–1275	7.27–7.84	v	CH_2 def
	1260–1180	7.94–8.48	v	CH_2 def
	1100–1075	9.09–9.30	s	C—O str
	860–760	11.63–13.16	v	CH_2 def

Thiophenes, 

Thiophenes[15, 16] absorb at 3100–3000 cm^{-1} (3.23–3.33 μm) due to the C—H stretching vibration and also have four bands of variable intensity in the region 1555–1200 cm^{-1} (6.43–8.33 μm) due to in-plane ring vibrations.

All monosubstituted thiophenes have two bands of variable, often medium-to-strong, intensity, one at 745–695 cm^{-1} (13.42–14.39 μm) and the other at 700–660 cm^{-1} (14.29–15.15 μm), possibly due to the out-of-plane bending of the =C—H group.

2-Monosubstituted thiophenes usually have two bands of variable intensity, one at 570–490 cm^{-1} (17.54–20.41 μm) and the other at 470–430 cm^{-1} (21.28–23.26 μm). For esters of thiophene-2-carboxylic acid, the former band is usually near 565 cm^{-1} (17.70 μm).[16]

3-Monosubstituted thiophenes have a band of medium intensity at 540–515 cm^{-1} (18.52–19.42 μm) and a band of variable intensity at 500–465 cm^{-1} (20.00–21.51 μm). Sometimes only one band is observed.

Most 2-nitro-5-substituted thiophenes have bands of variable intensity at 555–525 cm^{-1} (18.02–19.05 μm) and 490–445 cm^{-1} (20.41–22.47 μm), and usually one or two weak bands near 430 cm^{-1} (23.26 μm).

In general, mono-, di-, tri-, and tetrasubstituted thiophenes all have bands in the region 530–450 cm^{-1} (18.87–22.22 μm) due to the out-of-plane ring deformation.

Table 13.4. Thiophenes

Functional Groups	Region		Inten-	Comments
	cm^{-1}	μm	sity	
Thiophenes	3100–3000	3.23–3.33	m	=C—H str
	1555–1480	6.43–6.56	v	C=C in-plane vib
	1445–1390	6.92–7.19	v	C=C in-plane vib
	1375–1340	7.33–7.46	v	C=C in-plane vib
	1240–1195	8.07–8.37	v	C=C in-plane vib
	530–450	18.87–22.22	v	Ring def
Monosubstituted thiophenes	745–695	13.42–1439	v	=C—H out-of-plane def
	700–660	14.29–15.15	v	=C—H out-of-plane def
2-Monosubstituted thiophenes	1535–1515	6.52–6.60	v	C=C in-plane vib, see ref. 17
	1455–1430	6.87–6.99	v	C=C in-plane vib
	1360–1345	7.35–7.44	v	C=C in-plane vib
	570–490	17.54–20.41	v	Esters at ~565 cm^{-1}
	470–430	21.28–23.26	v	
2-Alkyl thiophenes	~1080	~9.26	w	C—H def
	1055–1030	9.48–9.71	w–m	
	940–905	10.64–11.05	m	Out-of-plane CH def
	865–840	11.56–11.90	m–s	Out-of-plane CH def
	845–800	11.83–12.50	m	Out-of-plane CH def
	~690	~14.50	m	Out-of-plane CH def
3-Monosubstituted thiophenes	~1530	~6.54	v	C=C in-plane vib
	~1410	~7.09	v	
	~1370	~7.30	v	
	540–515	18.52–19.42	m	Sometimes only one present
	500–465	20.00–21.51	v	
3-Alkyl substituted thiophenes	~1530	~6.54	v	C=C in-plane vib
	~1410	~7.09	v	C=C in-plane vib
	~1370	~7.30	v	C=C in-plane vib
	~1155	~8.66	w	C—H def
	1100–1070	9.09–9.35	w	C—H def
	895–850	11.17–11.76	m	
	795–745	12.58–13.42	s	Out-of-plane CH def
2,3-Disubstituted thiophenes	715–690	14.01–14.49	m	Out-of-plane C—H def
2,4-Disubstituted thiophenes	825–805	12.11–12.41	m	Out-of-plane C—H def
2,5-Dialkyl thiophenes	~795	~12.58	m–s	C—H def
2-Nitro-5-substituted thiophenes	555–525	18.02–19.05	v	
	490–445	20.41–22.47	v	
	~430	~23.26	w	One or two bands
3,4-disubstituted thiophenes	925–910	10.80–11.00	m	Out-of-plane C—H def
	860–835	11.63–11.98	m	Out-of-plane C—H def
	780–775	12.82–12.90	m	Out-of-plane C—H def
Selenophenes, mono- and dimethyl substituted	440–405	22.73–24.69		See ref. 18
Thiazoles, 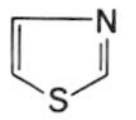	~1610	~6.21	v	See ref. 6
	1330–1305	6.45–6.64	m	
	~1380	~7.25	v	

Imidazoles, 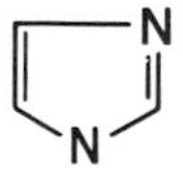

In general, azoles have three or four bands in the region 1670–1320 cm^{-1} (5.99–7.58 μm) due to C=C and C=N stretching vibrations. The intensities of these bands depend on the nature and positions of the substituents and on the position and nature of the ring heteroatoms.

In the solid phase, five-membered heteroatomic compounds with two or more nitrogen atoms in the ring have a broad absorption at 2800–2600 cm^{-1} (3.57–3.85 μm) due to the NH···N bond.

Imidazoles[5] have several bands of variable intensity in the range 1660–1450 cm^{-1} (6.02–6.90 μm) due to C=N and C=C stretching vibrations. Most 4-monosubstituted imidazoles have two strong bands, at 670–625 cm^{-1} (14.93–16.00 μm) and 630–605 cm^{-1} (15.87–16.53 μm). They also have two bands of medium intensity, at 445–355 cm^{-1} (22.47–28.17 μm) and 360–325 cm^{-1} (27.78–30.77 μm), although this last band is absent for some imidazoles. The first of these two bands is probably due to out-of-plane bending of the —N—H group.

4,5-Disubstituted imidazoles have two medium-to-strong bands, at 665–650 cm^{-1} (15.04–15.38 μm) and 645–610 cm^{-1} (15.50–16.39 μm). 1,4,5-Trisubstituted imidazoles have a medium-to-strong absorption at 660–650 cm^{-1} (15.15–15.63 μm) and a weak-to-medium band at 420–390 cm^{-1} (23.81–25.64 μm). A study of metal complexes with imidazole ligands can be found elsewhere.[19]

Table 13.5. Imidazoles

Functional Groups	Region		Inten-	Comments
	cm^{-1}	μm	sity	
Imidazoles, 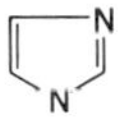	1660–1610	6.02–6.21	v	Imidazole I band
	1605–1585	6.23–6.31	w–m	Ring C=C str
	1560–1520	6.41–6.58	s	N=C—N str
4-Monosubstituted imidazoles, 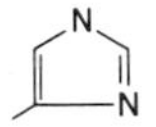	670–625	14.93–16.00	s	
	630–605	15.87–16.53	s	
	445–355	22.47–28.17	m	
	360–325	27.78–30.77	m	Not always present
4,5-Disubstituted imidazoles, 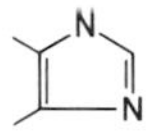	665–650	15.04–15.38	m–s	
	645–610	15.50–16.39	m–s	
1,4,5-Trisubstituted imidazoles, 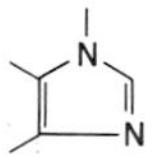	660–640	15.15–15.63	m–s	
	420–390	23.81–25.64	w–m	

Pyrazoles, 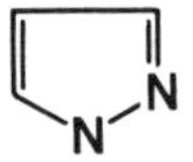

Due to tautomerism, positions 3 and 5 of pyrazoles[20, 21] are equivalent:

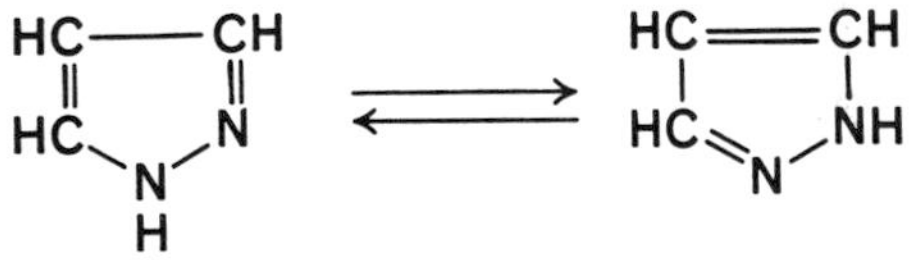

Some pyrazol-5-one derivatives[22] exist as a form in which the carbonyl group is no longer present, and indeed two such forms may exist:

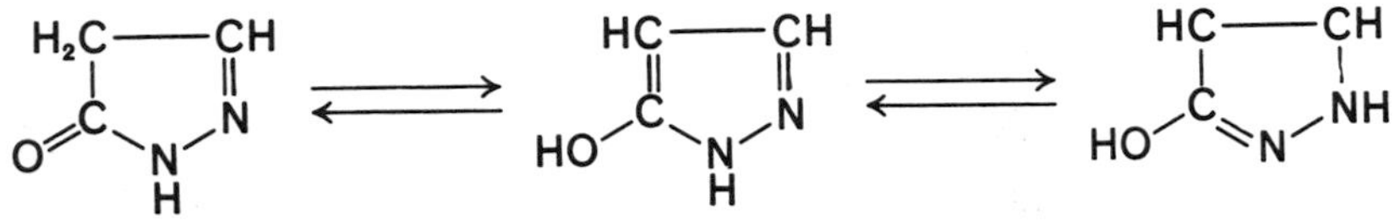

In the case of 4,4- and 1,2-disubstituted pyrazol-5-ones, the carbonyl group[24] is present and hence for these compounds an absorption band due to the carbonyl stretching vibration is observed.

5-Aminopyrazoles have a band of medium intensity near 1595 cm^{-1} (6.27 μm) and weaker bands near 1660 cm^{-1} (6.02 μm) and 1550 cm^{-1} (6.45 μm). All three bands have been attributed to ring vibrations.

Table 13.6. Pyrazoles

Functional Groups	Region cm^{-1}	Region μm	Intensity	Comments
N-Alkyl substituted pyrazoles	~1090	~9.17	m–s	See ref. 20
	~753	~13.28	m–s	
3-Alkyl substituted pyrazoles	~3175	~3.15	m	N—H str
	~937	~10.67	s	
	~770	~12.99	s	br
4-Alkyl substituted pyrazoles	1010–1000	9.90–10.00	s	
	~950	~10.53	s	
	~860	~11.63	s	
	~805	~12.42	s	
3-, 2,3-, 3,4-, 1,3,4-, and 2,3,4-substituted pyrazol-5-ones	3000–2200	3.33–4.55	m	br, O—H and N—H str
	1670–1450	5.99–6.90	w–m	Three or four bands due to C=C and C=N str
1,2,3-Trisubstituted pyrazol-5-ones	1675–1655	5.97–6.04	s	C=O str
3,4,4-Trisubstituted pyrazol-5-ones	~3150	~3.18	m	br, N—H str
	1705–1675	5.87–5.97	s	C=O str

References

1. R. A. Jones and A. G. Moritz, *Spectrochim. Acta*, 1965, **21,** 295
2. F. Millich and E. I. Becker, *J. Org. Chem.*, 1958, **23,** 1096
3. A. R. Katritzky and A. P. Ambler, *Physical Methods in Heterocyclic Chemistry*. Academic Press, New York, 1963, p. 161
4. A. I. Meyers, *J. Org. Chem.*, 1959, **24,** 1233
5. P. Bassignana *et al.*, *Spectrochim. Acta*, 1965, **21,** 605
6. J. Chouteau *et al.*, *Bull. Soc. Chim. France*, 1962, **18,** 1794
7. M. P. V. Mijovic and J. Walker, *J. Chem. Soc.*, **1961,** 3381
8. W. H. Washburn, *App. Spectrosc.*, 1964, **18,** 61
9. J. Baran, *Compt. Rend.*, 1959, **249,** 1096
10. J. H. Bayer *et al.*, *J. Amer. Chem. Soc.*, 1957, **79,** 1748
11. N. Baggett *et al.*, *J. Chem. Soc.*, **1960,** 4565
12. P. Grünager and F. Pozzi, *Gazz. Chim. Ital.*, 1959, **89,** 897
13. P. Grunager *et al.*, *Gazz. Chim. Ital.*, 1959, **89,** 913
14. A. R. Katritzky and J. M. Lagowski, *J. Chem. Soc.*, **1959,** 657
15. M. Rico *et al.*, *Spectrochim. Acta*, 1965, **21,** 689
16. A. Hidalgo, *J. Phys. Rad.*, 1955, **16,** 366
17. A. R. Katritzky and A. J. Boulton, *J. Chem. Soc.*, **1959,** 3500
18. N. A. Chumaevskii *et al.*, *Optics Spectrosc.*, 1959, **6** (1), 25
19. M. T. Forel *et al.*, *Colloq. Int. Cent. Nat. Rech. Sci.* 1970, **191,** 167
20. G. Zerbi and C. Alberti, *Spectrochim. Acta*, 1962, **18,** 407
21. A. A. Novikova and F. M. Shemyakin, *Khim.-Farm. Zh.*, 1968, **2** (10), 45
22. S. Refn, *Spectrochim. Acta*, 1961, **17,** 40
23. J. H. Lister, *Chemistry of Heterocyclic Compounds*, Vol. 24, Wiley, New York, 1971, p. 496
24. W. Freyer, *J. Prakt. Chem.*, 1977, **319,** 911

CHAPTER 14

Organic Nitrogen Compounds

Nitro- Compounds, $—NO_2$ [1, 2]

Saturated primary and secondary aliphatic nitro- compounds,[3–8] $—CH_2NO_2$ and $>CHNO_2$, have very strong bands at about 1550 cm^{-1} (6.45 μm) and 1385–1360 cm^{-1} (7.22–7.35 μm) which are due to the asymmetric and symmetric stretching vibrations respectively of the NO_2 group. Electron-withdrawing substituents adjacent to the nitro- group tend to increase the frequency of the asymmetric vibration and decrease that of the symmetric vibration.[7]

The band due to the C—N stretching vibration is of weak-to-medium intensity and occurs at 920–850 cm^{-1} (10.87–11.76 μm). Other groups have strong absorptions in this region which may obscure this band. Organic nitro- compounds have a very strong band at 655–605 cm^{-1} (15.27–16.53 μm) due to the deformation vibration of the NO_2 group. Primary aliphatic straight-chain nitro- compounds absorb strongly at 620–600 cm^{-1} (16.13–16.67 μm) and also have a medium-to-strong band at 490–465 cm^{-1} (20.41–21.51 μm), both bands being due to the NO_2 deformation vibration. Secondary nitroalkanes absorb at 630–610 cm^{-1} (15.87–16.39 μm) and 550–515 cm^{-1} (18.18–19.42 μm).

α,β-Unsaturated nitroalkenes absorb strongly at 1530–1510 cm^{-1} (6.54–6.62 μm) and 1360–1335 cm^{-1} (7.35–7.49 μm) due to the $—NO_2$ asymmetric and symmetric stretching vibrations. These bands are almost of equal intensity. The nitro- group does not appear to affect the position of the characteristic alkene C=C and C—H bands. However, the relative intensities of the bands due to the =C—H stretching and wagging vibrations are increased when the nitro- group is bonded to the same olefinic carbon as the hydrogen atom, the intensity of the band due to the C=C stretching vibration also being increased.

Aromatic nitro- compounds[9–12] have strong absorptions due to the asymmetric and symmetric stretching vibrations of the NO_2 group at 1570–1485 cm^{-1} (6.37–6.73 μm) and 1370–1320 cm^{-1} (7.30–7.58 μm) respectively. The intensity of this latter band is increased for electron-donating ring substituents. The former band is usually found in the range 1540–1515 cm^{-1} (6.49–6.60 μm). *Para*-substituted nitro- compounds whose substituent is a strongly electron-donating atom or group absorb at 1515–1485 cm^{-1} (6.60–6.73 μm), whereas those with electron-accepting groups absorb at 1570–1540 cm^{-1} (6.37–6.49 μm).

The asymmetric NO_2 stretching vibration of most singly-substituted aromatic *para-* nitro- compounds gives a band in the range 1535–1510 cm^{-1} (6.52–6.62 μm), exceptions to this being *p*-dinitrobenzene and some *p*-amino nitrobenzenes. Singly-substituted aromatic *meta-* nitro- compounds absorb in the range 1540–1525 cm^{-1} (6.49–6.59 μm) and nitro- compounds with small substituents in the *ortho-* position absorb at 1540–1515 cm^{-1} (6.49–6.60 μm). The band due to the asymmetric stretching vibration for nitro- groups forced out of the plane of the ring by bulky substituents in the *ortho-* positions is at 1565–1540 cm^{-1} (6.39–6.49 μm). Hydrogen bonding has little effect on the NO_2 asymmetric stretching vibration.[12]

The symmetric vibration of the NO_2 group for aromatic *para*-substituted nitro- compounds occurs at 1355–1335 cm^{-1} (7.38–7.49 μm) whereas for *meta*-compounds, and also ortho- compounds with small substituents, the range is 1355–1345 cm^{-1} (7.38–7.44 μm). In the case of bulky *ortho-* substituents, this band may be found as high as 1380 cm^{-1} (7.25 μm). In cases where strong hydrogen bonding occurs, this band may be found at about 1320 cm^{-1} (7.58 μm), an example being *o*-nitrophenol.

Aromatic nitro- compounds have a band of variable intensity in the region 580–520 cm^{-1} (17.24–19.23 μm) which is due to the in-plane deformation of the $—NO_2$ group.[30, 31] A strong band due to the C—N stretching vibration is observed at 865–835 cm^{-1} (11.56–11.98 μm) and a band is also sometimes observed at about 750 cm^{-1} (13.33 μm).

Due to the deformation vibration of the adjacent methylene group, primary nitroalkanes have a band of medium intensity near 1430 cm^{-1} (6.99 μm). In general, the band due to the symmetric deformation vibration of the methyl group is overlapped by that due to the NO_2 symmetric stretching vibration. However, in compounds where both the methyl and nitro- groups are attached to the same carbon atom, two well-separated bands are observed—one near 1385 cm^{-1} (7.22 μm) and the other near 1370 cm^{-1} (7.30 μm).

Alkali metal nitroparaffins[14] have a very strong absorption at 1605–1575 cm^{-1} (6.23–6.35 μm) due to the C=N stretching vibration, and a weak band near 1660 cm^{-1} (6.06 μm).

Table 14.1. Nitro- compounds

Functional Groups	Region		Inten-	
	cm^{-1}	μm	sity	Comments
Saturated primary and secondary aliphatic nitro- compounds, $-CH_2-NO_2$ and $>CH-NO_2$	1550–1545	6.45–6.47	vs	asym NO_2, str, see ref. 13, stronger than sym str
	1385–1360	7.22–7.35	vs	sym NO_2 str (CH_2 def also occurs in this region)
	1000–915	10.00–10.93	m	C—N str, *trans*- form
	920–850	10.87–11.76	m–s	br, C—N str, *gauche*- form
	655–605	15.27–16.53	vs	NO_2 def
Straight-chain primary nitroalkanes	620–600	16.13–16.67	s	sym NO_2 def (except nitromethane at ~649 cm^{-1})
	490–465	20.41–21.51	m–s	NO_2 rocking (except nitromethane at ~602 cm^{-1})
Secondary nitroalkanes	630–610	15.87–16.39	m–s	sym NO_2 def
	550–515	18.18–19.42	s	NO_2 rocking
Saturated tertiary aliphatic nitro- compounds, $\geqslant C-NO_2$	1550–1530	6.45–6.54	s	asym NO_2 str
	1360–1340	7.35–7.46	s	sym NO_2 str
Dinitroalkanes, $>C(NO_2)_2$	1590–1570	6.29–6.37	s	asym NO_2 str
	1340–1325	7.46–7.55	s	sym NO_2 str, may be split
α,β-Unsaturated nitro- compounds	1530–1510	6.54–6.62	s	asym NO_2 str
	1360–1335	7.35–7.49	s	sym NO_2 str
α-Halo-nitro-compounds	1580–1570	6.33–6.37	s	asym NO_2 str
	1355–1340	7.38–7.46	s	sym NO_2 str
α,α′-Dihalo-nitro- compounds compounds	1600–1575	6.25–6.35	s	asym NO_2 str
	1340–1320	7.46–7.58	s	sym NO_2 str
Aromatic nitro- compounds	1570–1485	6.37–6.73	s	asym NO_2 str
	1370–1320	7.30–7.58	s	sym NO_2 str
	865–835	11.56–11.98	s	C—N str
	~750	~13.33	s	Not always present
	580–520	17.24–19.23	v	In-plane bending of $-NO_2$ group
o-Amino nitro- aromatic compounds	1260–1210	7.94–8.26	s	sym NO_2 str
Nitroamines, $>N-NO_2$	1630–1530	6.14–6.54	s	asym NO_2 str, solids may be as low as 1500 cm^{-1}
	1315–1260	7.61–7.94	s	sym NO_2 str, solids may be as low as 1250 cm^{-1}
Nitrates, $-O-NO_2$	1635–1625	6.12–6.16	s	asym NO_2 str
	1285–1270	7.78–7.87	s	sym NO_2 str
Carbonitrates, $>C=NO_2^-$	1605–1575	6.23–6.35	s	C=N str, see ref. 15, low due to resonance
	1315–1205	7.60–8.30	s	asym NO_2 str
	1175–1040	8.56–9.62	s	sym NO_2 str
	735–700	13.61–14.28	m–s	NO_2 def
Nitrocycloalkanes (three-membered ring and larger)	1550–1535	6.45–6.51	s	asym NO_2 str
	1380–1355	7.25–7.38	s	sym NO_2 str

Nitroso- Compounds, —N=O,[16–19] (and Oximes, >C=N—OH)

In the solid and liquid phases, organic nitroso- compounds normally exist as dimers and may have *cis*- or *trans*- forms.

The fact that primary and secondary nitroso- compounds readily form oximes may present difficulties:

$$>CH-N=O \longrightarrow >C=N-OH$$

This reaction of nitroso- compounds, which in some cases occurs very easily due to either heat or light, may be used to identify bands associated with the nitroso- group by observing their disappearance from the spectrum. This conversion can easily be detected since nitroso- compounds are highly coloured and oximes are not.

Aliphatic nitroso- compounds in the solid phase have two strong absorptions when in the *cis*- form, one at 1425–1330 cm^{-1} (7.02–7.52 μm) and the other at 1345–1320 cm^{-1} (7.43–7.58 μm), whereas in the *trans*- form they have a band at 1290–1175 cm^{-1} (7.75–8.50 μm).

Aromatic nitroso- compounds, as dimers in the *cis*- form, absorb strongly at 1400–1390 cm^{-1} (7.14–7.19 μm) and at about 1410 cm^{-1} (7.10 μm), whereas in the *trans*- form a band at 1300–1250 cm^{-1} (7.69–8.00 μm) is observed.

As monomers,[20] which only occur in the gas phase and in dilute solution, aromatic nitroso- compounds absorb strongly at 1515–1480 cm^{-1} (6.06–6.75 μm) and aliphatic nitroso- compounds at 1590–1540 cm^{-1} (6.29–6.49 μm) due to the —N=O stretching vibration.

α-Halogenated nitroso- compounds absorb near 1620 cm^{-1} (6.17 μm).

The position of the band due to the N=O stretching vibration is affected by substituent groups in a very similar manner to that of the carbonyl band.

Nitroso- compounds usually have a band at 1180–1000 cm^{-1} (8.48–10.00 μm) and another at 865–750 cm^{-1} (11.56–13.33 μm), these being due to strong coupling of the C—N stretching vibration and the vibration of the carbon skeleton. The presence of chlorine atoms increases the intensity of these bands.

Oximes have a characteristic absorption at 3650–3500 cm^{-1} (2.74–2.86 μm) due to the O—H stretching vibration whose frequency is reduced, of course, in the presence of hydrogen bonding. A band which is weak, except for conjugated compounds, is observed at 1685–1650 cm^{-1} (5.94–6.06 μm) due to the C=N stretching vibration, the frequency of the band being increased in ring-strained situations. The band due to the N—O stretching vibration occurs at 960–930 cm^{-1} (10.42–10.75 μm).

Table 14.2. Organic nitroso- compound N—O stretching vibrations

Functional Groups	Region cm⁻¹	μm	Intensity	Comments
Cis- dimers, 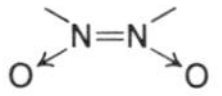				
Aliphatic compounds	1425–1330	7.02–7.52	s	
	1345–1320	7.43–7.58	s	
Aromatic compounds	~1410	~7.10	s	
	1400–1390	7.14–7.19	s	
Trans- dimers, 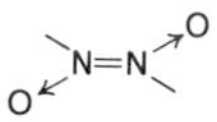				
Aliphatic compounds	1290–1175	7.75–8.50	s	
Aromatic compounds	1300–1250	7.69–8.00	s	
Monomers				
Aliphatic compounds	1590–1540	6.29–6.49	s	N=O str, usually at ~1550 cm^{-1}
α-Halogenated compounds	1620–1565	6.17–6.39	s	N=O str
Aromatic compounds	1515–1480	6.60–6.75	s	N=O str, usually at ~1500 cm^{-1}
Halogen-substituted aromatic compounds	1510–1485	6.62–6.73	s	N=O str

Table 14.3. Nitrosamines, >N—N=O

Functional Groups	Region cm⁻¹	μm	Intensity	Comments
Nitrosamines (vapour phase)	1500–1480	6.67–6.76	s	N=O str, monomer
Nitrosamines (dilute solution), see refs 21–23	~3200	~3.13	w	Overtone
	1460–1435	6.85–6.97	s	N=O str
	1065–1015	9.39–9.84	s	br, N—N str
	~660	~15.15	m–s	N—N—O def
Nitrosoamides, —N(NO)CO—	1535–1515	6.52–6.60	s	N=O str, see ref. 24
Alkyl thionitrites, R—S—N=O	~1535	~6.52	s	N=O str, multiple peaks

Table 14.4. Nitroamines, >N.NO_2, and nitroguanidines, —N=C(N—NO_2).N<

Functional Groups	Region cm⁻¹	μm	Intensity	Comments
Nitroamines	1315–1260	7.60–7.94	s	sym NO_2 str (see Table 14.1.)
	790–770	12.66–12.99	m	
Saturated aliphatic nitroamines	1585–1530	6.31–6.54	s	asym NO_2 str
Alkyl nitroguanidines	1640–1605	6.10–6.23	s	asym NO_2 str
Aryl nitroguanidines	1590–1575	6.29–6.35	s	asym NO_2 str
Aryl nitroureas	1590–1575	6.29–6.35	s	asym NO_2 str

Covalent Nitrates, —ONO_2

Organic nitrates[25] have strong absorptions due to the asymmetric and symmetric stretching vibrations of the NO_2 group which occur at 1640–1625 cm^{-1} (6.10–6.15 μm) and 1285–1270 cm^{-1} (7.78–7.87 μm) respectively. The symmetric NO_2 stretching vibration band of secondary alkyl nitrates and monocyclic nitrates consists of a doublet. The N—O stretching vibration also results in a very strong band, at 870–855 cm^{-1} (11.49–11.70 μm). Bands of weak-to-medium intensity are observed due to the NO_2 deformation vibrations at 760–755 cm^{-1} (13.10–13.25 μm) and 710–695 cm^{-1} (14.08–14.39 μm).

Inorganic nitrate salts[26] have a characteristic, sharp, weak-to-medium band in the region 840–720 cm^{-1} (11.90–13.89 μm) due to the bending vibration of the NO group.

Nitrato–metal complexes[27] absorb in the regions 1530–1480 cm^{-1} (6.54–6.76 μm) and 1290–1250 cm^{-1} (7.75–8.00 μm) due to the asymmetric and symmetric vibrations respectively of the NO_2 group.

Table 14.5. Organic (covalent) nitrates

Functional Groups	Region cm⁻¹	μm	Intensity	Comments
Nitrates, —ONO_2	1640–1625	6.10–6.15	vs	asym NO_2 str
	1285–1270	7.78–7.87	vs	sym NO_2 str
	870–855	11.49–11.70	vs	br, N—O str
	760–755	13.10–13.25	w–m	NO_2 out-of-plane def
	710–695	14.08–14.39	w–m	NO_2 def
Inorganic nitrate salts	1410–1350	7.09–7.41	vs	br, asym NO_3 str
	860–800	11.63–12.50	m	sh
	740–725	13.51–13.79	w	

Nitrites, —O—N=O

Nitrites[28,29] have very strong bands at 1680–1650 cm^{-1} (5.95–6.06 μm) and 1625–1610 cm^{-1} (6.16–6.21 μm) due to the N=O stretching vibration of the *trans-* and *cis-* forms respectively. The overtone band is of weak-to-medium intensity and occurs at 3360–3220 cm^{-1} (2.98–3.11 μm). Halogen substitution tends to increase these frequencies.

A strong absorption due to the N—O stretching vibration is observed at 815–750 cm^{-1} (12.27–13.33 μm) for the *trans-* form and at 850–810 cm^{-1} (11.76–12.35 μm) for the *cis-* form. Strong bands also occur at 690–615 cm^{-1} (14.49–16.26 μm) and 625–565 cm^{-1} (16.00–17.70 μm) for the *cis-* and *trans-* forms respectively, due to the deformation vibrations of the O—N=O group.

Table 14.6. Organic nitrites, —O—N=O

Functional Groups	Region cm⁻¹	μm	Intensity	Comments
Nitrite compounds	3360–3220	2.98–3.11	w–m	Overtones of N=O str
Nitrites, *cis*- form	1625–1610	6.16–6.21	vs	N=O str
	850–810	11.76–12.35	s	N—O str
	690–615	14.49–16.26	s	O—N=O def
Nitrites, *trans*- form	1680–1650	5.95–6.06	vs	N=O str
	815–750	12.27–13.33	vs	N—O str
	625–565	16.00–17.70	s	O—N=O def
Alkyl thionitrites, —S—N=O	~1535	~6.52	s	N=O str, multiple peaks
Inorganic nitrite salts	1275–1235	7.84–8.10	s	asym NO_2 str
	835–800	11.98–12.50	m	sh

Table 14.7. Amine oxides, $\geq N^+—O^-$

Functional Groups	Region cm⁻¹	μm	Intensity	Comments
Aliphatic *N*-oxides, $—N^+—O^-$	970–950	10.31–10.53	s	N—O str
Pyridine and pyrimidine *N*-oxides (non-polar solution)	1320–1230	7.58–8.13	m–s	N—O str, hydrogen bonding lowers frequency by 10–20 cm⁻¹, band position affected by ring substituents
	895–840	11.17–11.90	m	N—O def
Pyridine *N*-oxides	1190–1150	8.40–8.70	m–s	
Pyrazine *N*-oxides	1380–1280	7.25–7.81	m–s	N—O str, band position affected by ring substituents
	1040–990	9.62–10.10	m–s	
	~850	~11.76	m	N—O def
Nitrile oxides	1380–1290	7.25–7.75	s	N—O str
Oximes, >C=N—OH	960–930	10.42–10.75	s	N—O str

Table 14.8. Azoxy compounds, $—N=N^+—O^-—$

Functional Groups	Region cm⁻¹	μm	Intensity	Comments
Aliphatic azoxy compounds	1530–1495	6.54–6.69	m–s	Also band at 1345–1285 cm⁻¹
Aromatic azoxy compounds	1480–1450	6.76–6.90	m–s	asym N=N—O str
	1335–1315	7.49–7.60	m–s	sym N=N—O str

References

1. T. Y. Paperno and Y. V. Perekalin, *Spectra of Nitro Compounds*. Leningrad, Gas. Pedayag, 1974
2. M. Colette, *Ann. Sci. Univ. Besançon Chim.*, 1972, **9,** 3
3. R. N. Hazeldine, *J. Chem. Soc.*, **1953,** 2525
4. J. F. Brown, *J. Amer. Chem. Soc.*, 1955, **77,** 6341
5. Z. Eckstein *et al.*, *J. Chem. Soc.*, **1961,** 1370
6. F. Borek, *Naturwiss.*, 1963, **50,** 471
7. W. H. Lunn, *Spectrochim. Acta*, 1960, **16,** 1088
8. N. Jonathan, *J. Mol. Spectrosc.*, 1961, **7,** 105
9. C. J. W. Brooks and J. F. Morman, *J. Chem. Soc.*, **1961,** 3372
10. T. Kinugasa and R. Nakushina, *Nippon Kaguku Zasshi*, 1963, **84,** 365
11. C. P. Conduit, *J. Chem. Soc.*, **1959,** 3273
12. W. F. Baitinger *et al.*, *Tetrahedron*, 1964, **20,** 1635
13. A. S. Wexler, *App. Spectrosc. Rev.*, 1968, **1,** 29
14. A. G. Lee, *Spectrochim. Acta*, 1972, **28A,** 133
15. H. Feurer *et al.*, *Spectrochim. Acta*, 1963, **19,** 431
16. M. Colette, *Ann. Sci. Univ. Besançon Chim.*, 1971, **8,** 80
17. B. C. Gowenlock and W. Lüttke, *Quart. Rev.*, 1958, **12,** 321
18. L. J. Bellamy and R. L. Williams, *J. Chem. Soc.*, **1957,** 863
19. W. Lüttke, *Z. Elektrochem.*, 1957, **61,** 976
20. W. Lüttke, *Z. Elektrochem.*, 1957, **61,** 302
21. R. L. Williams *et al.*, *Spectrochim. Acta*, 1964, **20,** 225
22. P. Tarte, *Bull. Soc. Chim. Belges*, 1954, **63,** 525
23. C. E. Looney *et al.*, *J. Amer. Chem. Soc.*, 1957, **79,** 6136
24. E. H. White, *J. Amer. Chem. Soc.*, 1955, **77,** 6008
25. R. A. G. Carrington, *Spectrochim. Acta*, 1960, **16,** 1279
26. F. A. Miller and C. H. Wilkins, *Anal. Chem.*, 1952, **24,** 1253
27. E. Bannister and F. A. Cotton, *J. Chem. Soc.*, **1960,** 2276
28. P. Tarte, *J. Chem. Phys.*, 1952, **20,** 1570
29. R. N. Hazeldine and B. J. H. Mattinson, *J. Chem. Soc.*, **1955,** 4172
30. J. H. S. Green and D. J. Harrison, *Spectrochim. Acta*, 1970, **26A,** 1925
31. J. H. S. Green and H. A. Lauwers, *Spectrochim. Acta*, 1971, **27A,** 817

CHAPTER 15

Organic Halogen Compounds

Organic Halogen Compounds, $\geqslant C—X$ (where X = F, Cl, Br, I)

Strong characteristic absorptions due to the C—X stretching vibration are observed, the position of the band being influenced by neighbouring atoms or groups—the smaller the halide atom, the greater the influence of the neighbour. Different rotational isomers may often be identified since, in general, the *trans*-form absorbs at higher frequencies than the *gauche*- form. Bands of weak-to-medium intensity are also observed due to the overtones of the C—X stretching vibration.

Organic fluoride compounds

The band due to the C—F stretching vibration may be found over a wide frequency range, 1360–1000 cm^{-1} (7.35–10.00 μm),[1–6] since the vibration is easily influenced by adjacent atoms or groups. Monofluorinated compounds have a strong band at 1110–1000 cm^{-1} (9.01–10.00 μm) due to the C—F stretching vibration. With further fluorine substitution, two bands are observed due to the asymmetric and symmetric stretching vibrations, these occurring at higher frequencies.

Due to the strong coupling of the C—F and C—C stretching vibration, polyfluorinated compounds[2,4] have a series of very intense bands in the region 1360–1090 cm^{-1} (7.36–9.18 μm). A $—CF_3$ group[2] attached to an alkyl group absorbs strongly near 1290 cm^{-1} (7.75 μm), 1280 cm^{-1} (7.81 μm), 1265 cm^{-1} (7.91 μm), 1230 cm^{-1} (8.13 μm), and 1135 cm^{-1} (8.81 μm). Compounds with the group $CF_3CF_2—$ have a medium-intensity absorption in the region 1365–1325 cm^{-1} (7.33–7.55 μm) and a strong band at 745–730 cm^{-1} (13.42–13.70 μm) due to deformation vibrations. Compounds with $—CF_3$ on an aromatic ring have very strong bands near 1320 cm^{-1} (7.58 μm), 1180 cm^{-1} (8.47 μm), and 1140 cm^{-1} (8.77 μm).

The C—H stretching vibration of aliphatic groups with fluorine bonded to the carbon atom, such as $—CF_2H$ and $>CFH$, gives a band near 3000 cm^{-1} (3.33 μm).

Fluorine atoms directly attached to carbon double bonds have the effect of shifting the C=C stretching vibration to higher frequencies. For example, $—CF=CF_2$ absorbs at 1800–1780 cm^{-1} (5.56–5.62 μm) and $>C=CF_2$ at 1755–1735 cm^{-1} (5.70–5.76 μm).[6,7]

In general, C—F deformation vibrations give bands in the region 830–520 cm^{-1} (12.05–19.23 μm).

The difluoride hydrogen ion FHF^- has a very broad absorption in the region 1700–1400 cm^{-1} (5.88–7.14 μm) due to its asymmetric stretching vibrations and a band in the region 1260–1200 cm^{-1} (7.94–8.33 μm) due to its deformation vibrations.

Organic chlorine compounds

The C—Cl stretching vibrations[6,8–11] give generally strong bands in the region 760–505 cm^{-1} (13.10–19.80 μm). Compounds with more than one chlorine atom exhibit very strong bands due to the asymmetric and symmetric stretching modes. Vibrational coupling with other groups may result in a shift in the absorption to as high as 840 cm^{-1} (11.90 μm). For simple organic chlorine compounds, the C—Cl absorptions are in the region 750–700 cm^{-1} (13.33–14.29 μm) whereas for the *trans*- and *gauche*- forms they are near 650 cm^{-1} (15.38 μm),[8] the *trans*- form generally absorbing at higher frequencies.

In the liquid phase, since primary chloroalkanes exist as two or three isomers, two or three bands may be observed due to their C—Cl stretching vibrations. Primary chloro *n*-alkanes and α,ω-dichloro *n*-alkanes absorb strongly at 730–720 cm^{-1} (13.70–13.89 μm) and 655–645 cm^{-1} (15.27–15.50 μm), exceptions being the ethane and propane derivatives. In general, secondary chloroalkanes have a number of rotational isomers which therefore complicate the observed spectrum. For 2-chloroalkanes, strong bands are observed at 680–670 cm^{-1} (14.71–14.93 μm) and 615–610 cm^{-1} (16.26–16.39 μm), the latter band sometimes obscuring a further band which may be observed at about 625 cm^{-1} (16.00 μm).

Overtone bands of medium intensity resulting from the C—Cl stretching vibration are observed in the region 1510–1450 cm^{-1} (6.62–6.90 μm).

Organic bromine compounds

Bromine compounds[12] absorb strongly in the region 650–485 cm^{-1} (15.38–20.62 μm) due to the C—Br stretching vibrations, although when there is more than one bromine atom on the same carbon atom, two bands may be observed at higher frequencies.

Primary bromoalkanes of *n*-paraffins absorb strongly in the regions 645–635 cm^{-1} (15.50–15.75 μm) and 565–555 cm^{-1} (17.70–18.02 μm) due to the stretching vibration of the C—Br bond of the group $C—CH_2—CH_2Br$. Also, for *n*-bromoalkanes a band of variable intensity is observed at 440–430 cm^{-1} (22.73–23.26 μm), exceptions to this being the bromo- derivatives of ethane, propane, and *n*-tridecane. With the exception of small molecules, α,ω-dibromoalkanes have similar absorption regions to the monobromo *n*-alkanes except for the lower-frequency region where weak-to-medium intensity bands are observed at 490–480 cm^{-1} (20.41–20.83 μm) and 445–425 cm^{-1} (22.47–23.53 μm).

The spectra of *n*-alkyl bromides exhibit a similar dependence on conformation as those of the chlorides. It has been found that for the compounds in the series ethyl to *n*-decyl bromide, the C—Br stretching vibration gives a band at 645–635 cm^{-1} (15.50–15.75 μm) when the bromine atom is *trans*- to a carbon atom and at 565–555 cm^{-1} (17.70–18.02 μm) when *trans*- to a hydrogen atom.

Organic iodine compounds

Due to the large mass of the iodine atom, the C—I stretching vibration is coupled with skeletal vibrations. Also, a number of rotational isomers may exist thus affecting the position of the C—I band, which is found in the region 600–200 cm^{-1} (16.67–50.00 μm).[12] In general, primary iodo *n*-alkanes have strong absorptions at 600–585 cm^{-1} (16.67–17.09 μm) and 515–500 cm^{-1} (19.42–20.00 μm). It has been suggested that the former of these C—I stretching vibration bands is the result of the iodine atom being *trans*- to a carbon atom and the latter the result of it being *trans*- to a hydrogen atom. α,ω-Diiodoalkanes absorb in the same regions, strong bands usually being observed near 595 cm^{-1} (16.81 μm) and 500 cm^{-1} (20.00 μm).

Aromatic Halogen Compounds

Unlike aliphatic compounds, there appears to be no pure C—X stretching vibration band for aromatic halogen compounds.[5, 13–16] However, several X-sensitive bands[17] are observed, one of which occurs at about 1050 cm^{-1} (9.52 μm).

Aromatic fluoro- compounds[5] have medium-intensity bands in the region 1270–1100 cm^{-1} (7.87–9.09 μm), those with only one fluorine atom on the ring tending to absorb at about 1230 cm^{-1} (8.13 μm). Bands due to the C—H out-of-plane vibrations and other aromatic ring vibrations are also observed.

Due mainly to the bending of the ring–halogen bond, aromatic fluoro-compounds have a band of variable intensity at 420–375 cm^{-1} (23.81–26.77 μm), aromatic chloro- compounds have a band also of variable intensity (often medium-to-strong) at 390–270 cm^{-1} (25.64–37.04 μm), and aromatic bromo-compounds absorb strongly at 320–255 cm^{-1} (31.25–39.22 μm). These bands, as well as being observed for mono- and disubstituted benzenes, may also sometimes be observed, with different intensities, in polysubstituted aromatic compounds.

Most aromatic chloro- and bromo- compounds have strong absorptions at 760–395 cm^{-1} (13.10–25.32 μm) and 650–395 cm^{-1} (15.38–25.32 μm) respectively, which is due to a combination of vibrational modes. Monosubstituted benzenes, dihalogen-substituted benzenes, and compounds with electron-donor or methyl substituents in the *para*- position of halobenzenes all exhibit the former band.

Table 15.1. Organic fluorine compounds

Functional Groups	Region cm^{-1}	Region μm	Intensity	Comments
C—F	1400–1000	7.14–10.00	s	C—F str, general range
	830–520	12.05–19.23	s	C—F def, general range
Aliphatic monofluorinated compounds	1110–1000	9.01–10.00	vs	C—F str
	780–680	12.81–14.71	s	C—F def
Aliphatic difluorinated compounds	1250–1050	8.00–9.52	vs	Two bands, C—F str
Polyfluorinated alkanes	1360–1090	7.36–9.18	vs	A number of bands
$CF_3—CF_2—$	1365–1325	7.33–7.55	m–s	C—F str
	745–730	13.42–13.70	s	C—F def
$—CF_3$	1350–1120	7.41–8.93	vs	A number of bands
	780–680	12.82–14.71	s	
	680–590	14.71–16.95	s	asym CF_3 def
	600–540	16.67–18.52	s	sym CF_3 def } May be absent for α-unsaturated compounds
	555–505	18.02–19.80	s	asym CF_3 def }
	460–200	21.74–50.00	s	asym and sym rocking vib, two bands
Cyclic $—CF_2—$ (four- or five-membered ring)	1350–1140	7.41–8.77	s	
$>C=CF_2$	1755–1735	5.70–5.76	vs	C=C str
	525–505	19.05–19.80	m–s	Bending vib
	515–355	19.42–28.17	s	
	455–345	21.98–28.99	m–s	Rocking vib
$—CF=CF_2$	1800–1780	5.55–5.62	s	C=C str
	1340–1300	7.46–7.69	vs	C—F str
Trifluoromethyl aromatic compounds, $Ar—CF_3$	1330–1310	7.52–7.63	m–s	C—F str
	600–580	16.67–17.24	s	$—CF_3$ rocking vib

Table 15.2. Organic chlorine compounds

Functional Groups	Region cm^{-1}	μm	Inten-sity	Comments
C—Cl	760–505	13.10–19.80	s	C—Cl str, general range
	450–250	22.22–40.00	s	C—Cl def, general range
R—$(CH_2)_2$Cl and Cl—$(CH_2)_{n>3}$Cl	730–720	13.70–13.89	s	—CH_2Cl has a strong band at 1300–1240 cm^{-1} due to CH_2 wagging
	655–645	15.27–15.50	s	
R$(CH_2)_2$CH$(CH_3)_2$Cl	680–670	14.71–14.93	s	
	~625	~16.00	w–m	Easily overlooked
	615–610	16.26–16.39	s	
R$(CH_2)_2$CR′(CH_3)Cl	630–610	15.87–16.39	m–s	
(R′ = Me or Et)	580–560	17.24–17.86	m–s	
Polychlorinated compounds	800–700	12.50–14.29	vs	
>C=CCl_2	500–320	20.00–31.25	m	Bending vib (C=C str, ~1615 cm^{-1})
	265–235	37.74–42.55	w	
	260–180	38.46–55.56	s	Rocking vib
Chloroformates,	~690	~14.49	s	C—Cl str
RO—CO.Cl	485–470	20.62–21.28	s	C—Cl def
RS—CO.Cl	~580	~17.24	s	C—Cl str
	350–340	28.57–29.41	s	C—Cl def
>N—Cl	805–690	12.42–14.49		See ref. 18
Equatorial C—Cl	780–740	12.80–13.51	s	
Axial C—Cl	730–580	13.70–17.25	s	
Rotational configurations: chloroalkanes				
Primary chloroalkanes	730–720	13.70–13.89	s	Cl atom *trans* to C atom
	660–650	15.15–15.38	s	Cl atom *trans* to H atom
	685–680	14.60–14.71	s	Cl atom *trans* to H atom in branched alkane
Secondary chloroalkanes	~760	~13.10	m	Cl atom *trans* to two C atoms
	675–655	14.81–15.27	m–s	Cl atom *trans* to C and H atoms
	640–625	15.63–16.00	w	Cl atom *trans* to two H atoms in bent molecule
	615–605	16.26–16.53	s	Cl atom *trans* to two H atoms
Tertiary chloroalkanes	580–560	17.24–17.86	m–s	Cl atom *trans* to three H atoms
	635–610	15.75–16.39	m–s	Cl atom *trans* to one C and two H atoms

Table 15.3. Organic bromine compounds

Functional Groups	Region cm^{-1}	μm	Inten-sity	Comments
C—Br	650–485	15.38–20.62	s	C—Br str, general range
	300–140	33.33–71.43	m	C—Br def, general range
R—$(CH_2)_2$Br	645–635	15.50–15.75	s	C—Br str of C—$(CH_2)_2$Br, —CH_2Br has strong band near 1230 cm^{-1} due to CH_2 wagging
	565–555	17.70–18.02	s	
	440–430	22.73–23.26	v	
Br$(CH_2)_{n>3}$Br	645–635	15.50–15.75	s	C—Br str
	565–555	17.70–18.02	s	C—Br str
	490–480	20.41–20.83	w–m	
	445–425	22.47–23.53	w–m	
R—CH_2CHR′CH_2Br	650–645	15.38–15.50	s	C—Br str, *trans* form
(R′ = Me or Et)	620–610	16.13–16.39	s	C—Br str, *gauche* form
	515–500	19.42–20.00	v	
R—$(CH_2)_2$CH(CH_3)Br	620–605	16.13–16.53	s	
	590–575	16.95–17.39	m–w	
	540–530	18.52–18.87	s	
R—$(CH_2)_2$C$(CH_3)_2$Br	600–595	16.67–16.81	m–s	
	525–505	19.05–19.80	s	
>C=CBr_2	310–250	32.26–40.00	s	Bending vib
	185–135	54.05–74.07	m	
	160–120	62.50–83.33	s	Rocking vib
Equatorial C—Br	750–685	13.33–14.60	s	
Axial C—Br	690–550	14.50–18.20	s	

Table 15.4. Organic iodine compounds

Functional Groups	Region cm^{-1}	μm	Inten-sity	Comments
C—I	600–200	16.67–50.00	s	C—I str, general range
	300–50	33.33–200.00	v	C—I def, general range
R$(CH_2)_2$I	600–585	16.67–17.09	s	C—I str, —CH_2I have a strong band near 1170 cm^{-1} due to CH_2 wagging
	515–500	19.42–20.00	s	C—I str
I$(CH_2)_{n>3}$I	~595	~16.81	s	C—I str
	~500	~20.00	s	C—I str
Secondary iodoalkanes	~575	~17.39	s	
	550–520	18.18–19.23	s	
	490–480	20.41–20.83	s	
Tertiary iodides	580–560	17.24–17.86	s	C—I str
	510–485	19.61–20.62	m	C—I str
	485–465	20.62–21.51	s	
>C=CI_2	~200	~50.00		Bending vib
	~100	~100.00		
	~50	~200.00		Rocking vib
Equatorial C—I	~655	~15.27	s	Liquid phase
Axial C—I	~640	~15.63	s	Liquid phase

Table 15.5. Aromatic halogen compounds

Functional Groups	Region cm⁻¹	μm	Intensity	Comments
Aromatic halogen compounds (X = Cl, Br, I)	~1050	~9.52	m	X-sensitive band
Aromatic fluorine compounds	1270–1100	7.87–9.09	m	Approximate range, X-sensitive band
	680–520	14.71–19.23	m–s	Aromatic C—F str and ring def
	420–375	23.81–26.67	v	In-plane aromatic C—F bending
	340–240	29.41–41.67	s	Out-of-plane aromatic C—F bending
Aromatic chlorine compounds	1060–1035	9.43–9.66	m	*Ortho*-substituted benzenes } X-sensitive bands
	1080–1075	9.26–9.30	m	*Meta*-substituted benzenes } X-sensitive bands
	1100–1090	9.09–9.17	m	*Para*-substituted benzenes } X-sensitive bands
	760–395	13.10–25.32	s	Not always present
	500–370	20.00–27.03	m–s	Aromatic C—Cl str and ring def
	390–165	25.64–60.61	m–s	Out-of-plane aromatic C—Cl bending } Not always present
	330–230	30.30–43.48	m–s	In-plane aromatic C—Cl bending } Not always present
Aromatic bromine compounds	1045–1025	9.57–9.76	m	*Ortho*-substituted benzenes } X-sensitive bands
	1075–1065	9.30–9.39	m	*Meta*- and *para*-substituted benzenes } X-sensitive bands
	400–260	25.00–38.46	s	Aromatic C—Br str and ring def
	325–175	30.77–57.14	m–s	Out-of-plane aromatic C—Br def
	290–225	34.48–44.44	m–s	In-plane aromatic C—Br bending
Aromatic iodine compounds	1060–1055	9.43–9.48	m–s	X-sensitive band for *para*-substituted benzenes
	310–160	32.26–62.50	s	Out-of-plane aromatic C—I bending
	265–185	37.74–54.05		Aromatic C—I str and ring def
	~200	~50.00		In-plane aromatic C—I def

References

1. J. Murto *et al.*, *Spectrochim. Acta*, 1973, **29A,** 1121
2. O. Risgin and R. C. Taylor, *Spectrochim. Acta*, 1959, **15,** 1036
3. J. H. Simons (ed.), *Fluorocarbons and Related Compounds—Fluorocarbon Chemistry*, Vol. II. Academic Press, New York, 1954, p. 449
4. C. Y. Liang and S. Krimm, *J. Chem. Phys.*, 1956, **25,** 563
5. D. A. Long and D. Stecke, *Spectrochim. Acta*, 1963, **19,** 1947
6. N. C. Craig and D. A. Evans, *J. Amer. Chem. Soc.*, 1965, **87,** 4223
7. D. E. Mann *et al.*, *J. Chem. Phys.*, 1957, **27,** 51
8. J. J. Shipman *et al.*, *Spectrochim. Acta*, 1962, **18,** 1603
9. A. R. Katritzky, *Spectrochim. Acta*, 1960, **16,** 964
10. G. W. Chantry *et al.*, *Spectrochim. Acta*, 1966, **22,** 125
11. M. A. Ory, *Spectrochim. Acta*, 1960, **16,** 1488
12. F. F. Bentley *et al.*, *Spectrochim. Acta*, 1964, **20,** 105
13. G. Varsanyi *et al.*, *Spectrochim. Acta*, 1963, **19,** 669
14. T. R. Nanney *et al.*, *Spectrochim. Acta*, 1965, **21,** 1495
15. T. R. Nanney *et al.*, *Spectrochim. Acta*, 1966, **22,** 737
16. H. E. Shurvell *et al.*, *Spectrochim. Acta*, 1966, **22,** 333
17. E. F. Mooney, *Spectrochim. Acta*, 1964, **20,** 1021
18. R. C. Petterson *et al.*, *J. Org. Chem.*, 1960, **25,** 1595

CHAPTER 16

Sulphur and Selenium Compounds

Mercaptans, —SH

The band due to the S—H stretching vibration[1–7] is weak and may be missed in dilute solutions. It occurs in the region 2600–2550 cm^{-1} (3.85–3.92 μm) and is easily recognized since this is a region relatively free of other absorption bands. The N—H stretching vibrations of organic nitrogen compounds in the solid phase give a complex pattern of bands in this region whereas a single band is observed due to the S—H stretching vibration. Carboxylic acids also have bands in this region, forming a broad complex pattern due to the O—H stretching vibration. Aldehydes also may have weak, sharp bands in this region due to the aldehydic C—H stretching vibration, but usually a doublet is observed.

Hydrogen-bonding effects[2,7] are much smaller for the —S—H group than they are for the —O—H and >N—H groups. If dimers and monomers coexist, two S—H bands due to the S—H stretching vibration may be observed.

The C—S stretching vibration gives a weak band in the region 710–570 cm^{-1} (14.08–17.54 μm) (see the section dealing with sulphides).

Monothiocarboxylic acids,[5,6] —CO—SH, are a mixture of two forms:

$$-CO-SH \rightleftharpoons -CS-OH$$

and therefore exhibit bands due to S—H, O—H, C=O, and C=S vibrations.

C—S and S—S Vibrations: Organic Sulphides, >S, Mercaptans, —SH, Disulphides, —S—S—, and Polysulphides, —(—S—S—)$_n$—

In general, the assignment of the band due to the C—S stretching vibration in different compounds is difficult since it is of variable intensity and may be found over the wide region 1035–245 cm^{-1} (9.66–40.82 μm).

Both aliphatic[1] and aromatic[9] sulphides have a weak-to-medium band due to the C—S stretching vibration in the region 710–570 cm^{-1} (14.08–17.54 μm), primary sulphides absorbing at the higher-frequency end of the range and tertiary sulphides at the lower end.

Double-bond conjugation with the C—S bond, e.g. either vinyl or phenyl =C—S—, lowers the C—S stretching vibration frequency to about 590 cm^{-1} (16.95 μm) and increases its intensity significantly. For compounds in which the C—S group is adjacent to a C=O group, the C—S band is normally above 710 cm^{-1} (below 14.08 μm). The band due to the C—Cl stretching vibration also occurs in this region and may, in some cases, make interpretation more difficult. Thioethers absorb in the region 695–655 cm^{-1} (14.39–15.27 μm) due to the C—S—C stretching vibration.

The band due to the C—S—C bending vibration has been observed for a few sulphides and occurs at about 250 cm^{-1} (40.00 μm), the C—C—S band being near 325 cm^{-1} (30.77 μm).

Because of the symmetry of the S—S group, aliphatic and aryl disulphides[9] have a weak band which occurs at 520–500 cm^{-1} (19.23–20.00 μm) and 500–430 cm^{-1} (20.00–23.36 μm) respectively, due to the stretching of the S—S bond.

Table 16.1. Mercaptan S—H stretching and deformation vibrations

Functional Groups	Region cm^{-1}	Region μm	Intensity	Comments
Mercaptans, aliphatic thiols, and thiophenols (free)	2600–2550	3.85–3.92	w	S—H str, see ref. 1
Aryl mercaptans	2600–2450	3.85–4.08	w	S—H str, see refs 2–4, 7
Dithioacids, —CS—SH (free)	2600–2500	3.85–4.00	w	S—H str, sometimes a doublet, see ref. 6
Dithioacids (hydrogen-bonded)	2500–2400	4.00–4.17	w	br, S—H str
Compounds with —CO—SH (free)	2595–2560	3.85–3.91	w	S—H str
Trithiocarbonic acids (free)	2560–2550	3.91–3.92	w	S—H str
Organic compounds containing SeH (free)	2330–2280	4.29–4.39	w	S—H str, see ref. 10
Monothioacids, —CO—SH	840–830	11.90–12.05	m	S—H in-plane def
Dithioacids, —CS—SH	~860	~11.63	s	br, S—H in-plane def

Table 16.2. CH_3 and CH_2 vibration bands of organic sulphur compounds containing CH_3—S— and —CH_2S— groups

Functional Groups	Region cm^{-1}	Region μm	Intensity	Comments
CH_3—S—	2990–2955	3.34–3.39	m–s	asym CH_3 str
	2900–2865	3.45–3.49	m–s	sym CH_3 str
	1440–1415	6.94–7.06	m	asym CH_3 def
	1330–1290	7.52–7.76	m–s	sym CH_3 def
	1030–960	9.71–10.42	m	CH_3 rocking
—CH_2—S—	2950–2920	3.39–3.43	m	asym CH_2 str
	2880–2845	3.47–3.51	m	sym CH_2 str
	1435–1410	6.97–7.09	m	CH_2 def
	1270–1220	7.87–8.20	s	CH_2 wagging
—S—CH=CH_2	~1590	~6.29	m	C=C str
	~965	~10.36	s	C—H out-of-plane def
	~860	~11.63	s	CH_2 out-of-plane def

Table 16. 3. Organic sulphides, mercaptans, disulphides, and polysulphides: C—S and S—S stretching vibrations

Functional Groups	Region cm⁻¹	μm	Intensity	Comments
CH_3—S—	710–685	14.08–14.60	w–m	C—S str
R—CH_2—S—	660–630	15.15–15.87	w–m	C—S str (Increase in the length of the alkyl group(s) decreases the frequency)
R′R>CH—S—	630–600	15.87–16.67	w–m	C—S str
$R_1R_2R_3$C—S—	600–570	16.67–17.54	w–m	C—S str
Cyclohexyl sulphides	710–685	14.08–14.60	w–m	C—S str
Phenyl sulphides	715–670	13.99–14.93	w–m	C—S str
α,β-Unsaturated sulphides	~740	~13.51	v	C—S str
—CH_2—S—CH_2—	695–655	14.39–15.27	w–m	C—S—C str
—S—Ar	1110–1070	9.01–9.37	m	Ring vib with C—S interaction, X-sensitive band
Aliphatic disulphides	705–570	14.18–17.54	w	C—S str
	520–500	19.23–20.00	w	S—S str
Aromatic disulphides	500–430	20.00–23.26	w–m	S—S str
Polysulphides	500–470	20.00–21.28	w–m	S—S str
Mono- and disulphonyl chlorides	775–650	12.90–15.38	w–m	C—S str
Dithiolcarbonic acid esters, $(RS)_2C$=O	880–827	11.36–12.20	s	asym C—S str
	570–560	17.54–17.86	v	sym C—S str, review of thiol esters, see ref. 11
(RS) (ArS)C=O	~565	~17.70	v	C—S str
$(ArS)_2C$=O	~560	~17.86	s	C—S str
Thiolchloroformates, (RS)ClC=O	850–815	11.76–12.30	s	asym C—S str, often strongest band in spectrum
	~580	~17.24	m–s	sym C—S str
	~345	~28.99	s	C—Cl def
(ArS)ClC=O	~820	~12.20	s	asym C—S str, often strongest band in spectrum
	~595	~16.81	s	sym C—S str
Monothiol esters, —C(=O)—S—	1035–935	9.66–10.70	s	C—S str, see ref. 11; it has been suggested the C—S str for thiol acids and esters be assigned to a band near 625 cm⁻¹, see ref. 12
Monothiol acids, —C(=O)—S—H	~950	~10.53	s	C—S str, see ref. 11
Thioketals, $R_2C(S)_2CR_2$ (ring)	800–245	12.50–40.82	m–s	C—S str, a number of bands due to coupling
Xanthates, —O—C(=S)—S—	965–860	10.35–11.65	w–m	C—S str
Dithioacids, —C(=S)—SH	~580	~17.25	s	C—S str
Dixanthogens, —O—C(=S)—S—S—C(=S)—O—	965–860	10.35–11.65	w–m	C—S str
Thionitrites, —S—N=O	730–685	13.70–14.60	m–s	C—S str, see ref. 13

Table 16.3 (*cont.*)

Functional Groups	Region cm⁻¹	μm	Intensity	Comments
Thioacetals and trithiocarbonates, =C(SR)SR	900–800	11.11–12.50	m–s	asym S—C—S str
Ionic dithiolates, =C(S)S⁻	1050–900	9.52–11.11	m–s	asym S—C—S str
Ionic 1,1-dithiolates, =C(S⁻)S⁻	980–850	10.20–11.76	m–s	asym S—C—S str
Trialkyl arsine sulphides	~480	~20.83	vs	As—S str, band position dependent on size of alkyl groups

Compounds containing S=O: Organic Sulphoxides, >S=O, and Sulphites, —O—SO—O—

In a non-polar solvent such as carbon tetrachloride or *n*-hexane, sulphoxides[11, 14–19] have a strong absorption at 1070–1035 cm⁻¹ (9.35–9.66 μm) due to the stretching vibrations of the S=O group, while for solvents in which hydrogen bonding is possible, and for chloroform, the range is 1055–1010 cm⁻¹ (9.48–9.90 μm). In the case of strong intramolecular hydrogen bonding, the band due to the S=O stretching vibration of sulphoxides has been observed at about 995 cm⁻¹ (10.05 μm) with a very much weaker band being observed in the normal region.[20, 22]

In the solid phase, the S=O band appears 10–20 cm⁻¹ lower than as given above for the inert solvent and is broad, sometimes consisting of a number of peaks.[20–22] Conjugation has only a small effect on the position of the band. Dialkyl sulphites have a strong band due to this vibration at 1220–1170 cm⁻¹ (8.20–8.55 μm).

The position of the S=O band is dependent on the electronegativity of the attached group. Electronegative substituents tend to raise the frequency since they tend to stabilize the form S=O rather than S^+—O^-. Hence, the frequency of the S=O stretching vibration increases in the following order:

sulphoxides <	sulphinic acids <	sulphinic acid esters ~	sulphinyl chlorides <	sulphites
—S=O	—SO—OH	—SO—O—	—SO—Cl	—O—SO—O—

(For sulphites, there are two electronegative atoms adjacent to the S=O group.)

In general, organic compounds of the type $>SO$ may be distinguished from those of the type $-(SO_2)-$ which are not ionic in nature, i.e. $G-SO_2-G$ or $G-SO_3-G$, since the group $>SO$ has only one strong absorption in the region 1360–1100 cm^{-1} (7.35–9.09 μm) whereas sulphones, etc., have two (see section on sulphones).

Sulphoxides absorb in the region 730–690 cm^{-1} (13.70–14.49 μm) probably due to the stretching vibration of the C—S bond. A band of variable intensity at 395–360 cm^{-1} (25.32–27.78 μm) is also observed and has been assigned to the C—S=O deformation.

Sulphoxides may act as electron donors to either metals[25–27] or other molecules.[20–22]

For cyclic (six-membered ring) sulphoxides, the S=O group in the equatorial position absorbs at ~20 cm^{-1} higher than when in an axial position. Cyclic sulphites (five- to seven-membered rings) absorb at 1225–1200 cm^{-1} (8.16–8.33 μm).

Table 16.4. Organic sulphoxides, >S=O

Functional Groups	Region cm^{-1}	Region μm	Intensity	Comments
Sulphoxides, >S=O (in dilute CCl_4 solution)	1070–1035	9.35–9.66	vs	S=O str, halogen or oxygen atom bonded to S atom increases frequency
Sulphoxides	730–690	13.70–14.49	v	
	395–360	25.32–27.78	v	sym C—S—O def
Cyclic sulphoxides (six- and seven-membered rings) (in CCl_4 solution)	~1060	~9.43	s	S=O str, see ref. 17
Cyclic sulphoxides (four-membered rings) (in CCl_4 solution)	~1090	~9.17	s	S=O str
Aryl sulphoxides (in CCl_4 solution)	~1060	~9.43	s	See refs 18, 19; $CHCl_3$ solution spectra quite different
Methyl aryl sulphoxides Ar—SO—CH_3	535–495	18.69–20.20	s	C—S=O in-plane def
Thiosulphoxides, G_1—S.SO.G_2:				
G_1, G_2 = CH_3 and/or Ar	1110–1095	9.01–9.13	s	See ref. 24
G_1, G_2 = R and Ar	1090–1075	9.17–9.30	s	See ref. 24
G_1, G_2 = Ar	1115–1100	8.97–9.09	s	See ref. 24
Compounds of the type	345–255	28.99–39.22		sym S—O—S str, see ref. 35
R—S—O—S—R′	160–125	62.50–80.00		S—O—S bending
Dialkyl sulphites, $(RO)_2SO$	1220–1170	8.20–8.55	s	S=O str
	1050–850	9.52–11.76	s	Due to S—O—CH_2 group
	750–690	13.33–14.49	m	S—O str, two bands
Chloroalkyl sulphites	~1215	~8.23	s	S=O str
Sulphinic acid esters, —SO—O—	1140–1125	8.77–8.89	s	S=O str
Sulphinic acids, —SO—OH	2790–2340	3.58–4.27	w	O—H str (solid phase value)
	1090–990	9.17–10.10	vs	S=O str
	870–810	11.49–12.35	m–s	S—O str
Aryl sulphinic acids, Ar—SO—OH	~1100	~9.09	s	S=O str
Sulphinic anhydrides, R—SO_2—SO—R	~1100	~9.09	s	S=O str
Sulphinic acid salts, $RSO_2^-M^+$	~1030	~9.71	s	asym S⩦O str, stronger of the two bands
	~980	~10.20	s	sym S⩦O str
Alkyl sulphinyl chlorides, R—SO—Cl	~1135	~8.81	s	S=O str
Thionylamines, —N=S=O	1300–1230	7.69–8.13	s	asym N=S=O str, see ref. 23
	1180–1110	8.48–9.01	s	sym N=S=O str
Cyclic sulphites (five-, six-, and seven-membered rings)	1220–1210	8.20–8.26	s	S=O str

Organic Sulphones, $>SO_2$

In dilute solution in non-polar solvents, all organic sulphones[17, 28–33] have two very strong bands[34] due to the asymmetric and symmetric stretching vibrations[29] of the SO_2 group, at 1360–1290 cm^{-1} (7.41–7.75 μm) and 1170–1120 cm^{-1} (8.55–8.93 μm) respectively. In the solid phase, the band due to the asymmetric stretching vibration occurs 10–20 cm^{-1} lower than in dilute solution and usually appears to have a number of peaks whereas the band due to the symmetric stretching vibration usually consists of a single peak at 1180–1145 cm^{-1} (8.48–8.73 μm).

A number of sulphones have three components of the band due to the asymmetric SO_2 stretching vibration when in non-polar solvents such as carbon tetrachloride. In order of decreasing intensity, these bands occur at 1335–1315 cm^{-1} (7.49–7.61 μm), 1315–1305 cm^{-1} (7.61–7.66 μm), and 1305–1285 cm^{-1} (7.66–7.78 μm).

Conjugation does not alter the positions of the bands due to the SO_2 stretching vibration.

All sulphones have a characteristic medium-to-strong band at 610–545 cm^{-1} (16.39–18.35 μm) which is due to the scissor vibration of the $—SO_2$ group. Saturated aliphatic sulphones have a medium-intensity band at 525–495 cm^{-1} (19.05–20.20 μm) due to the wagging motion of the $—SO_2$ group.

Table 16.5. Organic sulphone SO_2 stretching and other vibrations

Functional Group	Region cm^{-1}	Region μm	Intensity	Comments
Sulphones (dilute solution)	1360–1290	7.41–7.75	vs	asym SO_2 str
	1170–1120	8.55–8.93	vs	sym SO_2 str
Dialkyl sulphones	1330–1305	7.52–7.67	vs	Straight-chain alkyl sulphones absorb at slightly higher frequencies than branched compounds
	1150–1135	8.70–8.81	vs	
Alkyl–aryl sulphones	1335–1325	7.49–7.54	vs	(For methylvinylsulphones, see ref. 94)
	1160–1150	8.62–8.70	vs	
Diaryl sulphones	1360–1335	7.35–7.49	vs	See ref. 30
	1170–1160	8.55–8.62	vs	
GSO_2CH_2COG	~1330	~7.52	vs	
	~1160	~8.62	vs	
Disulphones, $—SO_2—SO_2—$	1360–1280	7.35–7.78	vs	asym SO_2 str, see refs 9, 37
	1170–1120	8.55–8.93	vs	sym SO_2 str
Thiolsulphonates, $—SO_2—S—$	~1340	~7.46	vs	asym SO_2 str, see refs 28, 36, 37
	~1150	~8.70	vs	sym SO_2 str
Sulphinic acid anhydrides (sulphonyl sulphones), $—SO_2—SO—$	~1340	~7.46	vs	asym SO_2 str, see ref. 37
	~1140	~8.77	vs	sym SO_2 str
	~1100	~9.09	s	S=O str
Sulphones	610–545	16.39–18.35	m–s	SO_2 scissoring
Saturated aliphatic sulphones	525–495	19.05–20.20	m–s	SO_2 wagging

Sulphonyl Halides, $—SO_2—X$

The frequencies of the SO_2 stretching vibrations of sulphonyl fluorides and chlorides[28, 38–43] are higher than those of the sulphones due to the presence of the electronegative halogen atom.

Aliphatic sulphonyl chlorides[38, 39] absorb strongly at 1385–1360 cm^{-1} (7.22–7.35 μm) and 1190–1160 cm^{-1} (8.40–8.62 μm) due to the SO_2 asymmetric and symmetric stretching vibrations respectively. For aromatic sulphonyl chlorides,[40, 41] these ranges are extended to higher frequencies but the main difference observed is that the band due to the symmetric vibration forms a doublet.

Table 16.6. Sulphonyl halides

Functional Groups	Region cm^{-1}	Region μm	Intensity	Comments
Aliphatic sulphonyl chlorides	1385–1360	7.22–7.35	vs	asym SO_2 str
	1190–1160	8.40–8.62	vs	sym SO_2 str
	775–640	12.90–15.63	w	C—S str
	610–565	16.39–17.70	s	In-plane SO_2 def
	570–530	17.54–18.87	s	SO_2 rocking
Aromatic sulphonyl chlorides	1410–1385	7.09–7.22	s	asym SO_2 str
	1205–1175	8.30–8.51	s	sym SO_2 str, doublet
	~1090	~9.17	w	
Sulphonyl fluorides, $—SO_2—F$	1415–1395	7.07–7.17	s	asym SO_2 str
	1215–1200	8.23–8.33	s	sym SO_2 str
$RO.SO_2Cl$	1455–1405	6.87–7.12	s	asym SO_2 str
	1225–1205	8.17–8.30	s	sym SO_2 str
$RO.SO_2F$	1510–1445	6.62–6.92	s	asym SO_2 str
	1260–1230	7.94–8.13	s	sym SO_2 str

Sulphonamides, $—SO_2—N\langle$

In the solid phase, sulphonamides[28, 44–47] have strong bands due to their N—H stretching vibrations in the region 3390–3245 cm^{-1} (2.95–3.08 μm). In the unassociated state, these bands occur in the same region as for amines.

Also in the solid phase, sulphonamides have a very strong, broad absorption band at 1360–1315 cm^{-1} (7.35–7.61 μm) which generally consists of a number of peaks and is due to the asymmetric stretching vibration of the SO_2 group. In solution, this band is about 10–20 cm^{-1} higher and occurs at 1380–1325 cm^{-1} (7.25–7.55 μm). A very strong band due to the symmetric stretching vibrations of the SO_2 group occurs at 1180–1140 cm^{-1} (8.47–8.77 μm) when in the solid phase and at 1170–1150 cm^{-1} (8.55–8.70 μm) when in dilute solution (i.e. there is very little difference in the band position for this vibration). Due to the influence of the electronegative nitrogen atom, the frequencies of the SO_2 asymmetric and symmetric stretching vibrations are higher for sulphonamides than for sulphones. The positions of these bands are little influenced by molecular structure, i.e. whether the sulphonamides are aliphatic or aromatic.

A band of medium intensity is observed in the region 950–860 cm^{-1} (10.53–11.63 μm) and another band has been reported at 715–700 cm^{-1} (13.99–14.29 μm).

Table 16.7. Sulphonamides

Functional Groups	Region cm^{-1}	Region μm	Intensity	Comments
Primary sulphonamides, $—SO_2NH_2$ (hydrogen-bonded or solid phase)	3390–3245	2.95–3.08	s	Two bands due to asym and sym N—H str
N-Mono-substituted sulphonamides, $—SO_2NH—$ (hydrogen-bonded or solid phase)	3300–3250	3.03–3.08	s	N—H str, one band only
Sulphonamides, $—SO_2—N<$ (dilute solution)	1380–1325	7.25–7.55	vs	asym SO_2 str
	1170–1150	8.55–8.70	vs	sym SO_2 str
	950–860	10.53–11.63	m	
	715–700	13.99–14.29	w–m	
Sulphondiamides, $>N.SO_2.N<$	1340–1320	7.46–7.58	vs	asym SO_2 str, see ref. 47
	1145–1140	8.73–8.77	vs	sym SO_2 str

Covalent Sulphonates, $R—SO_2—OR'$ [28, 37]

Aliphatic sulphonates have a strong band in the region 1420–1330 cm^{-1} (7.04–7.52 μm) which may appear as a doublet and another strong band in the region 1200–1145 cm^{-1} (8.33–8.73 μm) which is usually found near 1175 cm^{-1} (8.51 μm). Aromatic esters of sulphonic acids have strong absorptions at 1380–1350 cm^{-1} (7.25–7.41 μm) and 1200–1190 cm^{-1} (8.33–8.40 μm).

Organic Sulphates, $—O—SO_2—O—$

Organic, covalent sulphates[28, 48, 49] have two strong bands, one at 1415–1380 cm^{-1} (7.07–7.25 μm) and the other at 1200–1185 cm^{-1} (8.33–8.44 μm), both of which are due to the stretching vibrations of the SO_2 group. As might be expected, electronegative substituents tend to increase the frequencies of the SO_2 stretching vibration. Studies of diaryl and alkylaryl sulphates have been published.[95]

Primary alkyl sulphate salts, $ROSO_2O^-M^+$,[50] have a very strong band at 1315–1220 cm^{-1} (7.61–8.20 μm) and a less intense band at 1140–1075 cm^{-1} (8.85–9.30 μm) due to the asymmetric and symmetric stretching vibrations respectively of the SO_2 group, whereas secondary alkyl sulphate salts have a very strong doublet at about 1270–1210 cm^{-1} (7.87–8.26 μm) and a strong band at 1075–1050 cm^{-1} (9.30–9.52 μm). The positions of these bands are influenced far more by different metal ions than by the nature of the alkyl group.

The asymmetric and symmetric S—O—C stretching vibration bands occur at about 875 cm^{-1} (11.43 μm) and 750 cm^{-1} (13.33 μm) respectively, the first band being of medium intensity and the second weak. These bands occur in a region where alkyl bands occur and may therefore be difficult to identify.

Sulphonic Acids, $—SO_3H$, and Salts, $SO_3^-M^+$

Small traces of water result in ionization of sulphonic acids, therefore extra care must be exercised if one is to observe covalent (non-ionized) sulphonic acids rather than the ionic (hydrated) form, $—SO_3^-H_3^+O$. The bands observed due to the SO_3 stretching vibration for both the anhydrous and hydrated form are strong and usually broad. In general, these two bands together form a broad absorption with two maxima and may thus be distinguished from the acid salts which have two separate bands.

The band due to the O—H stretching vibration of hydrated sulphonic acids is very broad and usually has several maxima, occurring in the region 2800–1650 cm^{-1} (3.60–6.06 μm). Sulphonic acid salts, of course, have no corresponding band. The broadness of the band due to the O—H stretching vibration may be used to distinguish between the hydrated and anhydrous forms of sulphonic acids.

The band due to the SO_3 asymmetric stretching vibration of sulphonic acid salts occurs at 1250–1140 cm^{-1} (8.00–8.77 μm), the position of the band being mainly dependent on the nature of the metal ion, not on whether the compound is alkyl or aryl. The band is usually broad with shoulders. The band due to the symmetric stretching vibration is sharper, also has shoulders and occurs at 1070–1030 cm^{-1} (9.35–9.70 μm). Ionic sulphates, which are a common impurity, have a very strong band in the region 1130–1080 cm^{-1} (8.85–9.26 μm). Substituted benzene and naphthalene sulphonic acid salts also have a band in this region which is not observed for alkyl acid salts.

Table 16.8. Compounds with the SO_2 group

Functional Groups	Region cm^{-1}	Region μm	Intensity	Comments
Covalent sulphates $(RO)_2SO_2$	1415–1380	7.07–7.25	s	asym SO_2 str
	1200–1185	8.33–8.44	s	sym SO_2 str
Primary alkyl sulphate salts, $RSO_4^- M^+$ (solid phase)	1315–1220	7.61–8.20	vs	asym SO_2 str, often doublet ~1250 and ~1220 cm^{-1} (aromatic compounds in same range)
	1140–1075	8.77–9.30	m	sym SO_2 str, aromatic compounds ~1040 cm^{-1}
	~1000	~10.00	m	Often split
	840–835	11.90–11.98	m	S—O—C str
	700–570	14.29–17.54	m–s	SO_3 bending, two bands
	440–410	22.73–24.39	w	SO_3 rocking
Secondary alkyl sulphate salts, $R_1R_2CHSO_4^- M^+$ (solid phase)	1270–1210	7.87–8.26	vs	asym SO_2 str, often doublet
	1075–1050	9.30–9.52	s	sym SO_2 str
	945–925	10.60–10.81	m	
	700–570	14.29–17.54	m–s	SO_3 bending, two bands
	440–410	22.73–24.39	w	SO_3 rocking
Covalent sulphonates, R—SO_2—OR	1420–1330	7.04–7.52	s	asym SO_2 str
	1200–1145	8.33–8.73	s	sym SO_2 str
Alkyl sulphonic acids (anhydrous), $RSO_2.OH$	~2900	~3.45	s	br, O—H str
	~2400	~4.15	w–m	br, O—H str
	1350–1340	7.41–7.46	s	asym SO_2 str
	1200–1100	8.33–9.10	s	sym SO_2 str
	1165–1150	8.59–8.70	s	br, S—O str
	1080–1040	9.26–9.62	w	
	910–890	10.99–11.24	s	S—O str
Alkyl sulphonic acids (hydrated), $RSO_3^- H_2O^+$	~2600	~3.85	m	Very broad band with three maxima, O—H str
	~2250	~4.45	m	
	~1680	~6.00	m	
	1230–1120	8.13–8.93	s	asym SO_3 str
	1120–1025	8.93–9.76	s	sym SO_3 str
Aryl sulphonic acids (solid phase)	~2760	~3.60	m–s	Broad band with shoulders, O—H str
	~2350	~4.25	m–s	
	~1345	~7.44	s	
	~1160	~8.62	s	
Aryl sulphonic acids (in inert solvent)	~3700	~2.70	v	sh, O—H str
	~2900	~3.45	v	
	~2500	~4.00	v	
Alkyl sulphonic acid sodium salts	1195–1175	8.37–8.51	vs	asym SO_3 str
	1065–1050	9.39–9.52	s	sym SO_3 str
Sulphonic acid salts, $SO_3^- M^+$	1250–1140	8.00–8.77	vs	asym SO_3 str
	1070–1030	9.35–9.70	s	sym SO_3 str
Sultones, 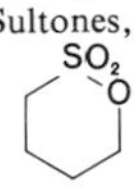	1385–1345	7.22–7.44	s	asym SO_2 str, often split
	1175–1165	8.51–8.58	s	sym SO_2 str, see ref. 28
Sulphate ion, SO_4^{2-}	1130–1080	8.85–9.26	s	SO_4 str
	~1000	~10.00	w	sh, not always present
Sulphite ion, SO_3^{2-}	990–910	10.10–10.99	s	
Bisulphate ion, HSO_4^-	1190–1160	8.40–8.62	s	asym SO_3 str
	1080–1000	9.26–10.00	s	sym SO_3 str
	880–840	11.36–11.90	m	Probably S—OH str

Thiocarbonyl Compounds, >C=S

The thiocarbonyl[51–59] absorption is not as strong as that due to the carbonyl C=O group, as might be expected since the sulphur atom is less electronegative than the oxygen atom and therefore the C=S group is less polar than the C=O group. In the case of compounds where the thiocarbonyl group is directly attached to a nitrogen atom, i.e. N—C=S,[53, 56, 60, 61, 64] the stretching vibration of the C=S portion is strongly coupled to that of the C—N part as a direct consequence of which several bands may, at least partly, be associated with the C=S stretching vibration. Hence thioamides,[53, 60, 61] thioureas, thiosemicarbazones, thiazoles, and dithio-oxamides have three absorption bands, in the regions 1570–1395 cm^{-1} (6.37–7.17 μm), 1420–1260 cm^{-1} (7.04–7.94 μm), and 1140–940 cm^{-1} (8.77–10.64 μm) which are in part due to the C=S stretching vibration.

The C=S stretching vibration for compounds where the thiocarbonyl group is not directly bonded to nitrogen gives rise to a band which is generally strong, often sharp, and occurs in the region 1230–1030 cm^{-1} (8.13–9.17 μm).

In general, the C=S band behaves in a similar manner to the carbonyl band. When a chlorine atom is directly bonded to the carbon of the C=S group, the band is observed at 1235–1225 cm^{-1} (8.10–8.16 μm).

Carbon disulphide, which is used as a solvent, absorbs strongly near 1510 cm^{-1} (6.62 μm) and 395 cm^{-1} (25.32 μm).

Table 16.9. Organic sulphur compounds containing C=S group

Functional Groups	Region cm^{-1}	μm	Intensity	Comments
Dialkyl thioketones, R—CS—R′	~1150	~8.70	s	C=S str, see ref. 54; normally dimerization makes the assignment of this band difficult, range has also been reported (see ref. 52) as 1270–1245 cm^{-1}
Diaryl thioketones, Ar—CS—Ar′	1225–1205	8.16–8.30	w–m	C=S str, see ref. 55
α,β-Unsaturated thioketones	1155–1140	8.66–8.77	s	C=S str
Dialkyl trithiocarbonates, (RS)$_2$C=S	1060–1050	9.43–9.52	s	C=S str
	~850	~11.76	s	asym S—C—S str
	~700	~14.29	w–m	Two bands for small alkyl groups
	~500	~20.00	w–m	Two bands, sym S—C—S str and C—S out-of-plane def
Thioncarbonates, (RO)$_2$C=S	1235–1210	8.10–8.26	s	C=S str, strong bands near 1200 cm^{-1} and 1100 cm^{-1} due to C—O str
Dithioacids, R—CS—SH	~1220	~8.20	s	C=S str
Dithioesters, R—CS—SR	1225–1185	8.16–8.44	s	C=S str, see ref. 53
	~870	~11.49	m–s	asym CS—S str
Thioacid fluorides, R—CS—F	1125–1075	8.90–9.30		
Thioacid chlorides	1235–1225	8.10–8.16	s	C=S str
	1100–1065	9.09–9.39		
RO—CSCH$_2$COOH and RS—CSCH$_2$COOH	~1050	~9.52	s	C=S str, see ref. 51
Xanthates	1200–1100	8.33–9.09	m–s	At least two bands
	1060–1040	9.43–9.62	vs	C=S str
Dixanthates, —O—C=S—S—	1200–1190	8.33–8.40	vs	asym C—O—C str, see refs 65–67
	1120–1100	8.93–9.09	m–s	sym C—O—C str
	1060–1000	9.43–10.00	vs	C=S str
Xanthate salts, R—O—CS—S$^-$	1100–1000	9.10–10.00	s	Have three strong bands in region 1250–1030 cm^{-1} (see above)
Oxyxanthates	~1580	~6.33	vs	br
	1115–1090	8.97–9.17	s	The only single strong band in region 1200–1000 cm^{-1}
	1050–1000	9.54–10.00	w	
	~695	~14.39	m–s	sh
Pyridthiones	1150–1100	8.70–9.09	s	C=S str, see refs 68, 69
Compounds of the type 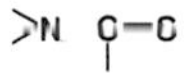	1570–1395	6.37–7.71	s	br } Due to strong coupling between C=S and C—N vibs
	1420–1260	7.04–7.94	v	
	1140–940	8.77–10.64	v	
	850–680	11.78–14.70		
Thioamides	1550–1500	6.45–6.67	s	Amide II band
Derivatives of 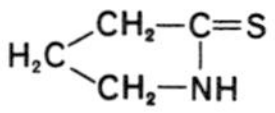 and 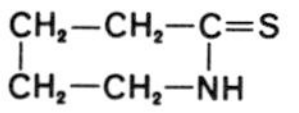	~1115	~8.70	s	C=S str (also seven-membered rings)

Table 16.9 (*cont.*)

Functional Groups	Region cm^{-1}	μm	Intensity	Comments
Cyclic thioureas (five- to seven-membered rings)	~1205	~8.30	s	Solid-phase spectra, also strong band at 1505–1570 cm^{-1}
P=S (solid phase)	865–655	11.56–15.27	m–s	
	730–550	13.70–18.18	v	
R$_3$P=S	770–685	12.99–14.60	m–s	
	595–530	16.81–18.87	v	
R$_3$As=S	490–470	20.41–21.28		As=S str
Methyl dithiocarbazic acids, salts and Ni^{II} and Cr^{III} coordination compounds	1040–960	9.62–10.42	s	asym CS$_2$ str, see ref. 64
	690–590	14.49–16.95	s	sym CS$_2$ str

Table 16.10. Other sulphur-containing compounds

Functional Groups	Region cm^{-1}	μm	Intensity	Comments
S—F	815–755	12.27–13.25	s	S—F str
S—O—CH$_2$—	1020–850	9.80–11.76	s	asym S—O—C str
	830–690	12.05–14.49	m	sym S—O—C str
Dialkyl thiolesters, R—CO—SR	1700–1690	5.88–5.59	vs	C=O str
Ar—CO—SR	1670–1665	5.99–6.00	vs	C=O str
Alkyl thiocarbonates, RS—CO—OR	1720–1700	5.81–5.88	vs	C=O str
ArS—CO—OAr	1740–1730	5.75–5.78	vs	C=O str
>C=C—S—	~1590	~6.29	m	C=C str, lower than expected, =C—H def band in normal position, see ref. 94
Thiooximes	~1620	~6.17	w–m	C=N str

Reviews

A review of the infrared spectra of sulphur compounds has been given by Billing.[92, 93] A review of the infrared spectra of gaseous diatomic sulphides is given by Barrow and Cousins.[8] The spectra of selenol and thiol esters have been reviewed by Ciurdaru and Denes.[11]

Organic Selenium Compounds

The infrared spectra of selenium compounds[10, 51, 60, 70–83] exhibit a great similarity to the corresponding sulphur analogues. This is hardly surprising—the change in mass, the (normally) weaker bonds formed by the selenium atom, and the slight variation in the bond angles account for the spectral differences.

A review of the spectra of selenium compounds has been published.[70]

Selenoamides, >N—CSe—

Selenoamides[60] do not have a band due solely to the C=Se stretching vibration because of the strong coupling of this vibration with the stretching vibration of the C—N bond. This type of behaviour has also been mentioned for thioamides. Two strong bands observed at 1500–1400 cm^{-1} (6.67–7.14 μm) and 700–600 cm^{-1} (14.29–16.67 μm) both have a contribution from the C=Se stretching vibration.

For metal complexes in which the metal ion is directly bound to the selenium atom, the former band position tends to higher frequencies whilst the latter tends to lower frequencies. Selenoamides have a medium-intensity band at 1200–1000 cm^{-1} (8.33–10.00 μm) which also has a contribution from the C=Se bond vibration. The spectra of other compounds[64, 82] with the N—C=Se group are similar to those of selenoamides.

The Se=O stretching vibration

The band due to the stretching vibration of the group Se=O is found over a wide range: 1000–800 cm^{-1} (10.00–12.50 μm).[75] As might be expected, the band is lower in frequency for metal complexes[76] than for the corresponding free selenoxide.

The P=Se stretching vibration

The band due to the P=Se stretching vibration[72–74] is of medium-to-strong intensity and occurs over the wide range 600–420 cm^{-1} (16.67–23.81 μm), more than one band often being observed.

The band generally occurs at higher frequencies for triaryl phosphine selenides, being at about 560 cm^{-1} (17.86 μm), than for the aliphatic compounds of this type, for which it occurs at about 425 cm^{-1} (23.53 μm). A similar difference is observed for the corresponding sulphur compounds. For metal complexes[84] of triaryl phosphine selenides, this absorption occurs at about 535 cm^{-1} (18.69 μm).

In the case of amide, ester, salt, and acid chloride derivatives of selenophosphoric acid, selenophosphonic acid, and selenophosphinic acid, the band is strong and in the range 600–500 cm^{-1} (16.67–20.00 μm).

Table 16.11. Organic selenium compounds

Functional Groups	Region cm^{-1}	Region μm	Intensity	Comments
Selenols, R—Se—H, and other compounds with Se—H	2330–2280	4.29–4.39	w	Se—H str, see refs 10, 78, 85
R—Se—D	~1680	~5.95	w	Se—D str
Selenides, R—Se—R	610–550	16.39–18.18	w–m	C—Se—C str, see ref. 79, 80
	585–505	17.09–19.80	w–m	C—Se—C str
Diselenides, R—Se—Se—R′	580–505	17.24–19.80	w–m	C—Se str, see ref. 81
	295–285	33.90–35.09	w	Se—Se str
Selenoxides, R—SeO—R and R—SeO_2—R	625–530	16.00–18.87	s	C—Se str, band weak for selenious acids and ions
Selenosemicarbazones, >N—CSe—N—N=C<	800–780	12.50–12.82	m	C=Se str
Coordinated carbon diselenide, e.g. $(phosphine)_2PtCSe_2$	~995	~10.05	s	C=Se str, carbon diselenide in CCl_4 absorbs strongly at ~1270 cm^{-1} and at ~310 cm^{-1}
Diselenocarbonates of the types RO—$CSeSeCH_2COOH$ and RS—$CSeSeCH_2COOH$	~940	~10.64	s	C=Se str, see ref. 51
Dialkyl triselenocarbonates	~800	~12.50	s	C=Se str, doubling of bands (see refs 71, 86) may be observed due to presence of different conformations
	~750	~13.33	s	asym Se—C—Se str
	~600	~16.67	w–m	Two bands
Phosphoniodiselenoformates, $R_3P^+CSeSe^-$	~900	~11.11	s	asym CSe_2 str, see ref. 87
Methyl diselenocarbazic acid, salts, and Ni^{II} and Cr^{III} coordination complexes	930–860	10.75–11.63	s	asym CSe_2 str, see ref. 88
	615–490	16.26–20.41	s	sym CSe_2 str
Selenoacetals, =C<(SeR)(SeR)	800–700	12.50–14.29	s	asym CSe_2 str
Ionic 1,1-ethylene diselenolates, =C<(Se^-)(Se^-)	870–750	11.49–13.33	s	asym =CSe_2 str
Ionic diselenocarbamates, =C<(Se)(Se^-)	950–800	10.53–12.50	s	asym =CSe_2 str
Cyanimidodiselenocarbonate alkali metal salts, CN—N=$C(SeH)_2$	~870	~11.49	s	asym CSe_2 str
Methyl aromatic selenothioesters, ArCSe.SMe	~980	~10.20	s	C=Se str
Aliphatic diselenocarboxylic acid esters, R—CSe.SeR	~780	~12.82	s	asym CSe.Se str
Selenoamides, ureas, and hydrazides	700–600	14.29–16.67	s	C=Se str, strongly coupled as with thioamides
Selenoamides	1500–1400	6.67–7.14	s	C=Se and C—N str, strongly coupled as with thioamides
	1200–1000	8.33–10.00	m	
	700–600	14.29–16.67	s	

Table 16.11 (*cont.*)

Functional Groups	Region cm^{-1}	μm	Intensity	Comments
Derivatives of 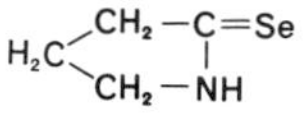 and 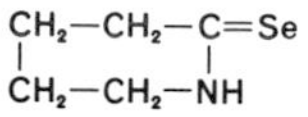	~1085	~9.21	s	C=Se str
Selenazoles, benzoselenazoles, and selenazolines	1570–1535	6.37–6.52	s	—N=C—Se str, see refs 89, 90
Selenazolines	1680–1650	5.95–6.06		C=N str
Benzoselenazoles	1610–1590	6.21–6.29		C=N str
Selenates, $(RO)_2SeO_2$	1040–1010	9.71–9.90	vs	asym O—Se—O str, see ref. 91
	960–930	10.42–10.75	vs	sym O—Se—O str
	700–600	14.29–16.67		Se—O—C str
Diselenates, $(RO)_2Se_2O_5$	700–600	14.29–16.67	vs	asym Se—O—Se str
	560–500	17.86–20.00	vs	sym Se—O—Se str
	~230	~43.48		Se—O—Se def
Selenones, R_2SeO_2	920–910	10.87–10.99	vs	asym O—Se—O str
	890–880	11.24–11.36	vs	sym O—Se—O str
	420–390	23.81–25.64		O—Se—O def
Selenites	~930	~10.75	s	Se=O str
Seleninic acids, R.SeO.OH	900–850	11.11–11.76	s	Se=O str
	700–680	14.29–14.71		Se—OH str
Seleninic acid anhydrides, R.SeO—O—SeO.R	900–850	11.11–11.76	s	Se=O str
	700–600	14.29–16.67		asym Se—O—Se str
	560–500	17.86–20.00		sym Se—O—Se str
	230–170	43.48–58.82		Se—O—Se def
Seleninic acid esters	900–850	11.11–11.76	s	Se=O str
Seleninyl halides	1005–930	9.96–10.75	s	Se=O str
Selenoxides, R_2SeO	840–800	11.90–12.50	s	Se=O str
Selenious amides	890–880	11.24–11.36	s	Se=O str
Dialkyl amino compounds of the type $(R_2N)_2Se$, $(R_2N)_2SeR_2$, $(R_2N)_2SeO$, $R_2N.SeO.Cl$	590–540	16.95–18.52		N—Se—N str, asym and sym str coincide
Triaryl phosphine selenides, Ar_3PSe	~560	~17.86	s	P=Se str, for metal complexes band occurs at 540–530 cm^{-1}
Trialkyl phosphine selenides, R_3PSe	~425	~23.53	s	P=Se str
$(EtO)_2.P{=}Se.SR$	~590	~16.95	s	P=Se str
Trialkyl arsine selenides, $R_3As{=}Se$	360–330	27.78–30.30		As=Se str, a doublet
Trialkyl stibine selenides, $R_3Sb{=}Se$	300–270	33.33–37.04	s	Sb=Se str

References

1. C. S. Hsu, *Spectrosc. Letters*, 1974, **7**, 439
2. J. G. David and H. E. Hallam, *Spectrochim. Acta*, 1965, **21**, 841
3. A. R. Cole *et al.*, *Spectrochim. Acta*, 1965, **21**, 1169
4. S. I. Miller and G. S. Krishnamurthy, *J. Org. Chem.*, 1962, **27**, 645
5. R. A. Nyquist and W. J. Potts, *Spectrochim. Acta*, 1959, **15**, 514
6. P. A. Tice and D. R. Powell, *Spectrochim. Acta*, 1965, **21**, 837
7. J. G. David and H. E. Hallam, *Trans. Faraday Soc.*, 1964, **60**, 2013
8. R. F. Barrow and C. Cousins, *Adv. High Temp. Chem.*, 1971, **4**, 161
9. J. Cymerman and J. B. Willis, *J. Chem. Soc.*, **1951**, 1332
10. N. Sharghi and I. Lalezari, *Spectrochim. Acta*, 1964, **20**, 237
11. G. Ciurdaru and V. I. Denes, *Stud. Cercet. Chim.*, 1971, **19**, 1029
12. G. A. Crowder, *App. Spectrosc.*, 1972, **26**, 486
13. R. J. Philippe and H. Moore, *Spectrochim. Acta*, 1961, **17**, 1004
14. T. Cairns *et al.*, *Spectrochim. Acta*, 1964, **20**, 31
15. T. Cairns *et al.*, *Spectrochim. Acta*, 1964, **20**, 159
16. D. Barnard *et al.*, *J. Chem. Soc.*, **1949**, 2442
17. W. Otting and F. A. Neugebauer, *Chem. Ber.*, 1962, **95**, 540
18. S. Pinchas *et al.*, *J. Chem. Soc.*, **1962**, 3968
19. G. Kresze *et al.*, *Spectrochim. Acta*, 1965, **21**, 1633
20. T. Granstad, *Spectrochim. Acta*, 1963, **19**, 829
21. R. H. Figueroa *et al.*, *Spectrochim. Acta*, 1966, **22**, 1563
22. R. H. Figueroa, *Spectrochim. Acta*, 1966, **22**, 1109
23. W. K. Glass and R. D. E. Pullin, *Trans. Faraday Soc.*, 1961, **57**, 546
24. S. Ghersetti and G. Modena, *Spectrochim. Acta*, 1963, **19**, 1809
25. M. F. Lappert and J. K. Smith, *J. Chem. Soc.*, **1961**, 3224
26. R. S. Drago and D. Meek, *J. Phys. Chem.*, 1961, **65**, 1446
27. R. Francis and F. A. Cotton, *J. Chem. Soc.*, **1961**, 2078
28. E. A. Robinson, *Canad. J. Chem.*, 1961, **39**, 247
29. L. J. Bellamy and R. L. Williams, *J. Chem. Soc.*, **1957**, 863
30. N. Marziano *et al.*, *Ann. Chim. Rome*, 1962, **52**, 121
31. S. Ghersetti and C. Zauli, *Ann. Chim. Rome*, 1963, **53**, 710
32. W. R. Fearirheller and J. E. Katon, *Spectrochim. Acta*, 1964, **20**, 1099
33. P. M. G. Bavin *et al.*, *Spectrochim. Acta*, 1960, **16**, 1312
34. A. S. Wexler, *App. Spectrosc. Rev.*, 1968, **1**, 29
35. R. J. Gillespie and E. A. Robinson, *Spectrochim. Acta*, 1963, **19**, 741
36. A. Simon and D. Kunath, *Zeit. Anorg. Chem.*, 1961, **308**, 21
37. M. Bredereck *et al.*, *Chem. Ber.*, 1960, **93**, 2736
38. G. Von Geiseler and K. O. Bindernagel, *Zeit. Elektrochem.* 1959, **63**, 1140
39. G. Von Geiseler and K. O. Bindernagel, *Zeit. Elektrochem.*, 1960, **64**, 421
40. G. Malewski and H. J. Weigmann, *Spectrochim. Acta*, 1962, **18**, 725
41. G. Malewski and H. J. Weigmann, *Zeit. Chem.*, 1964, **4**, 389
42. D. A. Long and R. T. Bailey, *Trans. Faraday Soc.*, 1963, **59**, 792
43. R. J. Gillespie and E. A. Robinson, *Canad. J. Chem.*, 1961, **38**, 2171
44. J. N. Baxter *et al.*, *J. Chem. Soc.*, **1955**, 669
45. A. R. Katritzky and R. A. Jones, *J. Chem. Soc.*, **1960**, 4497
46. E. von Merian, *Helv. Chim. Acta*, 1960, **49**, 1122
47. A. Vandi *et al.*, *J. Org. Chem.*, 1961, **26**, 1136
48. A. Simon *et al.*, *Chem. Ber.*, 1956, **89**, 1883, 2378, and 2384
49. S. Detoni and D. Hadzi, *Coll. Spectr. Inter. VI*, **1956**, 601
50. G. Chihara, *Chem. Pharm. Bull.*, 1960, **8**, 988
51. K. A. Jensen and U. Anthoni, *Acta Chem. Scand.*, 1970, **24**, 2055
52. C. Andrieu and Y. Mollier, *Spectrochim. Acta*, 1972, **28A**, 785
53. L. J. Bellamy and P. E. Rogash, *J. Chem. Soc.*, **1960**, 2218
54. R. Mayer *et al.*, *Ang. Chemie*, 1964, **76**, 157
55. N. Lozach and G. Guillouzo, *Bull. Soc. Chim. France*, **1957**, 1221
56. C. N. R. Rao and R. Venkataraghavan, *Spectrochim. Acta*, 1962, **18**, 541
57. C. N. R. Rao and R. Venkataraghavan, *Canad. J. Chem.*, 1961, **39**, 1757
58. H. E. Hallam and C. M. Jones, *Spectrochim. Acta*, 1969, **24A**, 1791

59. G. Keresztury and M. P. Marzocchi, *Spectrochim. Acta*, 1975, **31A,** 275
60. K. A. Jensen and P. M. Nielsen, *Acta Chem. Scand.*, 1966, **20,** 597
61. H. O. Desseyn *et al.*, *Spectrochim. Acta*, 1974, **30A,** 503
62. G. Borch *et al.*, *Spectrochim. Acta*, 1973, **29A,** 1109
63. K. Mergog *et al.*, *J. Mol. Struct.*, 1969, **3,** 339
64. U. Anthoni *et al.*, *Acta Chem. Scand.*, 1970, **29,** 959
65. F. G. Pearson and R. B. Stasiak, *App. Spectrosc.*, 1958, **12,** 116
66. L. M. Little *et al.*, *Canad. J. Chem.*, 1961, **39,** 745
67. M. L. Shankaranaryana and C. C. Patel, *Canad. J. Chem.*, 1961, **39,** 1633
68. A. R. Katritzky and R. A. Jones, *Spectrochim. Acta*, 1961, **17,** 64
69. A. R. Katritzky and R. A. Jones, *J. Chem. Soc.*, **1960,** 2947
70. K. A. Jensen *et al.* In *Organic Selenium Compounds—Their Chemistry and Biology* (D. L. Klayman and W. H. H. Günther, eds). Wiley, New York, 1973
71. L. Hendriksen *et al.*, *Spectrochim. Acta*, 1975, **31A,** 191
72. R. A. Zingaro, *Inorg. Chem.*, 1963, **2,** 192
73. K. A. Jensen and P. H. Nielsen, *Acta Chem. Scand.*, 1963, **17,** 1875
74. J. R. Durig *et al.*, *J. Mol. Spectrosc.*, 1968, **28,** 444
75. R. von Paetzold, *Zeit. Chem.*, 1964, **4,** 321
76. R. von Paetzold and G. Bochmann, *Zeit. Anorg. Allg. Chem.*, 1969, **368,** 202
77. V. Horn and R. von Paetzold, *Spectrochim. Acta*, 1974, **30A,** 1489
78. A. B. Harvey and M. K. Wilson, *J. Chem. Phys.*, 1966, **45,** 678
79. J. A. Allkins and P. J. Hendra, *Spectrochim. Acta*, 1967, **23A,** 1671
80. J. Shiro *et al.*, *Bull. Chem. Soc. Japan*, 1970, **43,** 612
81. G. Bergson, *Arkiv Kemi*, 1959, **13,** 11
82. B. A. Gingras *et al.*, *Canad. J. Chem.*, 1965, **43,** 1650
83. R. von Paetzold *et al.*, *Zeit. Anorg. Allg. Chem.*, 1967, **352,** 295
84. M. G. King and G. P. McQuillan, *J. Chem. Soc. A*, **1967,** 898
85. K. A. Jensen and L. Hendriksen, *Acta Chem. Scand.*, 1970, **24,** 3213
86. M. Dräger and G. Gatlow, *Chem. Ber.*, 1971, **104,** 1429
87. K. A. Jensen and P. H. Nielsen, *Acta Chem. Scand.*, 1963, **17,** 549
88. U. Anthoni *et al.*, *Acta Chem. Scand.*, 1970, **29,** 959
89. P. Bassignana *et al.*, *Spectrochim. Acta*, 1965, **21,** 605
90. R. V. Kendall and R. A. Olofsun, *J. Org. Chem.*, 1970, **35,** 806
91. R. von Paetzold and H. Amoulong, *Zeit. Chem.*, 1966, **6,** 29
92. F. A. Billing, *Intra-Science Chem. Reports*, 1967, **1,** 225
93. F. A. Billing and N. Kharasch, *Quart. Rpts Sulphur Chem.*, 1966, **1,** 189
94. B. Nagel and A. B. Remizov, *Zh. Obshsch. Khim.*, 1978, **45,** 1189
95. G. Paulson *et al.*, *Bio. Med. Mass Spectrom.*, 1978, **5,** 128

CHAPTER 17

Organic Phosphorus Compounds

P—H and P—C vibrations

The stretching vibration of the P—H group[6–8,10,11] gives rise to a sharp band of medium intensity in the region 2455–2265 cm^{-1} (4.07–4.42 μm). For aliphatic and aryl phosphines, this band occurs in a much narrower region: 2285–2265 cm^{-1} (4.38–4.42 μm).

The stretching vibration of the P—C bond gives a medium-to-strong band in the region 795–650 cm^{-1} (12.58–15.38 μm).

P—OH and P—O vibrations

Compounds with the P—OH group,[10] for which of course hydrogen bonding normally occurs, have two broad bands of weak-to-medium intensity, at 2700–2560 cm^{-1} (3.70–3.90 μm) and at 2300–2100 cm^{-1} (4.35–4.76 μm), which are due to the O—H stretching vibrations, and a medium-to-strong, broad band at 1040–910 cm^{-1} (9.62–10.99 μm) due to the P—O stretching vibration. However, since most phosphorus compounds absorb in this latter region, this band is of little value. Those compounds which also contain the P=O group[12,18,21,22] have a broad band near 1680 cm^{-1} (5.92 μm), e.g. dialkyl phosphoric acids, phosphorous acids. For phosphoric acids, the band near 2600 cm^{-1} (3.85 μm) is stronger than those near 2200 cm^{-1} (4.55 μm) and 1680 cm^{-1} (5.95 μm) whereas for phosphinic acids, the band near 1680 cm^{-1} is the strongest of the three and for phosphonic acids all three bands have about the same relative intensity.

Acid salts containing the P—OH group have broad bands in the regions 2725–2525 cm^{-1} (3.76–3.94 μm) and 2500–1600 cm^{-1} (4.00–6.25 μm).

P—O—C vibrations

For aliphatic compounds, the asymmetric stretching vibration of the P—O—C group[12,15,18,25] gives a very strong broad band, normally found in the region 1050–970 cm^{-1} (9.52–10.31 μm). In the case of pentavalent and trivalent methoxy compounds, this band is sharp and strong, occurring at 1090–1010 cm^{-1} (9.17–9.90 μm) and 1035–1015 cm^{-1} (9.67–9.85 μm) respectively, the characteristic symmetric methyl deformation band near 1380 cm^{-1} (7.25 μm) being absent in some cases. In general, the band due to the asymmetric stretching vibration of the P—O—C group of pentavalent phosphorus occurs at lower frequencies than that for the trivalent compound. Pentavalent ethoxy compounds have an additional strong band at 985–940 cm^{-1} (10.15–10.64 μm) which may be weak for higher alkoxy compounds. Methoxy and ethoxy compounds have a strong band at 830–740 cm^{-1} (12.05–13.51 μm) which is probably due to the symmetric stretching of the P—O—C group. However, this band is usually absent in other alkoxy compounds. Methoxy compounds have a weak, sharp band near 1190 cm^{-1} (8.40 μm). Other alkoxy phosphorus compounds have a medium intensity band near 1165 cm^{-1} (8.59 μm). For compounds which have only ethoxy groups (i.e. no other alkyl groups), two characteristic doublets are observed in the region 1500–1350 cm^{-1} (6.67–7.41 μm) due to the C—H deformation vibrations. For aromatic compounds, P—O—phenyl, the band due to the P—O—C asymmetric stretching vibration occurs at 995–855 cm^{-1} (10.05–11.70 μm).

P=O vibrations

The band due to the stretching vibration of the P=O group[7,13–24] is strong and in the region 1350–1150 cm^{-1} (7.41–8.70 μm). Due to the size of the phosphorus atom, the frequency of the P=O stretching vibration is almost independent of the type of compound in which the group occurs and of the size of the substituents. However, it is influenced by the number of electronegative substituents directly bonded to it, as well as being very sensitive to association effects,[23,24] for instance, a phase change results in a shift in band position of about 60 cm^{-1}. The P=O band may sometimes appear as a doublet,[14] the separation either being small, as for some triaryl phosphates, or as large as 50 cm^{-1}. This splitting is believed to be in some cases partly due to Fermi resonance and in others, such as some substituted triaryl phosphates, to rotational isomerism.

Pyrophosphates,[17] O=P—O—P=O, have only one P=O band unless the pyrophosphate is a non-symmetrical compound. Therefore, unlike carboxylic acid anhydrides, which in general have two bands that arise due to coupling between the C=O groups, no coupling appears to exist between the two P=O groups in pyrophosphates.

Phosphoric acids have extremely strong intermolecular hydrogen bonds which are present even in very dilute solution in inert solvents and result in the P=O band usually being about 50 cm^{-1} lower than for the corresponding ester.

The position of the band due to the P=O stretching vibration is dependent on the sum of the electronegativities of the attached groups. Electronegative groups tend to withdraw electrons from the phosphorus atom thus competing with the oxygen which would otherwise have a tendency to form $P^+—O^-$, therefore resulting in a stronger bond and hence in a higher vibration frequency. Similarly, hydrogen bonding tends to lower the frequency of the P=O stretching vibration and broaden the band. The frequency[7,13] of this band may be calculated for different compounds with reasonable accuracy from the relationship

$$\nu = 930 + 40 \sum \pi,$$

where π is the phosphorus inductive constant of a given substituent group. It should be noted that the frequency of the P=O stretching vibration has also been correlated with Taft σ values.[7,13]

Spectral changes for ⩾P=O compounds (a) where the P=O group acts as a good proton acceptor, (b) where coordination occurs, are given elsewhere.[26]

Other bands

The band due to the P=S stretching vibration[25,27–29] occurs at 865–655 cm^{-1} (11.56–15.27 μm) and is of medium-to-strong intensity. Also a band of variable intensity occurs at 730–550 cm^{-1} (13.70–18.18 μm), possibly due to the P—S bond stretching vibration. Like the phosphonyl group (P=O), the position of the P=S band is affected by the electronegativity of adjacent groups although this effect is not so marked as for the phosphonyl group since the P=S group has less ionic character. The P=S band may consist of a doublet due to the presence of rotational isomerism. Normally the band is difficult to identify since there are many other groups which have bands in the same region.

A band due to an aromatic ring vibration for compounds containing a phenyl—P bond occurs at 1455–1425 cm^{-1} (6.90–7.02 μm) and is of medium-to-strong intensity. This band is useful since it occurs in a region normally free from absorptions by phosphorus compounds.

Compounds containing the P—Cl bond[20] have a medium-to-strong absorption at 605–435 cm^{-1} (16.53–22.99 μm) due to the P—Cl stretching vibration. The position of the band due to the P—X (X = F or Cl) stretching vibration[30,31] is affected by the oxidation state of the phosphorus atom. In the presence of more than one halogen atom directly attached to the phosphorus atom, two peaks are observed due to the asymmetric and symmetric stretching vibrations respectively. Difluorides of the type $—P(O)F_2$ absorb at 930–895 cm^{-1} (10.75–11.17 μm) and 890–870 cm^{-1} (11.24–11.49 μm).

Reviews have been published dealing with the infrared spectra of organic phosphorus compounds,[1–8] as has a correlation chart[9] for inorganic phosphorus compounds.

Table 17.1. Organic phosphorus compounds. (The data given are, except where stated, for condensed phase spectra, i.e. liquids or solids, measured as nujol mulls or as discs)

Functional Groups	Region cm^{-1}	Region μm	Intensity	Comments
P—H vibrations:				
P—H	2455–2265	4.07–4.42	m	P—H str
	1150–965	8.70–10.36	w–m	P—H def
Alkyl phosphines, P—H	2285–2265	4.38–4.42	m	P—H str
	1100–1085	9.09–9.21	m	$P—H_2$ scissoring
	1065–1040	9.39–9.62	w–m	P—H def
	940–910	10.64–10.99	m	PH_2 wagging
Aryl phosphines, P—H	2285–2270	4.38–4.41	m	P—H str, see ref. 35
	1100–1085	9.09–9.21	m	P—H def
Phosphonates, $(GO)_2HP=O$	2455–2400	4.07–4.17	m	P—H str
Phosphine oxides, $G_2HP=O$	2340–2280	4.27–4.39	m	P—H str
	990–965	10.10–10.36	m–s	P—H wagging
$G_2HP=S$	950–910	10.53–10.99	m–s	P—H wagging
Phosphonates, $(RO)_2HP=O$	980–960	10.20–10.42	vs	Probably due to interaction between P—O—C stretching and P—H wagging
P—D	1795–1650	5.57–6.06	m	P—D str
	745–615	13.42–16.26	w–m	P—D bending
P—C and PC—H vibrations:				
P—C	795–650	12.58–15.38	m–s	P—C str
$P—CH_3$	1430–1390	6.99–7.19	m–s	asym CH_3 def
	1345–1275	7.49–7.85	m–s	sym CH_3 def
	980–840	10.20–11.90	s	CH_3 def, often doublet (for P^V compounds often at 935–870 cm^{-1}, for P^{III} at 905–860 cm^{-1}, for $⩾PHCH_3$ ~845 cm^{-1})
	790–770	12.66–12.99	s	P—C str
$⩾PHCH_3$	850–840	11.76–11.90	m–s	
$P(CH_3)_2$	960–835	10.42–10.70	m–s	Two or three bands
$(RO)_2PCH_3$	1285–1270	7.77–7.87	m–s	$P—CH_3$ bands
	870–865	11.49–11.56	s	
$CH_3(RO)HP=O$	1300–1295	7.69–7.72	m–s	$P—CH_3$ bands
	850–840	11.76–11.90	s	
$CH_3(RO)_2P=O$	1320–1305	7.58–7.66	m–s	$P—CH_3$ bands
	930–885	10.75–11.30	s	
$CH_3(RO)P(=O)—O^-$	1310–1280	7.63–7.81	m–s	$P—CH_3$ bands
	900–875	11.11–11.43	s	
$CH_3(RO)ClP=O$	1315–1300	7.60–7.69	m–s	$P—CH_3$ bands
	925–885	10.81–11.30	s	
$P—C_2H_5$	1285–1225	7.78–8.17	w	Doublet (P^{III} compounds also have medium intensity band at 1235–1205 cm^{-1})
$P—CH_2—P$	845–780	11.83–12.82	m–s	asym P—C—P vib
	770–720	12.99–13.89	m–s	sym P—C—P vib
$P—CH_2—$	1440–1405	6.94–7.12	m	CH_2 def
	780–760	12.82–13.16	s	P—C str
$P—CH_2—Ar$	795–740	12.58–13.51	s	P—C str
P—Ar	~3050	~3.33	m–w	C—H str
	~1600	~6.25	m–w	Aromatic ring in-plane str
	~1500	~6.67	m–w	
	1455–1425	6.90–7.02	m–s	

Table 17.1 (*cont.*)

Functional Groups	Region cm⁻¹	Region µm	Intensity	Comments
P—Ar (cont.)	1010–990	9.09–10.10	m–s	Interaction between aromatic ring vib and P—C str
	560–480	17.86–20.83	m–s	
P—Ph	1130–1090	8.85–9.71	s	
	750–680	13.33–14.71	s	
P—N—Ph	1425–1380	7.02–7.25	w–m	
P—O—H vibrations:				
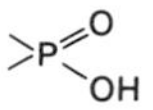	2725–2525	3.76–3.96	w–m	br, OH str, hydrogen-bonded
	2350–2080	4.26–4.81	w–m	br, may be doublet for aromatic phosphorus acids
	1740–1600	5.75–6.25	w–m	br, OH def
	1335–1080	7.55–9.26	s	P=O str
	1040–910	9.62–10.99	s	sh, P—O str
	540–450	18.52–22.22	w–m	Often a doublet
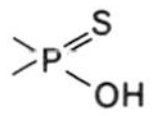	3100–3000	3.23–3.33	w	br, OH str
	2360–2200	4.24–4.55	w	br, OH str
	935–910	10.70–10.99	s	P—O str
	810–750	12.35–13.33	m–s	P=S str
	655–585	15.27–17.12	v	P=S str
P—O—C vibrations:				
P—O—R	1050–970	9.52–10.31	vs	asym P—O—C str (see ref. 15), (for phosphonium compounds, range extends to 1090 cm⁻¹)
	850–740	11.76–13.51	w–m	Sometimes very weak
P—O—CH_3	1190–1170	8.40–8.55	w–m	CH_3 def
	1090–1010	9.17–9.90	vs	asym P—O—C def
	830–740	12.05–13.51	s	sym P—O—C str (asym str ~1050 cm⁻¹)
P—O—C_2H_5	1165–1155	8.59–8.68	w–m	CH_3 rocking
	830–740	12.05 13.51	s	sym P—O—C str
P—O—CH_2R	1170–1100	8.55–9.09	w–m	Number of bands
Isopropyl—O—P	1190–1170	8.40–8.55	w	
	1150–1135	8.70–8.81	w	
	1115–1100	8.97 9.09	w	
P—O—Ar	1460–1445	6.85–6.92	w–m	
	1260–1160	7.94–8.62	s	sh, mainly O—C str
	995–915	10.05–10.93	vs	br, P—O—C str (pentavalent)
	875–855	11.43–11.70	s	P—O—C str (trivalent)
	790–740	12.66–13.51	s	sym P—O—C str
	625–570	16.00–17.54	s	P—O—Ar def
Alkyl phosphites $(RO)_3P$	1050–990	9.52–10.10	vs	P—O—C str
Aryl phosphites $(ArO)_3P$	1240–1190	8.07–8.40	vs	P—O—C str
Phosphites $(GO)_3P$	580–510	17.24–19.61	m	
	580–400	17.24–25.00	s	
	400–295	25.00–33.90	s	
Hydrogen phosphites	560–545	17.86–18.35	s	
	540–500	18.52–20.00	w–m	
P=O vibrations:				
P=O (unassociated)	1350–1175	7.41–8.51	vs	P=O str
P=O (associated)	1250–1150	8.00–8.70	vs	P=O str

Table 17.1 (*cont.*)

Functional Groups	Region cm⁻¹	Region µm	Intensity	Comments
Alkyl phosphates, $(RO)_3P=O$	1285–1255	7.78–7.97	vs	P=O str, see refs 13, 15
	1050–990	9.52–10.10	vs	P—O—C str
	595–520	16.81–19.23	m	br
	495–465	20.20–21.51	m	br
	430–415	23.26–24.10	w	
	395–360	25.32–27.78		
Aryl phosphates, $(ArO)_3P=O$	1315–1290	7.61–7.75	vs	P=O str
	1240–1190	8.07–8.40	vs	P—O—C str
	625–575	16.00–17.39	s	
	570–540	17.54–18.52	s	
	510–490	19.61–20.41		
	460–430	21.74–23.26	m–w	
Acid phosphates				
$(RO)_2(HO)P=O$	1250–1210	8.00–8.26	vs	P=O str, see ref. 32
	590–460	16.95–21.74	m	br
	400–380	25.00–26.32	w	Not observed for phosponates
$(ArO)_2(HO)P=O$	600–580	16.67–17.24	s	Sensitive to ring substitution
	565–535	17.70–18.69	s	
	515–500	19.42–20.00	s	
	490–470	20.41–21.28	s	
	400–380	25.00–26.32	w	Not observed for phosphonates
$(RO)(HO)_2P=O$	~1250	~8.00	vs	P=O str (aryl compounds ~1200 cm⁻¹)
Phosphonates, $G(RO)_2P=O$	1265–1230	7.91–8.13	vs	P=O str, see ref. 32
	800–750	12.50–13.33	w–m	P—O—C str
Alkyl phosphonates, $R(RO)_2P=O$	570–500	17.54–20.00	m	br
	490 410	20.41–24.39	m	br
	440–400	22.73–25.00	w	
Aryl phosphonates, $Ar(ArO)_2P=O$	620–600	16.13–16.67	m	See ref. 32
	535–515	18.69–19.42	s	
	500–480	20.00–20.83	vw	
	425 415	23.53–24.10	vw	
Dialkyl aryl phosphonates, $(RO)_2ArP=O$	585–565	17.09–17.70	s	See ref. 32
	530–520	18.87–19.23	s	
	435–420	22.99–23.81	w	
	320–310	31.25–32.26	w	
Hydrogen phosphonates				
$R'(RO)(HO)P=O$	1215–1170	8.23–8.55	vs	P=O str, see ref. 32
	570–540	17.54–18.52	m	br
	500–450	20.00–22.22	m	br
	320–300	31.25–33.33	w	
$Ar'(ArO)(OH)P=O$	1220–1205	8.20–8.30	vs	P=O str, see ref. 32
	605–570	16.53–17.54	s	
	550–535	18.18–18.69	s	
	495–485	20.20–20.62	m	
	~460	~21.74	m	
	430–420	23.26–23.81	m	
	370–350	27.03–28.57	w	
	315–290	31.75–34.48	w	
$(RO)(HO)HP=O$	1215–1200	8.23–8.33	vs	P=O str
$(RO)_2HP=O$	1265–1250	7.97–8.00	vs	P=O str
$(RO)_2FP=O$	1315–1290	7.61–7.75	vs	P=O str
$(ArO)_2FP=O$	1330–1325	7.52–7.55	vs	P=O str
$(RO)_2ClP=O$	1310–1280	7.63–7.81	vs	P=O str (CN substituted compounds at ~1290 cm⁻¹)

Table 17.1 (*cont.*)

Functional Groups	Region cm⁻¹	μm	Intensity	Comments
$(RO)_2(RS)P{=}O$	1270–1245	7.87–8.06	vs	P=O str
$(RO)_2(NH_2)P{=}O$	1250–1220	8.00–8.20	vs	P=O str, see ref. 34
$(ArO)_2(NH_2)P{=}O$	~1250	~8.00	vs	P=O str
$(RO)_2(NHR)P{=}O$	1260–1195	7.94–8.36	vs	P=O str
$(RO)_2(NR_2)P{=}O$	1275–1250	7.84–8.00	vs	P=O str
$R_2(R'O)P{=}O$	1220–1180	8.20–8.48	vs	P=O str
$R_2(HO)P{=}O$	1190–1140	8.40–8.77	vs	P=O str
$Ar_2(HO)P{=}O$	1205–1085	8.30–9.21	vs	P=O str
$R(HO)HP{=}O$	1175–1135	8.51–8.81	vs	P=O str
$R_3P{=}O$	1185–1150	8.44–8.70	vs	P=O str
$Ar_3P{=}O$	1190–1175	8.40–8.51	vs	P=O str
$R_2HP{=}O$	~1155	~8.66	vs	P=O str
$Ar_2HP{=}O$	1185–1170	8.44–8.55	vs	P=O str
$R_2ClP{=}O$	~1215	~8.23	vs	P=O str (for dichloro-, see ref. 36)
$Ar_2ClP{=}O$	~1235	~8.10	vs	P=O str
$G_2BrP{=}O$	~1250	~8.00	vs	P=O str
$(RS)_3P{=}O$	~1200	~8.33	vs	P=O str
$(ArS)_3P{=}O$	~1210	~8.26	vs	P=O str
$R_2(RS)P{=}O$	~1200	~8.33	vs	P=O str
$(RHN)_3P{=}O$	1230–1215	8.18–8.23	vs	P=O str
$(R_2N)_3P{=}O$	1245–1190	8.03–8.40	vs	P=O str
$R_2(NHR)P{=}O$	1180–1150	8.48–8.66	vs	P=O str
$R(NHR)_2P{=}O$	1220–1160	8.20–8.62	vs	P=O str
Pyrophosphates, $>P(=O)—O—P(=O)<$	1310–1205	7.63–8.30	vs	P=O str (see ref. 17); usually one band, two bands for unsymmetrical pyrophosphates
Alkyl pyrophosphates, $(RO)_2P(=O)—O—P(=O)(RO)_2$	1240–1205	8.07–7.63	vs	P=O str, usually one band
	1310–1280	7.63–7.81	vs	P=O str
Phosphonic anhydrides, $R(RO)P(=O)—O—P(=O)(RO)R$	1270–1250	7.87–8.00	vs	P=O str
	930–915	10.75–10.93	s	br, asym P—O—P str
P—O—P	1025–870	9.85–11.49	s	Usually broad, asym str often found in region 945–925 cm⁻¹ (a weak band also found near 700 cm⁻¹)
$>P(=O)—S—C$	615–555	16.26–18.18	m	
	575–510	17.39–19.61	m	
P=S vibrations:				
P=S	865–655	11.56–15.27	m–s	See refs 27–29
	730–550	13.70–18.18	v	
$(OH)P{=}S$	810–750	12.35–13.33	m	
	655–585	15.27–17.09	v	
$>P(=S)—OP<$	865–770	11.56–12.99	m	
	610–585	16.39–17.09	v	
$>P(X){=}S$ (X = F or Cl)	835–750	11.98–13.33	m–s	
	730–590	13.70–16.95	v	

Table 17.1 (*cont.*)

Functional Groups	Region cm⁻¹	μm	Intensity	Comments
$>P(=S)—S—C$	545–510	18.35–19.61	m–s	
	520–475	19.23–21.05	m	
$>P(=S)—N<$	860–725	11.63–13.79	m–s	
	715–570	13.99–17.54	v	
$—P(Cl)(=S)—N$	810–765	12.35–13.07	m–s	
	670–605	14.93–16.53	v	
P=Se	590–515	16.95–19.42	m	P=Se str, see refs 27–29
	535–420	18.69–23.81	m	
P=Te	470–400	21.28–25.00		P—Te str
$R_3P{=}S$	770–685	12.99–14.60	m	
	595–530	16.81–18.87	v	
$(RO)_3P{=}S$	845–800	11.83–12.50	m	Absent for tetrathionates
	665–600	15.04–16.67	v	($(RS)_3P{=}S$ ~690 cm⁻¹)
$R(RO)_2P{=}S$	805–770	12.42–12.99	m	
	650–595	15.38–16.81	v	
$R_2(RO)P{=}S$	795–770	12.58–12.99	m	
	610–580	16.39–17.24	v	
$(RO)_2(RS)P{=}S$	835–790	11.98–12.66	m	
	665–645	15.04–15.50	m–s	
$(RO)_2(RS)P{=}Se$	~590	~16.95	s	P=Se str
$(RO)_2(SH)P{=}S$	865–835	11.56–11.98	m	S—H bending vib
	780–730	12.82–13.70	m	
	660–650	15.15–15.38	m–s	
$R_2ClP{=}S$	775–750	12.90–13.33	m	
	625–590	16.00–16.95	m–s	
$RCl_2P{=}S$	780–775	12.82–12.90	s	($(RO)Cl_2P{=}S$ ~830 cm⁻¹)
	670–640	14.93–15.63	m–s	
$(R_2N)_3P{=}S$	840–790	11.90–12.66	m	
	715–690	13.99–14.49	m–s	
Metal phosphorodithioates, (MS) $(RO)_2P{=}S$ M = Zn, Cd, Ni	660–635	15.15–15.75	s	Probably due to P=S, see ref. 25
	555–535	18.02–18.69	s	Possibly due to P—S—M group
PN vibrations:				
P—N	1110–930	9.01–10.75	m–s	Probably asym P—N—C str, see ref. 20
	750–680	13.33–14.71	m–s	sym P—C—N str
$P^{III}—N$	1010–790	9.90–12.66	m–s	See ref. 33
$P—N—CH_3$	1320–1260	7.58–7.94	m	
	1205–1155	8.30–8.66	w–m	
	1080–1050	9.26–9.52	w–m	
	1010–935	9.90–10.70	s	
$P—N(C_2H_5)_2$	1225–1190	8.16–8.40		m–s for P^V, w for P^{III}
	1190–1155	8.40–8.66		m for P^V, s for P^{III}
	1110–1085	9.01–9.22	w–m	
	1075–1055	9.30–9.48	w–m	
	1050–1015	9.52–9.85	m–s	
	975–930	10.26–10.75	m–s	
	930–915	10.75–10.93	w	Not always observed
P=N (cyclic compounds)	1440–1170	6.94–8.55	vs	P=N str, see ref. 15
P=N (acyclic compounds)	1500–1230	6.67–8.13	s	P=N str

Table 17.1 (*cont.*)

Functional Groups	Region cm⁻¹	Region μm	Intensity	Comments
$(RO)_3P{=}N{-}Ar$, $R(RO)_2P{=}N{-}Ar$	1385–1325	7.22–7.55	s	P=N str
$R{-}NH{-}P(O)Cl_2$	~560	~17.86		
$-O{-}NH{-}P(O)Cl_2$	545–520	18.35–19.23		
$>N{-}P(S)Cl_2$	525–490	19.05–20.41		
$-O{-}O{-}PCl_2$	510–495	19.61–20.20		
	475–465	21.05–21.51		
$-O{-}O{-}P(O)Cl_2$	~590	~16.95		
$-O{-}O{-}P(S)Cl_2$	560–535	17.86–18.69		
$(RO)Cl_2P{=}O$	~570	~17.54		
$(RO)Cl_2P{=}S$	~535	~18.69	s	
Phosphinic acid, $R_2PO_2^-$, and phosphonous acid, $RHPO_2^-$	1200–1100	8.33–9.09	s	asym PO_2 str
	1075–1000	9.30–10.00	s	sym PO_2 str
$(RO)_2PO_2^-$ (salt)	1285–1120	9.78–8.93	s	asym PO_2^- str
	1120–1050	8.93–9.52	s	sym PO_2^- str
$R(RO)PO_2^-$	1245–1150	8.03–8.70	s	asym PO_2^- str
	1110–1050	9.01–9.52	s	sym PO_2^- str
RPO_3^{2-}	1125–970	8.89–10.31	s	asym PO_3^{2-} str
	1000–960	10.00–10.42	w–m	sym PO_3^{2-} str
$ROPO_3^{2-}$	1140–1055	8.77–9.48	s	asym PO_3^{2-} str
	995–945	10.05–10.58	w–m	sym PO_3^{2-} str
R_2POS^-	1140–1050	8.77–9.52	s	P—O str
	570–545	17.54–18.35	m	P—S str
$(RO)_2POS^-$, $R(RO)POS^-$	1215–1110	8.24–9.01	s	P—O str
	660–575	15.15–17.39	m	P—S str
Inorganic salts, PO_2^-	1300–1150	7.69–8.70	s	
Inorganic salts, PO_3^{2-}	1030–970	9.71–10.31	s	asym str
	990–920	10.10–10.87	m	sym str
Inorganic salts, PO_4^{3-}	1100–1000	9.09–10.00	s	
PS_2^-	585–545	17.09–18.35		
$P^{III}{-}F$	770–760	12.99–13.10	m–s	
$P^{V}{-}F$	930–805	10.75–12.42	m–s	
$R_2P(O)F$	835–805	11.98–12.42	m–s	P—F str, see ref. 30
$RP(O)F_2$	930–895	10.75–11.17	m–s	P—F str, see ref. 30
P—Cl	605–435	16.53–22.99	m–s	See refs 15, 31
P—Br	485–400	20.62–25.00	m–s	
PCl_2	590–485	16.95–20.62	s	
	545–420	18.35–23.81	m–s	
P—Cl where P is bonded to O, C, or F atom	565–440	17.70–22.73	m–s	
P—Cl where P is bonded to N or S atom	540–435	18.52–22.99	m–s	
P—S—H	550–525	18.18–19.05	m	
	525–490	19.05–20.41	m	
P—S—C	1050–970	9.52–10.31		Observed for aliphatic compounds
	565–550	17.70–18.18	m	
	490–440	20.41–22.73	m	
P—O—P	1025–870	9.85–11.49	s	Usually br, asym str, often found at 945–925 cm⁻¹ (weak band also found near 700 cm⁻¹), see ref. 15
P—S—P	~500	~20.00	m	
	495–460	20.20–21.74	m	

Table 17.1 (*cont.*)

Functional Groups	Region cm⁻¹	Region μm	Intensity	Comments
P—O—S	930–815	10.75–12.27		asym P—O—S str
	765–700	13.07–14.29		
P—O—Si	1070–855	9.35–11.70		
P—C=C	1645–1595	6.08–6.27	m	C=C str
P—C—C=C	1660–1630	6.02–6.14	m	C=C str
P—C≡N	2220–2180	4.51–4.59	m	C≡N str

References

1. E. Steger, *Z. Chem.*, 1972, **12,** 52
2. R. A. Nyquist and W. J. Potts. In *Analytical Chemistry of Phosphorus Compounds* (M. Halmann, ed.). Interscience, New York, 1972, pp. 189–293
3. J. R. Ferraro, *Prog. Infrared Spectrosc.*, 1964, **2,** 127
4. D. E. C. Corbridge, *Topics Phosphorus Chem.*, 1970, **6,** 235
5. L. C. Thomas, *Interpretation of the Infrared Spectra of Organo-phosphorus Compounds.* Heyden, London, 1974
6. D. E. C. Corbridge, *J. App. Chem.*, 1956, **6,** 456
7. L. C. Thomas and R. A. Chittenden, *Chem. Ind.*, **1961,** 1913
8. R. A. Chittenden and L. C. Thomas, *Spectrochim. Acta*, 1965, **21,** 861
9. D. E. C. Corbridge and E. J. Lowe, *J. Chem. Soc.*, **1954,** 493 and 4555
10. L. C. Thomas and K. P. Clark, *Nature*, 1963, **198,** 855
11. R. Wolf *et al.*, *Bull. Soc. Chim. France*, **1963,** 825
12. U. Dietze, *J. Prakt. Chem.*, 1974, **316,** 293
13. L. C. Thomas and R. A. Chittenden, *Spectrochim. Acta*, 1964, **20,** 467
14. R. A. Nyquist and W. W. Muelder, *Spectrochim. Acta*, 1966, **22,** 1563
15. E. M. Popov *et al.*, *Adv. Chem. Moscow*, 1961, **30,** 362
16. R. A. Jones and A. R. Katritzky, *J. Chem. Soc.*, **1960,** 4376
17. A. N. Lazarev and V. S. Akselrod, *Optics Spectrosc.*, 1960, **9,** 170
18. N. A. Slovochotova *et al.*, *Bull. Acad. Sci. URSS Ser. Chim.*, **1961,** 62
19. N. P. Greckin and R. R. Sagidullin, *Bull. Acad. Sci. URSS*, **1960,** 2135
20. R. A. McIvor and C. E. Hubley, *Canad. J. Chem.*, 1959, **37,** 869
21. D. F. Peppard *et al.*, *J. Inorg. Nuclear Chem.*, 1961, **16,** 246
22. J. R. Ferraro and C. M. Andrejasich, *J. Inorg. Nuclear Chem.*, 1964, **26,** 377
23. T. Gramstad, *Spectrochim. Acta*, 1964, **20,** 729
24. U. Blindheim and T. Gramstad, *Spectrochim. Acta*, 1965, **21,** 1073
25. J. Rockett, *App. Spectrosc.*, 1962, **16,** 39
26. D. M. L. Goodgame, *J. Chem. Soc.*, **1961,** 2298 and 3735
27. R. A. Chittenden and L. C. Thomas, *Spectrochim. Acta*, 1964, **20,** 1679
28. S. Husebye, *Acta Chem. Scand.*, 1965, **19,** 774
29. R. A. Zingaro and R. M. Hedges, *J. Phys. Chem.*, 1961, **65,** 1132
30. R. J. Schmutzler, *J. Inorg. Nuclear Chem.*, 1963, **25,** 335
31. R. R. Holmes *et al.*, *Spectrochim. Acta*,)1973, **29A,** 665
32. J. R. Ferraro *et al.*, *Spectrochim. Acta*, 1963, **19,** 811
33. P. R. Mathis *et al.*, *Spectrochim. Acta*, 1974, **30A,** 357
34. L. A. Strait and M. K. Hrenoff, *Spectrosc. Letters*, 1975, **8,** 165
35. M. I. Kabachnik, *Austral. J. Chem.*, 1975, **28,** 755
36. O. A. Raevskii *et al.*, *Izv. Akad. Nauk SSSR Ser. Khim.*, 1978, **3,** 614

CHAPTER 18

Organic Silicon Compounds

Due to the mass and size of the silicon atom, the infrared spectra of organosilicon compounds,[1,2] to a first approximation, consist of essentially independent group vibrations. In general, similar absorption bands to those of the corresponding carbon compounds are observed except that they are at lower frequencies and are usually more intense than their carbon analogues (due to the difference in electronegativity between carbon and silicon).

Si—H Vibrations

For organic silanes, a strong absorption band due to the Si—H stretching vibration[3–11] is found at 2250–2100 cm^{-1} (4.44–4.76 μm). In general, the frequency of this band tends to increase with increase in the electronegativity of the substituents on the silicon atom. It has been observed that as the number of hydrogen atoms directly bonded to the silicon atom decreases so does the frequency of the Si—H stretching vibration. Alkyl substituents on the silicon atom also tend to lower this frequency whereas aryl substituents tend to raise it.

The band due to the deformation vibration of the Si—H group occurs in the range 985–800 cm^{-1} (10.15–12.50 μm). The $—SiH_3$ group has two bands due to deformation vibrations in the region 945–910 cm^{-1} (10.58–10.99 μm), whereas deformation vibrations of the $>SiH_2$ group give rise to only one strong band in the region 945–920 cm^{-1} (10.58–10.87 μm) and $\geqslant$SiH has no strong band in this region.

Methyl–Silicon Compounds, $Si—CH_3$

Methyl groups attached to silicon atoms[7] have a characteristic, very sharp band at 1280–1250 cm^{-1} (7.81–8.00 μm) due to the symmetric deformation vibration of the CH_3 group. Electropositive groups or atoms (e.g. metals) directly bonded to the silicon atom make the band due to the symmetric CH_3 deformation vibration tend to the higher end of this range whereas for silanes and siloxanes the band occurs near the lower end. When there are three methyl groups attached to a silicon atom, the band due to the symmetric deformation often splits into two components of unequal intensity. The asymmetric deformation vibration of the CH_3 group results in a weak band near 1410 cm^{-1} (7.09 μm).

The frequencies of the stretching vibrations of the methyl group are not affected much by being bonded to a silicon atom rather than to a carbon atom.

The bands due to the methyl rocking vibration and the Si—C stretching vibration occur in the region 860–760 cm^{-1} (11.63–13.10 μm).

Ethyl–Silicon Compounds

Ethyl-substituted silicon compounds have a characteristic band of medium intensity at 1250–1220 cm^{-1} (8.00–8.20 μm) and two other useful bands at 1020–1000 cm^{-1} (9.80–10.00 μm) and 970–945 cm^{-1} (10.31–10.58 μm).

Alkyl–Silicon Compounds

The band due to the $SiCH_2R$ deformation vibration, which occurs at 1250–1175 cm^{-1} (8.00–8.51 μm), tends to decrease in intensity as the length of the aliphatic chain increases, the frequency of the vibration decreasing also. Obviously, as the chain length increases, the smaller becomes the influence of the silicon atom on the terminal C—H vibrations. Hence, the bands near 2950 cm^{-1} (3.39 μm), 1470 cm^{-1} (6.80 μm), and 1390 cm^{-1} (7.19 μm) increase in intensity with increase in the paraffin chain length.

Aryl–Silicon Compounds

The bands due to the aromatic ring vibration, which are normally found in the region 1600–1450 cm^{-1} (6.25–6.90 μm), are displaced to lower wavenumbers by about 20 cm^{-1} for phenyl–silicon compounds.[12] One of these bands, which is sharp and of medium intensity, is almost always found at 1430 cm^{-1} (6.99 μm). The band is broadened or altogether absent when the ring is substituted by an additional group.

Phenyl–silicon compounds have a strong, characteristic band at about 1100 cm^{-1} (9.09 μm) which often splits into two when two phenyl groups are attached to the one silicon atom, but appears as a single band in the case of three phenyl groups. In addition, phenyl–silicon compounds have two weak bands, one near 1030 cm^{-1} (9.71 μm) and the other near 1000 cm^{-1} (10.00 μm).

The band pattern normally observed in the overtone region 2000–1660 cm^{-1} (5.00–6.02 μm) cannot be relied upon for the determination of the substitution pattern, although it is satisfactory for a large number of aryl silanes.

Si—O Vibrations[13,17,18]

The band due to the asymmetric Si—O—Si stretching vibration is normally in the region 1100–1000 cm^{-1} (9.09–10.00 μm) and, due to the greater ionic character of the Si—O group, this band is much more intense than the corresponding C—O band for ether. The band pattern may be used to distinguish between cyclic and linear polysiloxanes.[13]

Due to the influence of ring strain, cyclic siloxane trimers absorb at 1020–1010 cm^{-1} (9.80–9.90 μm), which is about 60 cm^{-1} less than other cyclic siloxanes, whereas tetramers (which have less ring strain) absorb at 1090–1070 cm^{-1} (9.17–9.35 μm) along with higher cyclic siloxanes. It is difficult to distinguish between other cyclic siloxanes and the region of absorption overlaps, in fact, that of linear polysiloxanes.

Linear small-chain siloxanes tend to absorb at about 1050 cm^{-1} (9.52 μm) and with increase in molecular weight this band gradually broadens to occupy the region 1100–1000 cm^{-1} (9.09–10.00 μm). For long-chain polymers, a broad, strong band with maxima at about 1085 cm^{-1} (9.21 μm) and 1025 cm^{-1} (9.76 μm) is observed.[13]

Silicon–Nitrogen Compounds

The band due to the asymmetric Si—N—Si stretching vibration occurs at about 900 cm^{-1} (11.11 μm) and is of strong intensity, whereas, due to the influence of ring strain, cyclic disilazanes absorb at about 870 cm^{-1} (11.49 μm), cyclic trimers absorb at about 920 cm^{-1} (10.87 μm), and cyclic tetramers at about 940 cm^{-1} (10.64 μm). This behaviour is similar to that observed for siloxanes. A band which has been assigned to the N—H deformation vibration occurs at about 1150 cm^{-1} (8.70 μm) for cyclic trisilazanes and at about 1180 cm^{-1} (8.48 μm) for cyclic tetrasilazanes.

As might be expected, primary silyl amines[14] have two weak bands in the region 3580–3380 cm^{-1} (2.79–2.96 μm) due to the asymmetric and symmetric N—H stretching vibrations. Secondary silyl amine compounds have only one weak band, at about 3400 cm^{-1} (2.94 μm). Primary silyl amines also have a medium-to-strong intensity band at about 1530 cm^{-1} (6.54 μm) and linear secondary silyl amines have a medium-to-strong band at about 1175 cm^{-1} (8.51 μm) due to the N—H deformation vibration. This band is found about 30–40 cm^{-1} lower for cyclic silyl amines where the nitrogen atom forms part of the ring than for linear secondary silyl amines.

Silicon–Halide Compounds

Chloro-, bromo, and iodosilanes, in the presence of moisture, hydrolyse to form siloxanes and hydrogen halide, so that care must be exercised in handling samples.

Characteristic silicon–chloride stretching vibration bands are observed in the far infrared region below 600 cm^{-1} (above 16.67 μm).

Silicon compounds with more than one chlorine atom exhibit two bands due to the asymmetric and symmetric vibrations. The asymmetric band, which of course occurs at higher frequencies than the symmetric case, is generally the more intense of the two.

Hydroxyl–Silicon Compounds

The band due to the O—H stretching vibration occurs in the same region as that for alcohols, phenols, etc. However, the band due to the O—H deformation vibration occurs at 870–820 cm^{-1} (11.49–12.19 μm) when the hydroxyl group is bonded to a silicon atom, whereas it is near 1050 cm^{-1} (9.52 μm) when bonded to a carbon atom.

Table 18.1. Organic silicon compounds

Functional Groups	Region		Inten-	Comments
	cm^{-1}	μm	sity	
Silanes				
Si—H	2250–2100	4.44–4.76	s	Si—H str, general range
	985–800	10.15–12.50	s	Si—H def, general range
$-SiH_3$	2160–2140	4.63–4.67	s	Si—H str
	945–910	10.58–10.99	s	Two bands, asym and sym def
	680–540	14.71–18.52	s	Rocking vib
$>SiH_2$	2150–2115	4.65–4.73	s	Si—H str
	945–920	10.58–10.87	s	Si—H def
	895–840	11.17–11.90	s	Wagging vib
	745–625	13.42–16.00	w–m	Twisting vib
	600–460	16.67–21.74	m	sh, rocking vib
$\geqslant SiH$	2135–2095	4.68–4.77	s	Si—H str
	845–800	11.83–12.50	s	Wagging vib
Si—D	1690–1570	5.92–6.37	s	Si—D str
	710–665	14.08–15.04	s	Si—D def
R_4Si	1280–1240	7.81–8.07	s	sym Si—C bending vib
	850–800	11.76–12.50	s	Si—C rocking vib
	760–750	13.10–13.33	s	Si—C rocking vib
$Si(CH_3)_n$,	1280–1250	7.81–8.00	s	sh, sym CH_3 def
n = 1, 2, 3, or 4	860–760	11.63–13.61	s	Si—CH_3 rocking vib
$\geqslant SiCH_3$	~765	~13.07	s	Si—C rocking vib, see ref. 5
$> Si(CH_3)_2$	~855	~11.70	vs	Si—C rocking vib
	815–800	12.27–12.50	vs	Si—C rocking vib
$-Si(CH_3)_3$	~840	~11.90	vs	Si—C rocking vib
	~765	~13.07	vs	Si—C rocking vib
	330–240	30.30–41.67	w	$Si(CH_3)_3$ rocking vib
Si—C_2H_5	1250–1220	8.00–8.20	m	CH_2 wagging
	1020–1000	9.80–10.00	m	
	970–945	10.31–10.58	m	
Si—CH_2R	1250–1175	8.00–8.51	w–m	Long-chain aliphatics absorb at low-frequency end of range
	760–670	13.10–14.93	m	CH_2 rocking vib
Si—CH=CH_2	~1925	~5.20	w	Overtone
	1610–1590	6.21–6.29	m	C=C str, see ref 16
	1410–1390	7.09–7.19	s	CH_2 in-plane def
	1010–1000	9.90–10.00	m	*Trans* CH wagging vib
	980–940	10.20–10.64	s	CH_2 wagging vib
	580–515	17.24–19.42	w	Hydrogen out-of-plane def
Si—Ph	3080–3030	3.25–3.30	m	C—H str
	~1600	~6.25	m	Ring vib, usually stronger than band near 1430 cm^{-1}
	1480–1425	6.99–7.02	m	sh, ring vib
	1125–1090	8.89–9.17	vs	X-sensitive band
	~730	~13.70	s	Out-of-plane C—H vib
	700–690	14.29–14.49	s	Out-of-plane C—H vib
R_3SiPh	670–625	14.93–16.00	w	Ring in-plane bending vib
	490–445	20.41–22.47	s	Si—C—C out-of-plane bending vib
	405–345	24.69–28.99	w	Si—C str and ring in-plane def
	~290	~34.48	v	Si—Ph in-plane def
R_2SiPh_2	635–605	15.75–16.53	w	Ring in-plane bending
	495–470	20.20–21.28	s	S—C—C out-of-plane bending vib

Table 18.1 (*cont.*)

Functional Groups	Region		Inten-	Comments
	cm^{-1}	μm	sity	
	445–400	22.47–25.00	w	asym Si—C str
	380–305	26.32–32.79	w	sym Si—C str
$RSiPh_3$	625–605	16.00–16.53	w	Ring in-plane bending
	515–485	19.42–20.62	s	Si—C—C out-of-plane bending vib
	445–420	22.47–23.81	w	asym Si—Ph str
	~330	~30.30	v	sym Si—Ph str
Cyclopentamethylene dialkylsilanes	495–480	20.20–20.83	m–s	Probably due to heterocyclic ring, but cyclopentamethylene silane and the diphenyl derivative do not exhibit this band
Silanols				
Si—OH	3700–3200	2.70–3.13	m	May be br, O—H str
	955–835	10.47–11.98	s	Si—O str, for condensed-phase samples a br, m–w band occurs near 1030 cm^{-1} due to SiOH def
Silyl esters and ethers				
$RCOSiR_3$	~1620	~6.17	s	C=O str
Si—O—R	1110–1000	9.01–10.00	vs	asym Si—O—C str, at least one band, Si—O—Si also absorbs in this region
Si—O—CH_3	~2860	~3.50	m	sym CH_3 str
	~1190	~8.40	s	CH_3 rocking vib
	~1100	~9.09	vs	asym Si—O—C str
	850–800	11.76–12.50	s	sym Si—O—C str
$> Si(OCH_3)_2$	390–360	25.64–27.78	s	asym Si—O—C def
$-Si(OCH_3)_3$	480–440	20.83–22.73	s	asym Si—O—C def
	470 330	21.28–30.30	w	
Si—O—CH_2—	1190–1140	8.40–8.77	s	
	1100–1070	9.09–9.35	vs	asym Si—O—C str, usually a doublet
	990–945	10.10–10.64	s	sym Si—O—C str
$> Si(OC_2H_5)_2$	475–405	21.05–24.69	w	asym Si—O—C def
$-Si(OC_2H_5)_3$	500–440	20.00–22.73	s	
Si—O—Ar	1135–1090	8.81–9.17	vs	Several sh bands, probably Si—O—C str
	970–920	10.31–10.87	s	
Si—O—Si and Si—O—C	1090–1020	9.17–9.80	vs	Si—O str, two bands of almost equal intensity, siloxane chains absorb near 1085 and 1020 cm^{-1} increasing in intensity with increase in chain length, cyclic siloxanes have only one strong band although a second band is somtimes observed for tetramers and larger rings

Table 18.1 (*cont.*)

Functional Groups	Region cm⁻¹	Region μm	Intensity	Comments
Disiloxanes				
Si—O—Si	625–480	16.00–20.83	w	br, sym Si—O—Si str, band occurs at lower frequencies for substituted disiloxanes and linear polymeric siloxanes
Siloxanes				
—OSiCH$_3$ (end group)	850–840	11.76–11.90	s	Si—C str
—OSiC$_2$H$_5$ (linear polymer)	810–800	12.35–12.50	s	Si—C str
—OSiCH$_3$ (cyclic compounds)	820–780	12.20–12.82	s	Si—C str
Silyl amines				
Si—NH$_2$	3570–3475	2.80–2.88	m	NH$_2$ str
	3410–3390	2.93–2.95	m	NH$_2$ str
	1550–1530	6.45–6.54	m	NH$_2$ def
Si—NH—Si	~3400	~2.94	m	NH str
	~1175	~8.51	m	
	~935	~10.70	m	
Aminosilanes,	880–835	11.36–11.98	s	asym N—Si—N str
$H_2N-\overset{|}{\underset{|}{Si}}-NH_2$	800–785	12.50–12.74	m	sym N—Si—N str
Silicon halides				
⩾SiF	920–820	10.87–12.22	m	Si—F str (general ranges: Si—F str, 1000–800 cm⁻¹; Si—F def, 425–265 cm⁻¹)
>SiF$_2$	945–915	10.58–10.93	s	Probably asym str
	910–870	10.99–11.49	m	Probably sym str
—SiF$_3$	980–945	10.20–10.58	s	Probably asym str
	910–860	10.99–11.63	m	Probably sym str
⩾SiCl	550–470	18.18–21.28	s	Si—Cl str (Si—Cl def, 250 150 cm⁻¹—general range)
>SiCl$_2$	595–535	16.81 18.69	s	asym Si—Cl str
	540–460	18.52–21.74	m	sym Si—Cl str
—SiCl$_3$	625–570	16.00–17.54	s	asym Si—Cl str
	535 450	18.69–22.22	m	sym Si—Cl str
⩾SiBr	430–360	23.26–27.78	w	Si—Br str
>SiBr$_2$	460–425	21.74–23.53	w	asym Si—Br str
	395–330	25.32–30.30	w	sym Si—Br str
—SiBr$_3$	480–450	20.83–22.22	w	asym Si—Br str
	360 300	27.78–33.33	w	sym Si—Br str
⩾SiI	365–280	27.40–35.71	w	sym Si—I str
>SiI$_2$	390 330	25.64–30.30	w	Si—I str
	325–275	30.77–36.36	w	
—SiI$_3$	410–365	24.39–27.40	w	Si—I str
	280–220	35.71–45.45	w	
Other groups				
Si—Ph	1125–1090	8.89–9.17	s	See ref. 15
Ge—Ph	~1080	~9.26	s	See ref. 15
Sn—Ph	1080–1050	9.26–9.52	s	Usually 1065 cm⁻¹, see ref. 15
Pb—Ph	~1050	~9.52	s	See ref. 15

Table 18.1 (*cont.*)

Functional Groups	Region cm⁻¹	Region μm	Intensity	Comments
Organogermanium, Ge—O—Ge	900–700	11.11–14.29	s	asym Ge—O—Ge str; cyclic trimers ~850 cm⁻¹, cyclic tetramers ~860 cm⁻¹, linear polymers ~870 cm⁻¹
Organotin, Sn—O	780–580	12.82–17.24	s	br
Organolead, Pb—O	~625	~16.00		

References

1. K. Licht and P. Reich, *Literature Data for Infrared, Raman and N.M.R. Spectra of Silicon, Germanium, Tin and Lead Organic Compounds*. Dent. Verlag. Wiss., Berlin, 1971
2. A. L. Smith, *Spectrochim. Acta*, 1960, **16,** 87
3. G. J. Janz and Y. Mikawa, *Bull. Chem. Soc. Japan*, 1961, **34,** 1495
4. A. L. Smith and J. A. McHard, *Anal. Chem.*, 1959, **31,** 1174
5. H. G. Kuivila and P. L. Maxfield, *J. Organomet. Chem.*, 1967, **10,** 41
6. G. Kessler and H. Kriegsmann, *Z. Anorg. Allgem. Chem.*, 1966, **342,** 53
7. I. F. Kovalev, *Optic Spectrosc.*, 1960, **8,** 166
8. S. D. Gokhale and W. L. Jolly, *Inorg. Chem.*, 1964, **3,** 946
9. E. A. Groschwitz *et al.*, *J. Organomet. Chem.*, 1967, **9,** 421
10. A. L. Smith, *Spectrochim. Acta*, 1963, **19,** 849
11. R. N. Kinseley *et al.*, *Spectrochim. Acta*, 1959, **15,** 651
12. M. C. Harvey and W. H. Nebergall, *App. Spectrosc.*, 1962, **16,** 12
13. W. Noll, *Angew. Chem. Int. Ed.*, 1963, **2,** 73
14. W. Fink, *Helv. Chim. Acta*, 1962, **45,** 1081
15. F. J. Bajer and H. W. Post, *J. Org. Chem.*, 1962, **27,** 1422
16. V. F. Mironov and N. A. Chumaevskii, *Doklady Akad. Nauk SSSR*, 1962, **146,** 1117
17. P. Tarte *et al.*, *Spectrochim. Acta*, 1973, **29A,** 1017
18. J. Chiosnct *et al.*, *Spectrochim. Acta*, 1975, **31A,** 1023

CHAPTER 19
Boron Compounds

Boron compounds generally have intense bands, as, for example, those due to the B—H,[1,2] B—halogen,[3–8] B—O, and B—N[9,10] groups. The position and intensity of certain bands give information not only on the boron-containing group itself but frequently also on its environment. The bands due to certain boron-containing groups often appear as doublets, this being due to the presence of two naturally-abundant isotopes of boron.

Bands due to the B—H stretching vibrations[1,2] occur at 2640–2450 cm^{-1} (3.79–4.08 μm) for the groups BH and BH_2 in which the hydrogen atom is free. No isotope band-splitting is observed for compounds containing a single, free (exo) B—H group whereas it does occur for free (exo) BH_2 groups (in gas-phase spectra).

The B—H stretching vibration of samples enriched in ^{10}B is at slightly higher frequencies than for samples with the naturally-occurring ratio of boron isotopes.

In some cases, the band due to the B—H stretching vibration of borane–amine complexes exhibits isotope-splitting. For alkyl boranes, the band tends to lower frequencies with increasing substitution.

Compounds with the B···H···B bridge have a series of weak-to-medium intensity bands in the region 2140–1710 cm^{-1} (4.67–5.85 μm) and a strong band at 1610–1540 cm^{-1} (6.21–6.49 μm). The band due to the B—H stretching vibration of compounds for which the boron atom has a complete octet of electrons occurs in the range 2400–2200 cm^{-1} (4.17–4.55 μm).

The asymmetric and symmetric methyl deformation vibrations of B—CH_3 occur at 1460–1405 cm^{-1} (6.85–7.12 μm) and 1330–1280 cm^{-1} (7.52–7.81 μm) respectively.

Compounds with the B—aryl group have a strong, sharp band, due to the ring vibration, at 1440–1430 cm^{-1} (6.94–6.99 μm). Compounds with a B—phenyl group have a strong band at about 760 cm^{-1} (13.16 μm) due to the ring CH wagging motion.

A review of the infrared spectra of inorganic boron compounds has been published.[17]

Table 19.1. Boron compounds

Functional Groups	Region cm⁻¹	Region μm	Intensity	Comments
B—H ((free) hydrogens)	2565–2480	3.90–4.03	m–s	B—H str
	1180–1110	8.48–9.01	s	B—H in-plane def
	920–900	10.87–11.11	m–w	Out-of-plane bending
Alkyl diboranes $B—H_2$ ((free) hydrogens)	2640–2570	3.79–3.89	m–s	sym $B—H_2$ str
	2535–2485	3.95–4.02	m–s	asym $B—H_2$ str
	1205–1140	8.30–8.77	m–s	Sometimes br, $B—H_2$ def
	975–920	10.26–10.87	m	$B—H_2$ wagging
Alkyl diboranes B···H···B (bridged hydrogen)	2140–2080	4.67–4.81	w–m	sym in-phase motion of H atom
	1990–1850	5.03–5.41	w	sym out-of-phase motion of H atom, several bands
	1800–1710	5.56–5.85	w–m	asym out-of-phase motion of H atom
	1610–1540	6.21–6.49	vs	asym in-phase motion of H atom
Borazines, borazoles	3500–3400	2.86–2.94	m	N—H str, see refs 11–14
	2580–2450	3.88–4.08	m	B—H str
	1465–1330	6.83–7.52	s	B—N str
	700–680	14.29–14.71	m	B—N out-of-plane def, doublet
Boron hydride salts and amine–borane complexes (with boron octet complete)	2400–2200	4.17–4.55	m	B—H str
Borane $B—H_3$ (in complexes)	2380–2315	4.20–4.32	s	asym B—H str
	2285–2265	4.38–4.42	s	sym B—H str
	~1165	~8.58	s	$B—H_3$ def
BH_4^- ion	2310–2195	4.33–4.56	s	B—H str, two bands (one due to Fermi resonance)
B—OH, boric acid, boronic acids (solid phase)	3300–3200	3.03–3.13	m	br, O—H str
B—OH, aryl boronic acids	~1000	~10.00	m	br } not present in anhydrides
	800–700	12.50–14.29	m	br } not present in anhydrides
1,1-Dialkyl diboranes and trialkyl boranes	1185–1100	8.44–9.09	s	asym C—B—C str (isotope splitting large, ~10–30 cm⁻¹)
	845–770	11.83–12.98	m	sym C—B—C str
$B—CH_3$	1460–1405	6.85–7.12	m	asym CH_3 def
	1330–1280	7.52–7.81	m	sym CH_3 def
B—R	1270–620	7.87–16.13	v	B—C str, isotopic splitting sometimes observed. For BR_3 compounds, one strong band due to asym C—B—C str and one weak band (sometimes absent) due to sym C—B—C str
Monomethyl boranes	1010–835	9.90–11.98	m	B—C str (isotopic splitting observed)
Di- and trimethyl boranes	1240–1140	8.07–8.77	vs	asym C—B—C str
	720–675	12.20–14.18	m–w	sym C—B—C str, infrared-inactive for symmetrical compounds
Alkyl boranes (other than methyl)	1135–1110	8.81–9.01	m	asym C—B—C str (isotopic splitting of ~20 cm⁻¹)
	675–620	14.82–16.22	w	sym C—B—C str, often absent
B—Ar	1440–1430	6.94–6.99	m–s	sh, ring vib
	1280–1250	7.81–8.00	s	X-sensitive band

Table 19.1 (*cont.*)

Functional Groups	Region cm⁻¹	Region μm	Intensity	Comments
B—Ar (cont.)	~760	~13.16	s	Ring C—H out-of-plane def, for phenyl compounds only, doublet if more than one phenyl group on boron atom (~20 cm⁻¹ separation)
bis-(Alkyl amino) phenyl boron compounds, $PhB\begin{smallmatrix}\diagup NHR\\ \diagdown NHR\end{smallmatrix}$	1450–1440	6.90–6.94	s	B—C str
bis-Phenyl boron compounds, $Ph_2B—$	1260–1250	7.94–8.00	s	B—C str
Aryl boron dihalides, $ArBX_2$ (X = halide)	1270–1215	7.87–8.23	s	B—C str, isotopic splitting present
B—O, borates, boronates, boronites, boronic anhydrides, boronic acids, borinic acids	1380–1310	7.25–7.63	s	B—O str, weak band when boron octet complete, e.g. compounds with a nitrogen coordinated to the boron
Trialkyl borates, $B(RO)_3$	1350–1310	7.41–7.63	vs	br, also have strong band at 1070–1040 cm⁻¹ probably due to C—O str
Dialkyl phenyl boronates, $(RO)_2BPh$	1435–1425	6.97–7.02	m–s	B—C str
	1330–1310	7.52–7.63	s	asym C—O—B—O—C str
	1180–1120	8.48–8.93	s	sym C—O—B—O—C str
	675–600	14.81–16.67	m–s	B—O def, isotopic splitting present
Boronites, R_2BOR	1350–1310	7.41–7.63	s	B—O str
B-alkoxyl borazoles, 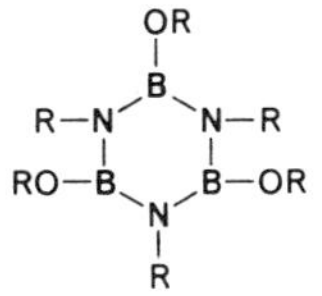	1500–1435	6.67–6.97	s	B—N str
	1330–1310	7.52–7.63	m–s	B—O str
Alkyl and aryl metaborates, 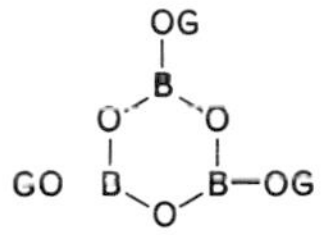	1380–1335	7.25–7.49	s	B—O str
	1225–1080	8.16–9.26	s	C—O str, higher frequencies for aryl compounds, lower for *n*-alkyl compounds
Haloboroxines, 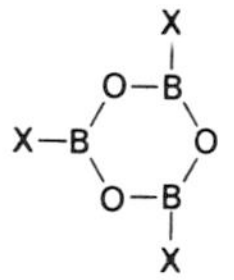 (X = halogen)	1470–1180	6.80–8.48	s	B—O str, isotopic splitting present
Fluoroboroxine	~970	~10.31	m	asym B—F str (sym B—F str infrared-inactive)
Chloroboroxine	~760	~13.16	s	asym B—Cl str, isotopic splitting present, see ref. 16

Table 19.1 (*cont.*)

Functional Groups	Region cm^{-1}	Region μm	Intensity	Comments
Aryl boronic acid esters, 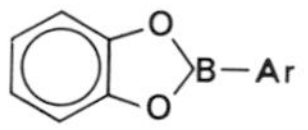	1360–1330	7.35–7.52	s	asym B—O str, isotopic splitting present
	1240–1235	8.07–8.10	s	asym C—O str
	1075–1065	9.30–9.39	s	sym B—O str, isotopic splitting present
	1030–1020	9.71–9.80	m	sym C—O str
Boronic acid anhydrides 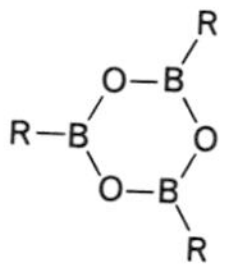	1390–1355	7.19–7.38	s	B—O str
	1255–1145	7.97–8.74	m	B—C str, isotopic splitting present
Metallic orthoborates $M_x(BO_3)_y$	1280–1200	7.82–8.34	s	br, asym B—O str, isotopic splitting present
	~900	~11.11	w	sym B—O str, often absent, see ref. 15
BX_3 (X = F, Cl) (complexes of acids, esters, and ethers)	725–610	13.79–16.39	s	br, B···O str, isotopic splitting present
Covalent boron–nitrogen compounds	1550–1330	6.45–7.52	s	B—N str (general range), isotopic splitting present
Amine–borane complexes	780–680	12.82–14.71	m–s	B···N str (general range), isotopic splitting present, see ref. 18
N-Alkyl B-halo borazoles, 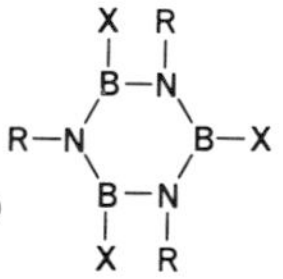 (X = halogen)	1510–1400	6.62–7.14	s	B—N str
	720–635	13.89–15.75		B—N def
N-Alkyl B-chloro borazoles	1090–960	9.17–10.42	s	B—Cl str
N-Alkyl B-bromo borazoles	1075–950	9.30–10.53	s	B—Br str
N-Alkyl amino borazoles, 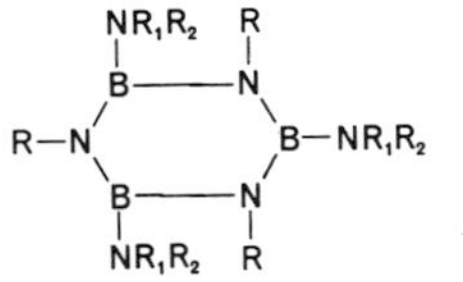	1520–1490	6.58–6.71	s	B—N str, see refs 12, 14
N-Methyl B-aryl borazoles, 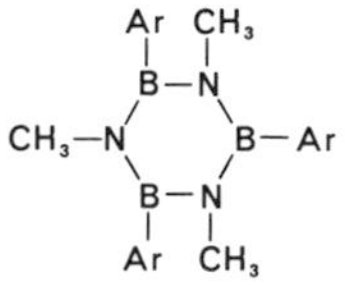	1470–1440	6.80–6.94	s	B—N str, isotopic splitting—^{10}B shoulder present, see ref. 13
	750–720	13.33–13.89		B—N def

Table 19.1 (*cont.*)

Functional Groups	Region cm^{-1}	Region μm	Intensity	Comments
N-Alkyl B-aryl borazoles	1430–1410	6.99–7.09	s	B—N str, isotopic splitting—^{10}B shoulder present, see ref. 13
	750–720	13.33–13.89		B—N def
N-Aryl B-methyl borazoles	1400–1375	7.14–7.27	s	B—N str, isotopic splitting—^{10}B shoulder present, see ref. 13
Alkyl borazenes, $(CH_3)_2B—NR_1R_2$	1550–1330	6.45–7.52	s	B—N str
bis-Dimethylamino boranes, $—B[N(CH_3)_2]_2$	1550–1500	6.45–6.67	s	asym B—N str, isotopic splitting present
	1415–1375	7.07–7.32	s	sym B—N str, isotopic splitting present
Boron fluorine compounds	1500–840	6.67–11.90	v	B—F str (general range), usually strong, isotopic splitting present
Boron difluorides, XBF_2 (in boron trihalides)	1500–1410	6.67–7.09	s	asym B—F str
	1300–1200	7.69–8.33	s	sym B—F str (for BF_3 this vib is infrared-inactive (Raman ~885 cm^{-1})), see ref. 3
Boron monofluorides, X_2BF (in boron trihalides)	1360–1300	7.35–7.69	s	B—F str, see ref. 3
BF_3 complexes	1260–1125	7.94–8.89	s	asym B—F str; Isotopic splitting present, band may be split further
	1030–800	9.71–12.50	s	sym B—F str, see refs 5, 6, 8, 19; Isotopic splitting present, band may be split further
Tetrafluoroborate ion, BF_4^-	~1030	~9.71	vs	asym B—F str, shoulder at ~1060 cm^{-1} (sym B—F str infrared-inactive), see ref. 5
Chlorotrifluoroborate ion, $ClBF_3^-$	1080–1025	9.26–9.76	s	asym B—F str, doublet
	890–840	11.24–11.90	w	sym B—F str, doublet
Boron chlorine compounds	1090–290	9.17–34.48	v	B—Cl str (general range), isotopic splitting present, higher frequency end of range for trichloroborazoles, lower end for BCl_3 complexes
Boron dichlorides (in boron trihalides)	1030–950	9.71–10.53	s	asym B—Cl str, isotopic splitting present (for BCl_3, band at ~955 cm^{-1})
	920–470	10.87–21.28	s	sym B—Cl str, isotopic splitting present (vib infrared-inactive for BCl_3)
Boron monochlorides (in boron trihalides)	955–690	10.47–14.49	s	B—Cl str
Alkyl aryl chloroboronites	1220–1195	8.20–8.36	s	Probably asym C—B—C str
	910–890	10.99–11.24	s	B—Cl str
BCl_3 in complexes	785–660	12.74–15.15	s	asym B—Cl str, isotopic splitting present, see ref. 8
	540–290	18.52–34.48	s	sym B—Cl str
	290–200	34.48–50.00	m–s	B—Cl def, several bands
Tetrachloroborate ion, BCl_4^-	760–645	13.16–15.50	s	br, asym B—Cl str, several peaks, (sym B—Cl str vib infrared-inactive)

Table 19.1 (*cont.*)

Functional Groups	Region cm⁻¹	μm	Intensity	Comments
Aryl boron dichlorides	970–915	10.31–10.93	s	asym BCl_2 str, isotopic splitting present
	645–630	15.50–15.87	s	sym BCl_2 str
	585–550	17.09–18.18	w	BCl_2 out-of-plane def
	~340	~29.41	s	BCl_2 rocking vib
	~230	~43.48	s	BCl_2 scissoring vib
	~130	~76.92	s	BCl_2 torsional vib
Boron–bromine compounds	1080–240	9.26–41.67	v	B—Br str (general range), isotopic splitting often present, higher frequency end of range for bromoboronazoles, lower end for BBr_3 complexes
Boron dibromides (in boron trihalides)	910–820	10.99–12.20	s	asym B—Br str, (BBr_3 band at ~820 cm⁻¹ with shoulder at ~855 cm⁻¹ due to isotopic splitting)
	420–275	23.81–36.36	s	sym B—Br_3 str, (infrared-inactive for BBr_3 (Raman at ~280 cm⁻¹)
BBr_3 in amine complexes	~700	~14.29	s	asym B—Br str, isotopic splitting sometimes present, see ref. 8
	~250	~40.00	s	sym B—Br str, isotopic splitting sometimes present
	~200	~50.00	s	asym B—Br def
	~175	~57.14	s	
	~125	~80.00	vs	B—Br rocking vib
Tetrabromoborate ion, BBr_4^-	~600	~16.67	s	asym B—Br str, isotopic splitting present
	~240	~41.67	w	sym B—Br str
	~165	~60.61	m	B—Br def
Aryl boron dibromides	890–865	11.24–11.56	s	asym BBr_2 str, isotopic splitting present
	~620	~16.13	s	sym BBr_2 str
	~525	~19.05		Out-of-plane BBr_2 vib
	~270	~37.04	s	BBr_2 rocking vib
	~160	~62.50	m	BBr_2 scissoring vib
Thio orthoborate esters (symmetrical), 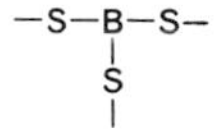	955–905	10.47–11.05	s	asym B—S str, several peaks due to isotopic splitting (sym B—S str infrared-inactive)

References

1. W. Gerrard, *Organic Boron Compounds*. Academic Press, New York, 1961, p. 223
2. W. J. Lenhmann and I. Shapiro, *Spectrochim. Acta*, 1961, **17,** 396
3. L. P. Lindemann and M. R. Wilson, *J. Chem. Phys.*, 1956, **24,** 242
4. R. L. Amster and R. C. Taylor, *Spectro. Acta*, 1964, **20,** 1487
5. T. C. Waddington and F. Klanberg, *J. Chem. Soc.*, **1960,** 2339
6. M. Taillander *et al.*, *J. Mol. Struct.*, 1968, **2,** 437
7. W. Kynaston *et al.*, *J. Chem. Soc.*, **1960,** 1772
8. P. G. Davies *et al.*, *Inorg. Nuclear Chem. Letters*, 1967, **3,** 249
9. A. Meller, *Organometall. Chem. Rev.*, 1967, **2,** 1
10. W. Sawodny and J. Goubeau, *Zeit. Phys. Chem. N. F.*, 1965, **44,** 227
11. J. M. Butcher *et al.*, *Spectrochim. Acta*, 1962, **18,** 1487
12. D. W. Aubrey *et al.*, *J. Chem. Soc.*, **1961,** 1931
13. J. E. Burch *et al.*, *Spectrochim. Acta*, 1963, **19,** 889
14. W. Gerrard *et al.*, *Spectrochim. Acta*, 1962, **18,** 149
15. A. Mitchell, *Trans. Faraday Soc.*, 1968, **62,** 530
16. B. Latimer and J. R. Devlin, *Spectrochim. Acta*, 1965, **21,** 1437
17. H. J. Becher and F. Thevenot, *Zeit. Anorg. Allg. Chem.*, 1974, **410,** 274
18. M. T. Forel *et al.*, *Colloq. Int. Cent. Rech. Sci.*, 1970, **191,** 167

CHAPTER 20

The Near Infrared Region

The near infrared region, 14 000–4000 cm^{-1} (0.7–2.5 μm) is more akin to the ultraviolet and visible regions than the normal infrared region and hence longer path-length cells are employed. This means that the cells are easy to clean and more robust. Also, being made of glass with quartz windows, or of silica, they are not attacked by water.

One of the most useful solvents employed is carbon tetrachloride, as it has no absorptions in this region whatsoever.

In general, bands in the near infrared region are due to the overtones or combinations of fundamental bands occurring in the region 3500–1600 cm^{-1} (2.8–6.2 μm). Therefore, qualitatively, this spectral region is not as characteristic as the 'fingerprint' region. Although a fair amount of investigation still needs to be done in this region, it is obvious that the straightforward compilation of spectra will not, in the main, yield the type of qualitative information to which we are accustomed in the normal infrared range. Since intensity measurements are reliable and relatively easy to make, both band position and accurate values of intensity are usually quoted in the literature.

The near infrared region has been found extremely useful in the assignment of particular groups containing hydrogen.

Since the bands in this region are combination or overtone bands, the path-length of the sample must be increased in order to examine successfully the higher frequency part of the range.

Carbon–Hydrogen Groups

Strong combination bands associated with C—H groups occur in the region 5000–4000 cm^{-1} (2.00–2.50 μm) and first and second overtones of the C—H stretching vibration are observed at 6250–5550 cm^{-1} (1.60–1.80 μm) and 9090–8200 cm^{-1} (1.10–1.22 μm) respectively. Methyl groups absorb in the region 8375–8360 cm^{-1} (~1.195 μm), methylene groups at 8255–8220 cm^{-1} (1.211–1.217 μm), and compounds containing aromatic C—H bonds at 8740–8670 cm^{-1} (1.144–1.153 μm). Aldehydes have a characteristic band in the region 4760–4520 cm^{-1} (2.10–2.21 μm) which probably arises from a combination of the C=O and C—H stretching vibrations. Aromatic aldehydes have characteristic

Chart 6* The absorption ranges and corresponding intensities in terms of average molar absorptivity in the near infrared region.

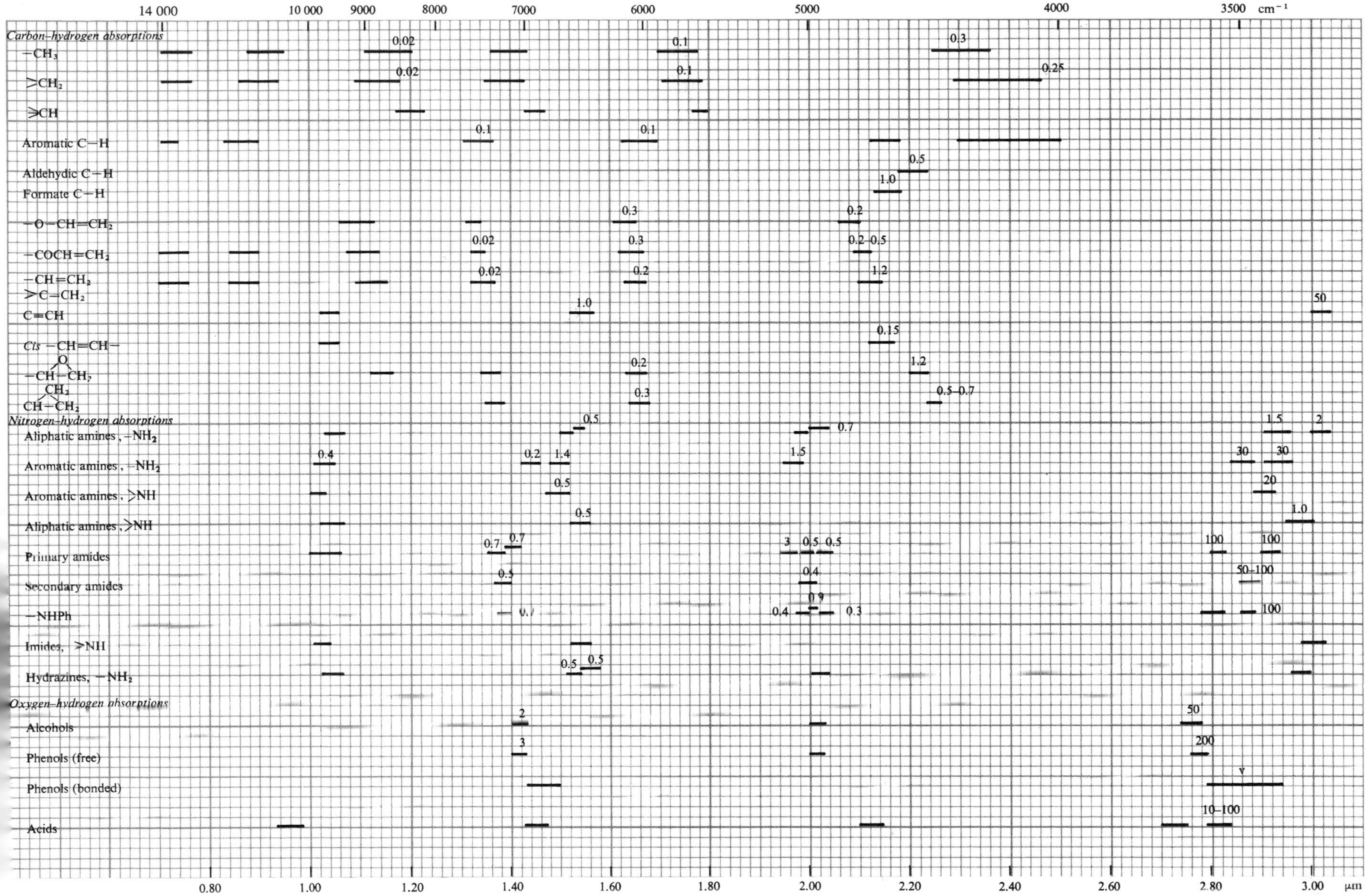
14 000
10 000
9000
8000
7000
6000
5000
4000
3500 cm⁻¹
Carbon–hydrogen absorptions
–CH₃
>CH₂
>CH
Aromatic C–H
Aldehydic C–H
Formate C–H
–O–CH=CH₂
–COCH=CH₂
–CH=CH₂
>C=CH₂
C≡CH
Cis –CH=CH–
Nitrogen–hydrogen absorptions
Aliphatic amines, –NH₂
Aromatic amines, –NH₂
Aromatic amines, >NH
Aliphatic amines, >NH
Primary amides
Secondary amides
–NHPh
Imides, >NH
Hydrazines, –NH₂
Oxygen–hydrogen absorptions
Alcohols
Phenols (free)
Phenols (bonded)
Acids
0.80
1.00
1.20
1.40
1.60
1.80
2.00
2.20
2.40
2.60
2.80
3.00 μm

* Based with some additions, on that of R. F. Goddu and D. A. Delker, *Anal. Chem.*, **32**, 140.

Chart 6 (*continued*)

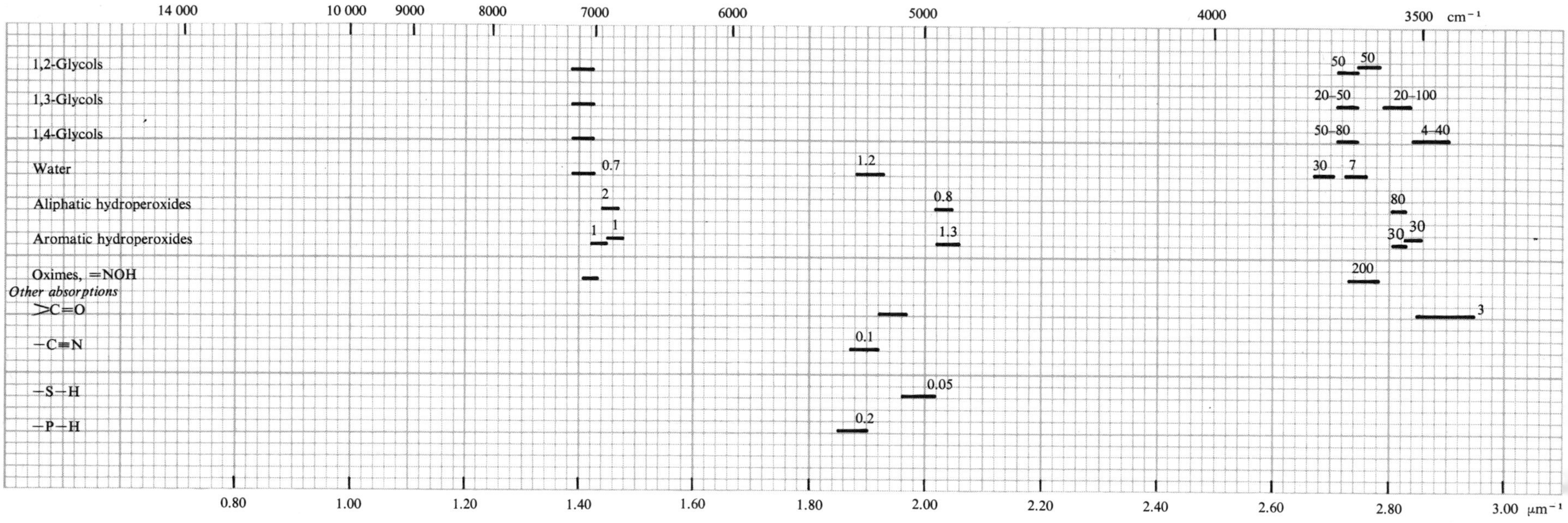
14 000
10 000
9000
8000
7000
6000
5000
4000
3500 cm⁻¹
1,2-Glycols
1,3-Glycols
1,4-Glycols
Water
Aliphatic hydroperoxides
Aromatic hydroperoxides
Oximes, =NOH
Other absorptions
>C=O
−C≡N
−S−H
−P−H
50
50
20–50
20–100
50–80
4–40
0.7
1.2
30
7
2
0.8
80
1
1
1.3
30
30
200
3
0.1
0.05
0.2
0.80
1.00
1.20
1.40
1.60
1.80
2.00
2.20
2.40
2.60
2.80
3.00 μm⁻¹

bands near 4525 cm^{-1} (2.21 μm), 4445 cm^{-1} (2.25 μm), and 8000 cm^{-1} (1.25 μm).

Terminal epoxide groups have absorption bands near 6060 cm^{-1} (1.65 μm) and 4550 cm^{-1} (2.20 μm), these positions being similar to those of terminal methylene groups discussed below. However, the epoxide bands are not so complicated and are much more intense. Cyclopropanes also have similar absorptions at 6160–6060 cm^{-1} (1.62–1.65 μm) and 4500–4400 cm^{-1} (2.22–2.27 μm).

Terminal methylene groups, $>C{=}CH_2$, absorb near 6200 cm^{-1} (1.613 μm) and near 4750 cm^{-1} (2.11 μm). The terminal methylene groups of vinyl ethers, $—O—CH{=}CH_2$, absorb near 6190 cm^{-1} (1.616 μm) and those of α,β-unsaturated ketones, $—CO—CH{=}CH_2$, near 6175 cm^{-1} (1.619 μm) whereas for unsaturated hydrocarbons this absorption occurs near 6135 cm^{-1} (1.630 μm).

cis-Alkenes, —CH=CH—, have at least three bands in the near infrared region, one of which is near 4650 cm^{-1} (2.15 μm), whereas the *trans*- isomers have no strong absorptions in the near infrared region.

For terminal methyne groups, —C≡CH, the band due to the C—H stretching vibration, as discussed previously, occurs near 3330 cm^{-1} (3.00 μm), the overtone of this band being found near 6535 cm^{-1} (1.53 μm). Both of these bands are sharp and may easily be distinguished from the absorptions of amino groups, which also occur at these positions, since the C—H overtone band has almost twice the molar absorptivity of the N—H absorption.

Oxygen–Hydrogen Groups

In dilute carbon tetrachloride solution, primary alcohols absorb near 3635 cm^{-1} (2.751 μm), secondary alcohols near 3625 cm^{-1} (2.759 μm), and tertiary alcohols near 3615 cm^{-1} (2.766 μm). Aryl and unsaturated alcohols, in which the hydroxyl group may interact with the π-electrons of the system, normally have their maximum intensity absorptions near 3615 cm^{-1} (2.766 μm) with a shoulder near 3635 cm^{-1} (2.751 μm), in dilute solution spectra. The greater the interaction, the smaller the intensity of the shoulder.

Carboxylic acids, depending on their degree of association, have several bands in the region 3700–3330 cm^{-1} (2.70–3.00 μm). Even in dilute solutions, carboxylic acids exist in a high proportion as dimers. However, resulting from the fundamental stretching vibration of the OH groups of monomers, combination and overtone bands are observed near 3570 cm^{-1} (2.80 μm), 4750 cm^{-1} (2.11 μm), and 6900 cm^{-1} (1.45 μm).

Hydroperoxides absorb near 4800 cm^{-1} (2.08 μm) and 6850 cm^{-1} (1.46 μm).

Carbonyl Groups

Carbonyl groups have an overtone band, due to the C=O stretching vibration, in the region 3600–3330 cm^{-1} (2.78–3.00 μm). This band may easily be distinguished from those due to N—H and O—H groups, which may also occur in this region, due to its comparatively low intensity. The position of this carbonyl overtone band follows the pattern observed for the position of bands due to the C=O stretching vibration—that is, in general, esters absorb at higher frequencies than aliphatic ketones which in turn absorb at higher frequencies than aromatic ketones.

Nitrogen–Hydrogen Groups

Primary, secondary, and tertiary amines may be distinguished on examination of the spectra of their dilute solutions in this region. The fundamental N—H vibrations have been discussed previously in the section dealing with amines. Primary amines have two bands in the region 3500–3300 cm^{-1} (2.86–3.03 μm) due to their fundamental N—H stretching vibrations. In the first overtone region, 7000–6500 cm^{-1} (1.43–1.54 μm), they have two bands and there is, in addition, a single band near 10 000 cm^{-1} (1.00 μm). Secondary amines have single bands in each of these regions and since tertiary amines have no NH group they do not, of course, absorb at all in these regions.

Primary amines also have a band resulting from the combination of the N—H bending and stretching modes which appears near 5000 cm^{-1} (2.00 μm) whereas secondary amines do not.

Alkyl and aryl amines may also be easily distinguished as the latter have, in general, the more intense absorptions in the near infrared region.

Primary amine hydrohalides have broad bands at 4600–4500 cm^{-1} (2.17–2.22 μm) which may be the result of the combination of NH_3 bending and CH_2 stretching vibrations.

CHAPTER 21

Inorganic Compounds

The infrared study of inorganic compounds presents some difficulties in that the use of conventional organic solvents is not always possible and the use of aqueous solutions is generally precluded by energy considerations. Therefore, the use of solids (as powders) in dispersive sampling techniques is extensive. Polyethylene cells may be used in the far infrared region.

Unfortunately, the infrared spectra of inorganic substances may not always be reproducible for a given sample since the extensive grinding necessary for the sampling techniques employed may result in (a) decomposition of the sample, (b) the crystal lattices being strained, (c) polymorphic changes, (d) varying degrees of hydration (or solvation), and indeed differences in particle size may also result in spectral changes.[1,2] Bands due to water occur very frequently in the spectra of inorganic compounds and this needs to be borne in mind.

Compared with the spectra of organic compounds, those of inorganic compounds often consist of a relatively small number of broad bands, the exceptions being the spectra of organometallic compounds.

A number of books[3–8] and useful reviews of a general[14] or specific nature may be found in the literature. References are given which deal with ionic crystals,[15] ions and ion pairs,[16–20] complexes,[18,21–24] donor–acceptor complexes,[25] carbonyl compounds,[26–29] sulphate and other oxyanions of Group VI,[30] tetracyanides of Pt, Pd, and Ni,[31] complex oxides and nitrides (spinel and garnet structures),[32] inorganic hydrazides,[33] and far infrared.[34]

Ions

The spectra of ionic solids composed of monoatomic ions, such as sodium chloride and potassium bromide, consist of broad bands. The only vibrations which can occur are the vibrations of individual ions and these are dependent on their nearest neighbouring ions. Vibrations of this type are referred to as lattice vibrations. With increase in the atomic weights of the ions involved, these vibrations occur at lower frequencies.

Absorption by polyatomic ions may be due to (a) internal vibrations of the ion, (b) torsional oscillations of water or other solvation molecules, (c) lattice vibrations.[15–20]

The internal vibrations of the ions are independent of the sample phase and of the associated ion(s) and dependent only on the atomic structure of the polyatomic ion. These vibrations are similar to those occurring in organic substances and are characteristic of the particular ion. For example, carbonate or sulphate ions have characteristic vibration frequencies which are very nearly independent of the cation.

Torsional oscillations result from the water or other solvent molecules being restricted in their rotational motion, hence resulting in bands due to these torsional vibrations which may complicate the infrared spectrum.

Lattice vibrations for inorganic compounds are due to the translational and rotational motion of molecules or ions within the crystalline lattice and normally result in absorptions below 300 cm^{-1} (above 33.33 μm).

The sulphate ion SO_4^{2-} has two characteristic bands—a very strong broad band at 1130–1080 cm^{-1} (8.85–9.26 μm) and a less intense band at 680–580 cm^{-1} (14.71–17.24 μm).

The nitrate ion has a very strong absorption at 1410–1340 cm^{-1} (7.09–7.46 μm) and a sharp, less intense band at 860–800 cm^{-1} (11.63–12.50 μm).

Small shifts in band position may be observed for different cations. The various radii and charges of different cations alter the electrical environment of polyatomic anions and hence affect their vibrational frequencies. Obviously, different crystalline arrangements may result when the cation is altered. Normally, with increase in mass of the cation there is a shift to lower frequency.

The characteristic bands of particular polyatomic ions are given in Table 21.1.

Coordination Complexes[3,8,10–13,18,21–29]

When a ligand coordinates to a metal atom M, new modes of vibration, not present in the free state, may become infrared active. For example, for a coordinated water molecule, rocking, twisting, and wagging modes become possible and, of course, this is in addition to the M—O stretching vibration.[52] The ammonia molecule exhibits bands in its spectrum which result from rocking and M—N stretching vibrations not observed in the free molecule. This general behaviour is true of other ligands, e.g. NO_2, PH_3.

In general, the frequencies of these new bands not only depend on the ligand involved but are also sensitive to the nature of the metal atom—its size, charge, etc.

In true aquo-complexes,[3,35,36] the water molecule is firmly bound to the metal atom by means of a partial covalent bond and is known as coordinated water. However, in some other cases, the metal–oxygen bond may be almost ionic in nature and in these hydrates the water molecule may be considered as crystal or lattice water. Since this type of water molecule is trapped, certain rotational and vibrational motions become partially hindered by environmental interactions and may in fact become infrared- (and Raman-) active. These bands are observed in the region 600–300 cm^{-1} (16.67–33.33 μm). The positions of the bands are sensitive to the anions present since hydrogen bonding also occurs.

Chart 7 The positions and intensities of the absorption bands due to ions and certain common inorganic groups. (Note the change of scale at 2000 cm^{-1} (5.0 μm)).

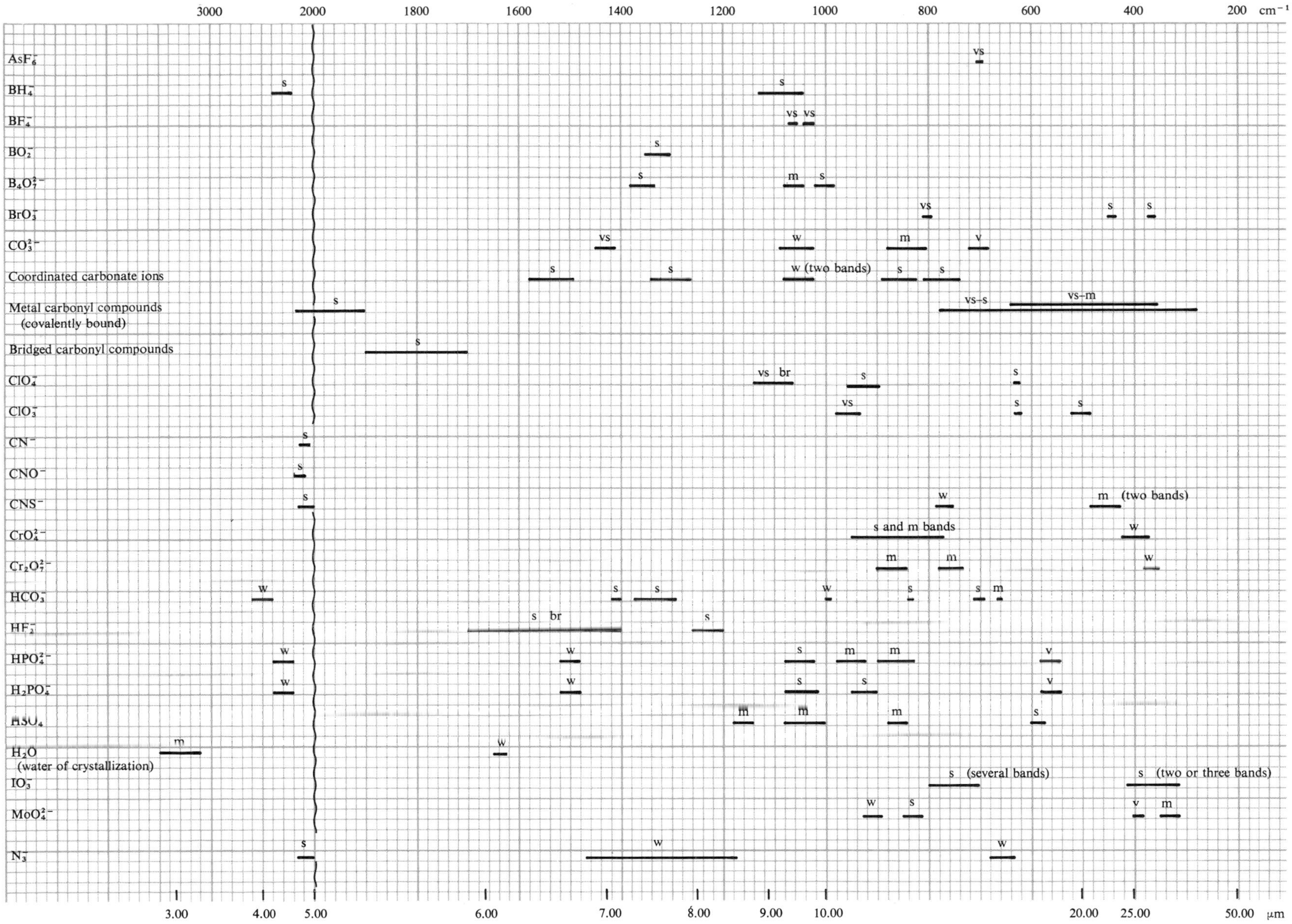
3000
2000
1800
1600
1400
1200
1000
800
600
400
200 cm^{-1}
AsF_6^-
BH_4^-
BF_4^-
BO_2^-
$B_4O_7^{2-}$
BrO_3^-
CO_3^{2-}
Coordinated carbonate ions
Metal carbonyl compounds (covalently bound)
Bridged carbonyl compounds
ClO_4^-
ClO_3^-
CN^-
CNO^-
CNS^-
CrO_4^{2-}
$Cr_2O_7^{2-}$
HCO_3^-
HF_2^-
HPO_4^{2-}
$H_2PO_4^-$
HSO_4^-
H_2O (water of crystallization)
IO_3^-
MoO_4^{2-}
N_3^-
w (two bands)
vs–s
vs–m
vs br
m (two bands)
s and m bands
s br
s (several bands)
s (two or three bands)
3.00
4.00
5.00
6.00
7.00
8.00
9.00
10.00
20.00
25.00
50.00 μm

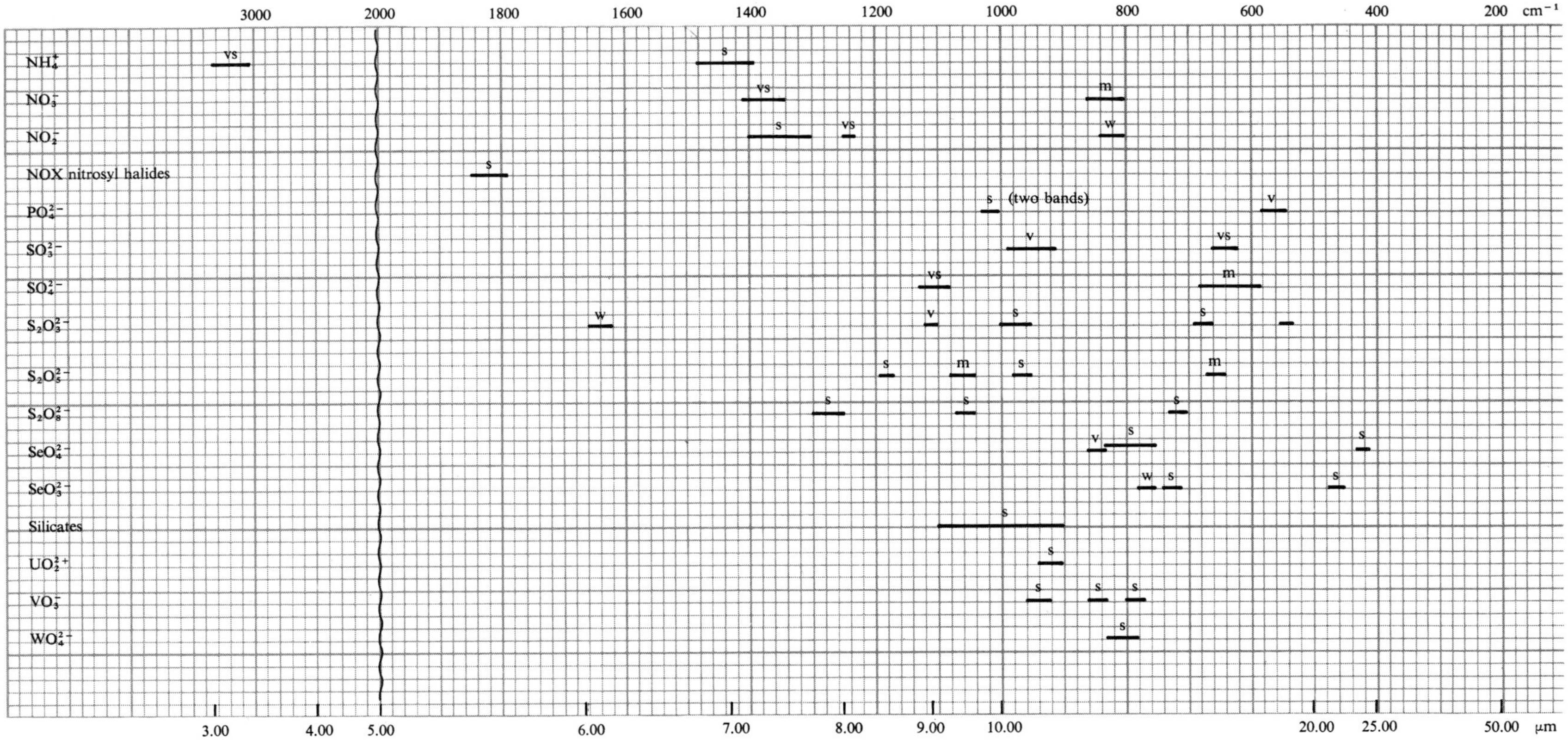
3000
2000
1800
1600
1400
1200
1000
800
600
400
200 cm⁻¹
NH_4^+
NO_3^-
NO_2^-
NOX nitrosyl halides
PO_4^{3-}
SO_3^{2-}
SO_4^{2-}
$S_2O_3^{2-}$
$S_2O_5^{2-}$
$S_2O_8^{2-}$
SeO_4^{2-}
SeO_3^{2-}
Silicates
UO_2^{2+}
VO_3^-
WO_4^{2-}
vs
s
m
w
v
(two bands)
3.00
4.00
5.00
6.00
7.00
8.00
9.00
10.00
20.00
25.00
50.00 μm

In a similar manner, vibrational modes of coordinated water molecules such as wagging, twisting, and rocking (which cannot occur for lattice water molecules) may become infrared-active, the resulting bands occurring in the region 880–650 cm^{-1} (11.36–15.38 μm).

In addition to the above bands which may become infrared-active, bands due to asymmetric and symmetric H—O—H stretching vibrations are observed in the region 3550–3200 cm^{-1} (2.82–3.13 μm) and bands due to the H—O—H bending vibrations occur in the region 1630–1600 cm^{-1} (6.13–6.25 μm).

As with a ligand, when a free ion becomes covalently bound there is a decrease in its structural symmetry. This results in the removal of the degeneracy of some vibrations hence resulting in new bands being observed in the infrared (and (Raman) region. Hence, infrared spectroscopy may be used to distinguish between ionic and covalent bonds in coordination complexes.

In general, the positions of the absorption bands for a particular ligand depend on the metal atom(s) to which coordination occurs. However, the position may also be dependent on the crystalline environment.

Coordination of Free Ions having Tetrahedral Symmetry

Typical of this class of ion are sulphates, perchlorates, and phosphates.

The sulphate ion, for instance, may coordinate to a metal atom as a unidentate ligand M—OSO_3, as a chelating bidentate ligand

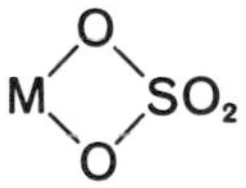
O
M
SO₂
O

or as a bridging bidentate ligand M—O—SO_2—O—M.

Free ions with tetrahedral symmetry, T_d, have four fundamental vibrations, only two of which are infrared-active (one stretching mode and one bending mode). For unidentate coordination, the symmetry is reduced to C_{3v}, each of the bands for the free ion being split into two bands and, in addition, the two previously only Raman-active vibrations now becoming infrared-active. Therefore, three bands due to stretching vibrations and three due to bending vibrations are expected. For bidentate coordination, the symmetry is reduced to C_{2v} and each of the bands due to the two modes of vibration of the free ion is now split into three, so that, taking into account the bands which were inactive for the free ion, four bands due to stretching vibrations and four due to bending vibrations are observed. The above applies equally to all other ions with tetrahedral symmetry.

Coordination of Free Ions having Trigonal–Planar Symmetry

Typical of this class of ion are carbonates and nitrates. These ions may form complexes as a unidentate or bidentate ligand. The free ion, which has D_{3h} symmetry, has one stretching and one in-plane deformation vibration which are infrared-active, the bands each being split into two in the case of both unidentate and bidentate coordination. In addition, a band due to the symmetric stretching vibration, which previously appeared only in the Raman spectrum, appears in the infrared spectrum, this vibration now being infrared-active. Unidentate coordination may be distinguished from bidentate coordination since the separation of two of the bands due to stretching vibrations is larger for the latter.

Coordinate Bond Vibration Modes

Bands due to the stretching or deforming of the coordinate bond are generally found at the low-frequency end of the infrared range, both the heavy metal atom and the nature of the coordinate bond being responsible for this.

Coordination complexes frequently contain metal–oxygen or metal–nitrogen bonds, but the absorption bands associated with these bonds are normally difficult to assign empirically since their position is not only dependent on the metal but also on the ligand and, in addition, coupling with other vibration modes often occurs.

By comparing the spectrum of the free ligand with that of the complex, metal–ligand vibrations may often be identified although, since some ligand vibrations may become infrared-active on forming the complex (as explained in the previous section), it is not uncommon for no clear assignments to be made by this comparison.

Metal–ligand vibrations[52] may sometimes be identified by changing the central metal atom or its valency state. This technique is useful when a series of complexes with the same structure is being studied.

Isotopic substitution of the ligand in order to observe isotopic shifts in spectral bands may also be used for the study of metal ligand vibrations, although caution must be exercised as shifts in other ligand bands are also observed. Isotopic substitution of the metal atom[32] is preferable, if possible, since only bands due to vibrations involving the metal atom will be shifted. The magnitude of the isotopic shift is usually small (2–10 cm^{-1}), depending, of course, on the relative mass difference.

Similar difficulties to those mentioned above arise in assignments of bands due to metal–carbon vibrations. Cyano- complexes[31, 49, 55] fall into this category and are extremely important and also common. The back donation of electrons by the metal atom to the ligand may complicate matters further by altering the character of the M—C bond. Carbonyl complexes[13, 26–29] also fall into this category.

Metal Halides

Metal halides of the MX_6 and MX_4 type, the former being octahedral and the latter either square–planar or tetrahedral, should have only one stretching vibration band. However, more than one may be observed if the symmetry is lowered. This may be either due to molecular distortion due to stereochemical considerations or due to interactions with neighbouring molecules in a particular crystalline environment. The effect of molecular distortion is much larger than that due to the crystalline phase, the latter often being too small to be observed. Broadening of the band(s) may also be observed for chlorine and bromine due to the presence of different naturally-occurring isotopes.

The bands due to metal–halogen vibrations occur in the following regions:

	F	Cl	Br	I
M—X str (cm^{-1})	840–550	540–220	430–200	360–150
M—X def (cm^{-1})	490–250	200–100	130–50	100–30

The position of a metal–halogen absorption band is dependent on (a) the strength of the bond, (b) the mass of the metal atom, and (c) the valence state of the metal atom. Other factors being equal, the frequency decreases with increase in mass of the metal atom.

The spectra of Group II metal dihalides[38] and of Group III–V metal fluorides[51, 53] have been reviewed.

Halogen atoms may act as bridging ligands between two metal atoms.[39] Bands due to both bridging and terminal halogens are observed for binuclear complexes. The metal–halogen–metal bond angle and the degree of interaction between bond stretching vibration modes affect the position of the observed bridging halogen bands. Planar binuclear complexes of the type

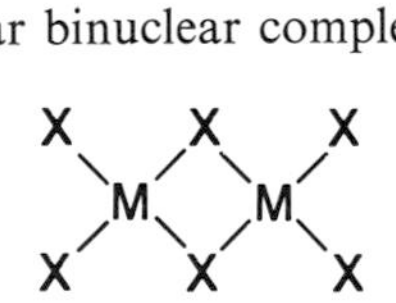

exhibit two bands due to bridging halogen–metal stretching vibrations and one due to the end-halogen stretching vibration.

Metal–π-Bond Vibrations

The infrared spectra of organometallic coordination compounds containing olefins and similar ligands show great similarity to the spectra of the free ligand. For example, in the case of ethylene coordinated to platinumII, most of the bands are at the positions observed for free ethylene except that due to the C=C stretching vibration which is found near 1525 cm^{-1} (6.56 μm) instead of near 1625 cm^{-1} (6.15 μm) and that due to the CH_2 twisting vibration which occurs near 730 cm^{-1} (13.70 μm) instead of near 1005 cm^{-1} (9.95 μm). The metal–olefin stretching vibration for platinum and palladium complexes results in a band at about 400 cm^{-1} (25.00 μm).

For platinum–acetylene complexes, the C≡C stretching vibration band is found at about 2000 cm^{-1} (5.00 μm) instead of near 2230 cm^{-1} (4.48 μm) as for free acetylene.

The cyclopentadiene ligand[40–43] may be bound to a metal (a) ionically, e.g. KC_5H_5, (b) by means of σ-bonds, e.g. $Mg(C_5H_5)_2$ and $Pb(C_5H_5)_2$, and (c) by means of π-bonds, e.g. $Fe(C_5H_5)_2$ and $Co(C_5H_5)_2$ (sandwich-type complexes).

In the first case, (a), four bands are observed in the infrared spectra, these being due to C—H stretching, ring deformation, C—H deformation, and C—H out-of-plane vibrations. In case (b), seven infrared bands are expected—due to two C—H stretching, two ring deformation, one C—H deformation, and two out-of-plane C—H bending vibrations. In case (c), in addition to the seven infrared bands of (b), other bands due to ring–metal vibrations are observed—asymmetric ring tilting, asymmetric metal–ring stretching, and metal–ring deformation vibrations.

Ferrocenes have several characteristic absorptions: a typical band due to the C—H stretching vibration near 3075 cm^{-1} (3.25 μm), a band of medium intensity near 1440 cm^{-1} (6.94 μm) due to the C—C stretching vibration, and strong bands near 1110 cm^{-1} (9.01 μm) and 1005 cm^{-1} (9.95 μm) due to asymmetric ring in-plane and C—H out-of-plane vibrations respectively. The latter two bands are absent in the case of disubstituted ferrocenes. Cyclopentadienyl complexes also have from three to six weak bands, possibly overtones, in the region 1750–1615 cm^{-1} (5.71–6.19 μm). Almost all ferrocenes have two strong bands at 515–465 cm^{-1} (19.42–21.51 μm) which for solid samples are normally sharp and for liquids, broad.

In general, compounds with a heavy metal atom coordinated between two parallel cyclopentadiene or benzene[44, 45] rings (or other suitable ligands such as cyclo-octadiene and norbornadiene) have strong bands at 530–375 cm^{-1} (18.87–26.67 μm) due to the asymmetric ring tilting motion, at 460–305 cm^{-1} (21.74–32.79 μm) due to the heavy metal atom moving perpendicular to the two rings in an asymmetric stretching vibration, and at 185–125 cm^{-1} (54.05–80.00 μm) due to the metal atom moving parallel to the two rings. This last band is not always observed and in benzene sandwich compounds, the former two bands are very sensitive to the nature of the metal atom.

Phthalocyanine[47] has bands at 620 cm^{-1} (16.13 μm), 616 cm^{-1} (16.23 μm), and 557 cm^{-1} (17.95 μm). Metal phthalocyanine compounds absorb near 645 cm^{-1} (15.50 μm), this band often appearing as a double peak, and near 555 cm^{-1} (18.02 μm). A band of medium-to-weak intensity is also observed at about 435 cm^{-1} (22.99 μm).

Metal Carbonyl Compounds[13, 26–29, 50]

The carbonyl groups of metal carbonyl compounds absorb strongly at 2170–1700 cm^{-1} (4.61–5.88 μm) due to the CO stretching vibration. Bridging carbonyl compounds, in which a carbonyl group is associated with two metal atoms, absorb at the lower frequencies in this range—1900–1700 cm^{-1} (5.26–5.88 μm). Most other carbonyl groups absorb above 1900 cm^{-1} (below 5.26 μm) except in the case of complexes with strong electron-donor ligands.

The in-plane bending vibration of the M—CO group of metal carbonyl compounds gives rise to a band of very strong or strong intensity at 775–275 cm^{-1} (12.90–36.36 μm) and the stretching vibration of the M—C group of these compounds causes a band of very strong-to-medium intensity at 640–340 cm^{-1} (15.63–29.41 μm).

Metal–Cyano Compounds

In these complexes,[55] the C≡N group may act as a terminal or bridging group. Terminal C≡N groups absorb near 2070 cm^{-1} (4.83 μm) whereas bridging C≡N groups absorb near 2130 cm^{-1} (4.69 μm). Inorganic cyanides also absorb in the ranges 570–180 cm^{-1} (17.54–55.55 μm) and 450–295 cm^{-1} (22.22–33.90 μm).

Metal–Urea Compounds

Coordination of the urea molecule to a metal atom may occur through either the oxygen or the nitrogen atoms. The electronic structure of urea may be considered as a hybrid of the three resonance structures:

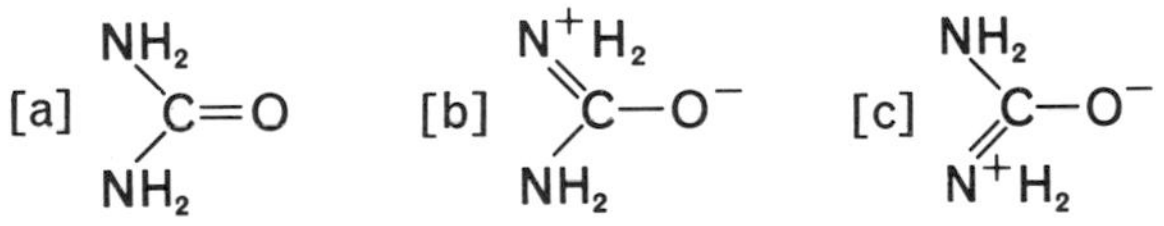

If coordination through the oxygen atom occurs, the contribution by structure (a) will be small. Therefore, coordination through the oxygen atom tends to decrease the frequency of the CO stretching vibration and increase that of the C—N stretching vibration relative to that observed for 'free' urea, for which the bands due to the CO and CN stretching vibrations are observed at 1683 cm^{-1} (5.94 μm) and 1471 cm^{-1} (6.80 μm) respectively. Coordination through the nitrogen atom tends to have the reverse effect. Similar observations apply to thiourea. Hence the donor atom may be determined and linkage isomers distinguished by infrared spectroscopy.

Metal–Acetylacetonato Compounds

Acetylacetone may coordinate[48] to a metal atom through the oxygen atoms:

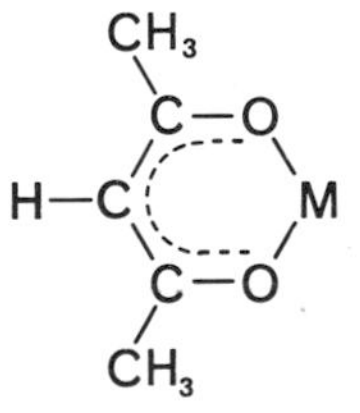

In this type of complex, the 'carbonyl' stretching vibration occurs at lower frequencies, usually 1605–1560 cm^{-1} (6.23–6.41 μm), than in the free acetylacetone, for which this band occurs at 1640 cm^{-1} (6.10 μm). A second strong band is observed near 1380 cm^{-1} (7.25 μm). The band due to the C—H stretching vibration tends to higher frequencies because of the new benzene-type environment in which it is found.

Acetylacetonates have two bands at 600–490 cm^{-1} (16.67–20.41 μm) and may also absorb near 430 cm^{-1} (23.26 μm) and at 390–290 cm^{-1} (25.64–34.48 μm).

Acetylacetone may also form coordination complexes by bonding through its γ-carbon atom. In this type of complex, the carbonyl band position is the same as for free acetylacetone. The bands due to the carbon–carbon stretching vibration are also in different positions for the two types of complex.

Coordination through the carbon–carbon double bond may also occur in some complexes.

Oxalato complexes have a number of bands in the region 590–290 cm^{-1} (16.95–34.48 μm).

Metal Oxides and Other Inorganic Oxides

Many simple metal oxides do not absorb in the region 4000–650 cm^{-1} (2.50–15.38 μm). However, oxides with more than one oxygen atom bound to a single metal atom usually absorb in the region 1020–970 cm^{-1} (9.80–10.31 μm) and, in general, metal oxides containing the group M=O absorb at 1100–825 cm^{-1} (9.09–12.12 μm).

Different polymorphic forms can be distinguished in the region 700–300 cm^{-1} (14.29–33.33 μm). Cubic crystalline forms of rare earth oxides have a characteristic band at 570–530 cm^{-1} (17.54–18.87 μm).

Ti—O—Ti absorbs in the range 900–700 cm^{-1} (11.11–14.29 μm) and Ti—O—Si at 950–900 cm^{-1} (10.53–11.11 μm). The stretching frequencies of silicate, Si—O, borate, B—O, metaphosphate, P—O, and germanate, Ge—O, bonds are 1100–900 cm^{-1} (9.09–11.11 μm), 1380–1310 cm^{-1} (7.25–7.63 μm), 1300–1140 cm^{-1} (7.69–8.77 μm), and 930–840 cm^{-1} (10.75–11.90 μm) respectively.

Structural Isomerism

Cis–trans *isomerism*

Infrared spectroscopy may be used to distinguish between *cis* and *trans* isomers of compounds. The structural symmetry of the molecule is used to determine the point group, the vibrational selection rules then being applied to determine which vibration bands are observed.

Nitro- (—NO_2) and nitrito- (—ONO) complexes

Linkage isomerism is possible in the case of metal complexes containing the unit NO_2. Coordination to the metal atom may occur through the nitrogen atom, resulting in a nitro- complex, or through an oxygen atom, resulting in a nitrito- complex.

Nitro- complexes exhibit bands due to asymmetric and symmetric —NO_2 stretching vibrations and, in addition, one due to a NO_2 deformation vibration. The nitrito- complexes exhibit bands due to asymmetric and symmetric —ONO stretching vibrations.

Nitro- groups in metal coordination complexes may exist as bridging or as end groups. Terminal nitro- groups absorb at 1440–1390 cm^{-1} (6.94–7.19 μm) and 1340–1315 cm^{-1} (7.46–7.61 μm) due to the asymmetric and symmetric stretching vibrations respectively of the NO_2 group. However, nitro- groups acting as bridging units between two metal atoms absorb at 1485–1470 cm^{-1} (6.73–6.80 μm) and at about 1200 cm^{-1} (8.33 μm), these bands being broader than those for terminal nitro- groups.

Thiocyanato- (—SCN) and isothiocyanato- (—NCS) complexes

The thiocyanate ion may act as an ambidentate ligand, i.e. bonding may occur either through the nitrogen or through the sulphur atom.[49] The bonding mode may easily be distinguished by examining the band due to the C—S stretching vibration which occurs at 730–690 cm^{-1} (13.70–14.49 μm) when the bonding occurs through the sulphur atom and at 860–780 cm^{-1} (11.63–12.82 μm) when it is through the nitrogen atom.

The C≡N stretching vibration of thiocyanato- complexes (sulphur-bound, i.e. M—SCN) gives rise to a sharp band at about 2100 cm^{-1} (4.76 μm), whereas for isothiocyanato- complexes (i.e. nitrogen bound) the resulting band is broad and below 2100 cm^{-1} (above 4.76 μm). In addition, a deformation vibration gives a band at 440–400 cm^{-1} (22.73–25.00 μm) for thiocyanato- complexes, which appears at 490–450 cm^{-1} (20.41–22.22 μm) for isothiocyanato- complexes.

Thiocyanates acting as bridging groups in platinum and palladium complexes absorb in the region 2185–2150 cm^{-1} (4.58–4.65 μm).

Table 21.1. Free inorganic ions and coordinated ions

Functional Group or Ion	Region cm^{-1}	Region μm	Intensity	Comments
AsO_4^{3-}	~800	~12.50	s	
AsF_6^-	705–690	14.18–14.49	vs	
BH_4^-	2400–2200	4.17–4.55	s	Two bands
	1130–1040	8.85–9.62	s	
BN^-	~1390	~7.19	s	
	~810	~12.35	w	
BF_4^-	~1060	~9.43	vs	
	~1030	~9.71	vs	
BO_2^-	1350–1300	7.41–7.69	s	
$B_4O_7^{2-}$	1380–1330	7.25–7.52	s	
	1080–1040	9.26–9.62	m	
	1020–980	9.80–10.20	s	
Bromate, BrO_3^-	810–790	12.35–12.66	vs	
	450–430	22.22–23.26	s	sh
	370–355	27.03–28.17	s	sh
Carbonate, CO_3^{2-}	1450–1410	6.90–7.09	vs	
	1090–1020	9.17–9.80	w	
	880–800	11.36–12.50	m	
	720–680	13.89–14.71	v	Not always present
Coordinated carbonate	1580–1490	6.33–6.71	s	
	1340–1260	7.46–7.94	s	
	1085–1020	9.22–9.80	w	Two bands
	890–820	11.24–12.20	s	
	810–735	12.35–13.61	s	
Perchlorate, ClO_4^-	1140–1060	8.77–9.43	vs	br
	~935	~10.70	s	
	630–620	15.87–16.13	s	sh
Chlorate, ClO_3^-	980–930	10.20–10.75	vs	
	630–615	15.87–16.26	s	sh
	510–480	19.61–20.83	s	sh
Carbonyl, CO (covalently bound)	2170–1900	4.61–5.26	s	
	775–275	12.90–36.36	s–vs	
	640–350	15.63–28.57	m–vs	
Bridged carbonyl	1900–1700	5.26–5.88	s	
Cyanide, cyanate, and thiocyanate: CN^-, CNO^-, and SCN^-	2200–2000	4.55–5.00	s	CN^- has no bands below 700 cm^{-1}, SCN^- has sharp doublet at 485–425 cm^{-1} (20.62–23.53 μm), separation ~30 cm^{-1}
Chromate, CrO_4^{2-}	950–770	10.53–12.99	s	Several bands, not all strong (except for complexes—all strong)
	420–370	23.81–27.03	w	Usually doublet
Dichromate, $Cr_2O_7^{2-}$	900–840	11.11–11.90	m	
	780–730	12.82–13.70	m	
	380–350	26.32–28.57	w	
$Fe(CN)_6^{4-}$	600–580	16.67–17.24		
Bicarbonate, HCO_3^-	1420–1400	7.04–7.14	s	
	1370–1290	7.30–7.75	s	
	1000–990	10.00–10.10	w	
	840–830	11.90–12.05	s	
	710–690	14.08–14.49	s	
	665–655	15.04–15.27	m	

Table 21.1 (*cont.*)

Functional Group or Ion	Region cm⁻¹	μm	Intensity	Comments
HF_2^-	1700–1400	5.88–7.14	s	Very br, max ~1450 cm^{-1} (6.90 μm)
	1260–1200	7.94–8.33	s	
Dibasic phosphate, HPO_4^{2-}	2400–2200	4.17–4.55	w	
	~1700	~5.88	w	
	1080–1020	9.26–9.80	s	
	980–925	10.20–10.81	m	
	900–825	11.11–12.12	m	
	580–540	17.24–18.52	v	Not always present
$H_2PO_4^-$	2400–2200	4.17–4.55	w	
	~1700	~5.88	w	
	1080–1020	9.26–9.80	s	
	950–900	10.53–11.11	s	
	580–540	17.24–18.52	v	Not always present
Bisulphate, HSO_4^-	1190–1160	8.40–8.62	s	
	1080–1000	9.26–10.00	s	
	880–840	11.36–11.90	m	
	600–570	16.67–17.54	s	
H_2O (water of crystallization)	3500–3200	2.86–3.13	m	
	1650–1625	6.06–6.15	w	
Iodate, IO_3^-	800–700	12.50–14.29	s	Several bands
	415–310	24.10–32.26	s	Two or three sh bands
Iodates (covalently bonded or coordinated)	810–755	12.35–13.25	m–s	
	795–715	12.58–13.99	s	
	690–630	14.49–15.87	s	
	480–420	20.83–23.81	m	
MnO_4^-	920–890	10.47–11.24	vs	br
	850–840	11.76–11.90	m	
	400–380	25.00–26.32	v	Often weak doublet
MoO_4^{2-}	930–890	10.75–11.24	w	
	850–810	11.76–12.35	s	
	400–380	25.00–26.32	v	
	350–310	28.57–32.26	m	
Azide, N_3^-	2170–2030	4.61–4.93	s	
	1375–1175	7.27–8.51	w	
	680–630	14.71–15.87	w	
Ammonium, NH_4^+	3335–3030	3.00–3.30	vs	
	1485–1390	6.73–7.19	s	(No bands below 700 cm^{-1})
Nitrate, NO_3^-	1410–1340	7.09–7.46	vs	
	860–800	11.63–12.50	m	sh (no bands below 700 cm^{-1})
Nitrite, NO_2^-	1400–1300	7.14–7.69	s	Two bands for nitrite complexes
	1250–1230	8.00–8.13	vs	
	840–800	11.90–12.50	w	(No bands below 700 cm^{-1})
Nitric oxide (monomer)	~1883	~5.31		(*Cis* dimer, ~1860 and ~1765 cm^{-1}; *trans* dimer, ~1740 cm^{-1})
NO^+	2370–2230	4.22–4.48	s	(Nitric acid, ~2220 cm^{-1})
NO^+ (coordinated M—NO)	1940–1630	5.15–6.13	s	
Nitrosyl halides NOX	1850–1790	5.41–5.59	s	
OsO_4^{2-}	350–300	28.57–33.33	v	
PF_6^-	~915	~10.93	m	
	850–840	11.76–11.90	vs	
	~555	~18.02		
Phosphate, PO_4^{3-}	1030–1000	9.71–10.00	s	
	580–540	17.24–18.52	v	Not always present
RuO_4^{2-}	350–300	28.57–33.33	v	
Thiocyanate, SCN^-	(See cyanate)			
Thiosulphate, $S_2O_3^{2-}$	1660–1620	6.02–6.17	w	
	1120–1100	8.93–9.09	v	Usually strong
	1000–950	10.00–10.53	s	
	690–660	14.49–15.15	s	
	550–530	18.18–18.87		
$S_2O_5^{2-}$	1190–1170	8.40–8.55	s	
	1080–1040	9.26–9.62	m	
	980–950	10.20–10.53	m–s	
	670–640	14.93–15.63	m	
$S_2O_8^{2-}$	1300–1250	7.69–8.00	s	
	1070–1040	9.35–9.62	s	sh
	730–700	13.70–14.29	s	
Sulphate, SO_4^{2-}	1130–1080	8.85–9.26	vs	
	680–580	14.71–17.24	m	
Sulphite, SO_3^{2-}	990–910	10.10–10.99	v	Often strong
	660–615	15.15–16.26	vs	sh
Selenate, SeO_4^{2-}	860–830	11.63–12.05	v	Often strong
	830–750	12.05–13.33	s	
	430–410	23.26–24.39	s	
Selenites, SeO_3^{2-}	780–750	12.82–13.33	w	
	740–710	13.51–14.08	s	
	~460	~21.74	s	
SiF_6^{2-}	~725	~13.79	s	
Silicate	1100–900	9.09–11.11	s	
SiO_3^{2-}	1030–960	9.71–10.42	s	
	790–750	12.66–13.33	w	
	500–450	20.00–22.22	s	
Metasilicate	750–730	13.33–13.70	w	
	470–460	21.28–21.74	s	
UO_2^+	940–900	10.64–11.11	s	
VO_3^-	960–920	10.42–10.87	s	
	860–830	11.63–12.05	s	
	800–770	12.50–12.99	s	
WO_4^2	830–780	12.05–12.82	s	

Table 21.2. Cyclopentadienyl–metal complexes

Functional Groups	Region cm⁻¹	μm	Intensity	Comments
Cyclopentadienyl complexes	3110–3025	3.22–3.31	m	C—H str
	~1440	~6.94	m	C=C str
	1115–1090	8.97–9.17	m–s	sh, C=C str
	1010–990	9.90–10.10	s	sh, C—H in-plane def
	830–700	12.05–14.29	vs	br, C—H out-of-plane def
	515–485	19.42–20.62	s	asym ring tilting vib (not always present)
	495–465	20.20–21.51	s	asym ring–metal str

Table 21.3. Metal hydrides

Functional Groups	Region cm^{-1}	μm	Intensity	Comments
Sn—H	1855–1800	5.39–5.56	m	Sn—H str
	~570	~17.54	s	Sn—H def
M—H (M = Pt, Ir, Ru, Os, Re)	2200–1890	4.55–5.29	m	Highest frequencies for Pt, lowest for Re; for H *trans* to a halogen, frequency ~100 cm^{-1} higher than when *trans* to a phosphine
Ge—H	2160–2010	4.63–4.98	m	Ge—H str
	~720	~13.89	s	Ge—H def
AlH_3 and AlH_4	1810–1675	5.52–5.97	m–s	Al—H str

Table 21.4. Metal–carbon stretching vibrations (and others)

Functional Groups	Region cm^{-1}	μm	Intensity	Comments
Alkyl metal compounds				
Ge—R	645–510	15.50–19.61	m–s	Ge—C str
Sn—R	570–450	17.54–22.22	m–s	Sn—C str
Pb—R	480–420	20.83–23.81	m–s	Pb—C str
Ga—R	620–535	16.13–18.69	m–s	Ga—C str
In—R	570–465	17.54–21.51	m–s	In—C str
Sb—R	600–475	16.67–21.05	m–s	Sb—C str
Metal carbonyl compounds	775–275	12.90–36.36	s–vs	In-plane bending of M—CO
	640–340	15.63–29.41	m–vs	M—C str
Triphenyl compounds,	460–440	21.74–22.73	m–s	X-sensitive band
Ph_3MA (M = Sn or Ge; A = H, Cl, Br, I)	375–235	26.67–42.55	m	M—A str
Tetraphenyl compounds, Ph_4M (M = Ge, Sn, Pb)	480–440	20.83–22.73	s	Doublet, X-sensitive band
Metal–ammine	510–275	19.61–36.36	w–m	M—N str, triplet
	330–190	30.30–52.63	s	In-plane bending of N—M—N
Borate complexes	580–550	17.24–18.18	w–m	

References

1. G. Duyckaerts, *Analyst*, 1959, **84,** 201
2. A. W. Baker, *J. Phys. Chem.*, 1957, **61,** 450
3. K. Nakemoto, *Infrared Spectra of Inorganic and Coordination Compounds.* Wiley, New York, 1970
4. K. E. Lawson, *Infrared Absorption of Inorganic Substances.* Reinhold, Princeton, N.J., 1961
5. I. A. Cadsen, *Infrared Spectra of Mineral and Related Inorganic Compounds.* Butterworth, London, 1975
6. S. D. Ross, *Inorganic Infrared and Raman Spectra.* McGraw-Hill, London, 1972
7. R. A. Nyquist and R. O. Kagel, *Infrared Spectra of Inorganic Compounds (3800–45 cm^{-1}).* Academic Press, New York, 1971
8. I. Yarwood (ed.), *Spectroscopy and Structure of Molecular Complexes.* Plenum, New York, 1973
9. A. J. Downs, *Essays in Structural Chemistry.* Plenum, New York, 1971, pp. 404 and 433
10. J. Lewis and R. G. Lewis, *Modern Coordination Chemistry*, Interscience, New York, 1960
11. M. Tsutsui (ed.), *Characterization of Organometallic Compounds*, Interscience, New York, 1971,
12. A. E. Martell (ed.), *Coordination Chemistry*, Vol. 1. Interscience, New York, 1971, pp. 134–185
13. A. S. Braterman, *Metal Carbonyl Spectra*, Academic Press, New York, 1975
14. I. R. Beattie, *Chem. Soc. Rev.*, 1975, **4,** 107
15. I. Motohiko *et al.*, *Kagaku No Ryoiki*, 1973, **27,** 769
16. F. Miller and C. H. Wilkins, *Anal. Chem.*, 1952, **24,** 1253
17. S. R. Yoganarsimhan and C. N. R. Rao, *Chemist-Analyst*, 1962, **51,** 21
18. J. W. Brasch *et al.*, *App. Spectrosc. Rev.*, 1968, **1,** 187
19. W. F. Edgell, in *Ions, Ion Pairs Organic Reactions*, Vol. 1 (M. Szwarc, ed.). Interscience, New York, 1972, pp. 153–176
20. A. I. Popov, *Pure App. Chem.*, 1975, **41,** 275
21. M. Goldstein, *Spectrosc. Prop. Inorg. Organomet. Compds*, 1974, **7,** 320
22. G. Davidson, *Spectrosc. Prop. Inorg. Organomet. Compds*, 1974, **7,** 377
23. R. J. H. Clerk, *Rec. Chem. Progress*, 1965, **26,** 269
24. K. E. Stine and W. F. Ulrich, *Purif. Inorg. Org. Mater.*, **1969,** 47
25. M. T. Forel *et al.*, *Colloq. Int. Cent. Nat. Rech. Sci.*, 1970, **191,** 167
26. R. Poilblanc *et al.*, *Rev. Inst. Fr. Pet.*, 1974, **29,** 387
27. S. C. Tripathi, *J. Sci. Ind. Res.*, 1974, **33,** 570
28. M. Bigorgna, *J. Organomet. Chem.*, 1975, **94,** 161
29. S. F. A. Kettle and I. Paul, *Adv. Organomet. Chem.*, 1972, **10,** 199
30. S. D. Ross, *Infrared Spectra Miner*,. **1974,** 423
31. S. Jerome-Lerutte, *Struct. Bonding (Berlin)*, 1972, **10,** 153
32. G. A. Kurbatov *et al.*, *Tr. Vses. Gos. Inst. Nauch-Issled Prockt. Rab. Ogneupor. Prom.*, 1971, **43,** 120
33. R. I. Machkhoshvili *et al.*, *Issled Obl. Khim. Kompleksn. Prastykh. Soedin Nek. Perekhodnykh*, 1974, **2,** 37
34. F. A. Miller *et al.*, *Spectrochim. Acta*, 1960, **16,** 135
35. L. Nakagawa and T. Shimanouch, *Spectrochim. Acta*, 1964, **20,** 429
36. J. D. S. Goulden and D. J. Manning, *Spectrochim. Acta*, 1967, **23A,** 2249
37. K. Nakamoto, *Angew. Chem. Int. Ed.*, 1972, **11,** 666
38. I. Eliezer and A. Reyer, *Coord. Chem. Rev.*, 1972, **9,** 189
39. R. J. H. Clark and C. S. Williams, *Inorg. Chem.*, 1965, **4,** 350
40. F. Rocquet *et al.*, *Spectrochim. Acta*, 1973, **29A,** 1101

41. L. S. Myants *et al.*, *Optics Spectrosc.*, 1962, **13,** 177
42. E. R. Lippincott and R. D. Nelson, *Spectrochim. Acta*, 1958, **10,** 307
43. W. K. Winter *et al.*, *Spectrochim. Acta*, 1959, **15,** 1085
44. H. P. Fritz *et al.*, *Spectrochim. Acta*, 1061, **17,** 1068
45. R. G. Snyder, *Spectrochim. Acta*, 1959, **15,** 807
46. H. P. Fritz and J. Mauchot, *Spectrochim. Acta*, 1962, **18,** 171
47. A. N. Sidorov and I. P. Kotlyar, *Optics Spectrosc.*, 1961, **11,** 92
48. Y. Kawaski, *Spectrochim. Acta*, 1966, **22,** 1571
49. A. Sabatini and I. Bertini, *Inorg. Chem.*, 1965, **4,** 959
50. E. L. Burrows *et al.*, *J. Chem. Soc. Dalton Trans.*, 1975, **22,** 2353
51. R. L. Davidovich and V. I. Kostin, *Atlas of Long Wavelength Infrared Absorption Spectra of Complex Group III–V Metal and Uranyl Fluorides.* Nauka, Moscow, 1977
52. I. M. Cheremisina, *Zh. Strukt. Khim.*, 1978, **19,** 336
53. R. L. Davidovich *et al.*, *Atlas of Infrared Absorption Spectra and X-Ray Measurement Data for Complex Group IV and V Metal Fluorides*, Nauka, Moscow, 1972
54. G. Socrates, *J. Inorg. Nucl. Chem.*, 1969, **31,** 1667
55. M. W. Adlard and G. Socrates, *J. Inorg. Nucl. Chem.*, 1972, **34,** 2339

APPENDIX
Further Reading

L. J. Bellamy, *The Infrared Spectra of Complex Molecules*. Methuen, London, 1960
L. J. Bellamy, *Advances in Infrared Group Frequencies*. Methuen, London, 1968
H. A. Szymanski and R. E. Erickson, *Infrared Band Hand Book*, Vols I and II. Plenum, New York, 1970
H. A. Szymanski, *Interpreted Infrared Spectra*. Plenum, New York, 1964
R. M. Silverstein and G. C. Bassler, *Spectrometric Identification of Organic Compounds* Wiley, New York, 1967
T. Shimanouchi, 'Tables of molecular vibrational frequencies consolidated', *Nat. Stand. Ref. Data Ser.*, 1972, NSRDS–NBS–39
K. D. Moller and W. G. Rothschild, *Far Infrared Spectroscopy*. Wiley, New York, 1971
A. Finch, P. N. Gates, K. Radcliffe, F. N. Dickson, and F. F. Bentley, *Chemical Applications of Far-infrared Spectroscopy*. Academic Press, New York, 1970
S. S. Mitra and S. Nudelman (eds), *Far Infrared Properties of Solids. Proceedings of NATO Advanced Study Institute 1968*. Plenum, New York, 1970
C. S. Blackwell and R. C. Lord, 'Far-infrared spectroscopy of four-membered ring compounds', *Vib. Spectra Struct.*, 1972, **1,** 1
J. Loane, 'Far-infrared spectroscopy of five-membered ring compounds', *Vib. Spectra Struct.*, 1972, **1,** 25
A. R. Katritzky and P. J. Taylor, 'Infrared spectroscopy of heterocyclic compounds'. In *Physical Methods in Heterocyclic Chemistry*, Vol. 4 (A. R. Katritzky, ed.). Academic Press, New York, 1974, p. 265
S. Pinchas with I. Laulight, *Infrared Spectra of Labelled Compounds*. Academic Press, New York, 1971
'Tables of infrared and Raman fundamental vibration frequencies', *J. Phys. Chem. Ref. Data*, 1973, **2,** 121 and 225
R. G. White, *Handbook of Industrial Infrared Analysis*. Plenum, New York, 1964.
G. Herzberg, *The Spectra and Structures of Simple Free Radicals*. Cornell University Press, Ithaca, N.Y., 1971
E. A. Glebovokaya, *Use of Infrared Spectroscopy in Petroleum Geochemistry*. Nedra, Leningrad, 1971
P. M. A. Sherwood, *Vibrational Spectroscopy of Solids*. Cambridge University Press, London, 1972
G. Turrel, *Infrared and Raman Spectra of Crystals*. Academic Press, New York, 1972
H. E. Hallam (ed.), *Vibrational Spectroscopy of Trapped Species*. Wiley, New York, 1973
L. H. Little, A. V. Kiselev, and V. I. Lygin, *Infrared Spectra of Absorbed Species*. Academic Press, New York, 1966
A. V. Kiselev and V. I. Lygin, *Infrared Spectra of Surface Compounds and Absorbed Substances*. Nauka, Moscow, 1973
L. S. Mayants and B. S. Averbukh, *Theory and Calculation of Intensities in the Infrared of Molecules*. Nauka, Moscow, 1971
A. V. Karakin and G. A. Kirventsova, *State of Water in Organic and Inorganic Compounds*. Nauka, Moscow, 1973
V. C. Farmer (ed.), *The Infrared Spectra of Minerals*. Mineral Society, London, 1974
V. C. Farmer, 'Infrared of anhydrous mineral oxides', *Infrared Spectra Miner.*, 1974, 183
N. Neuroth, 'Infrared spectroscopy of glass', *Fachausschussber. Dtsch. Glasstech. Ges.*, 1974, **70,** 141
R. E. Hester, 'Infrared spectra of molten salts', *Adv. Molten Salt. Chem.*, 1971, **1,** 1
D. E. Irish, 'Infrared spectra of fused salts', *Ionic Interactions*, 1971, **2,** 187
C. Karr, 'Far infrared spectroscopy of minerals and inorganic compounds, review since 1930', *Progr. Nucl. Energy Ser.* **9**, 1972, **11,** 109
G. A. Kurbatov *et al.*, 'Use of infrared spectroscopy in study of complex metal oxides and nitrides (spinel and garnet structures)', *Tr. Vses. Gos. Inst. Nauch-Issled. Prockt. Rab. Ogneupor. Prom.*, 1971, No. **43,** 120
I. S. Shamina and I. K. Kuchkaeva, 'Infrared spectra of metal hydroxides', *Issled. Obl. Khim. Istochnikov Toka*, 1971, No. **2,** 16
H. J. Becher and F. Thevenot, 'Infrared spectra of inorganic boron compounds', *Zeit. Anorg. Allg. Chem.*, 1974, **410,** 274
A. G. Vlasov *et al.*, *Infrared Spectra of Inorganic Glasses and Crystals*. Khimiya, Leningrad, 1972.
K. Nakamoto, *Amer. Chem. Soc. Monogr.*, 1971, No. 168 (Coord. Chem. VI)
J. L. Ryan, 'Infrared and Raman spectra of actinide compounds'. In *International Review of Science: Inorganic Chemistry, Series One*, Vol. 7, *Lanthanides, Actinides* (K. W. Bagnall, ed.), Butterworth, London, 1972, pp. 323–367
D. O. Hummel, *Infrared Spectra of Polymers in the Medium and Long Wavelength Regions*. Interscience, New York, 1966
R. Zbinden, *Infrared Spectra of High Polymers*. Academic Press, New York, 1964
C. J. Heinniker, *Infrared Analysis of Industrial Polymers*. Academic Press, New York, 1967
F. J. Boerio and J. L. Koenig, 'Infrared spectroscopy of synthetic and natural polymers', *J. Macromol. Sci. Rev. Macromol. Chem.*, 1972, **7,** 209
R. R. Hampton, 'Infrared spectroscopy of rubbers', *Rubber Chem. Technol.*, 1972, **45,** 546
A. Elliott, 'Infrared spectra of polymers', *Adv. Spectrosc.*, 1959, **1,** 214
V. Fawcett and D. A. Long, 'Infrared spectra of polymers', *Mol. Spectrosc.*, 1973, **1,** 352
S. Krimm, 'Infrared spectra of polymers', *Fortschn Hochpolym. Forsch.*, 1960, **2,** 51
C. Clark and M. Chianta, 'Bibliography of infrared spectra of biochemical interest', *Ann. NY Acad. Sci.*, 1957, **69,** 205
N. K. Freeman, 'Infrared spectra of serum lipids', *Ann. NY Acad. Sci.*, 1957, **69,** 131.
N. K. Freeman, 'Infrared spectra of blood lipids', *Blood Lipids Lipoprotein Quant. Compos. Metab.*, 1972, 113
H. P. Schwarz *et al.*, 'Infrared spectra of tissue lipids', *Ann. NY Acad. Sci.*, 1957, **69,** 38
F. S. Parker, 'Infrared spectra of carbohydrates, lipids, proteins, polypeptides, steroids and porphyrins', *App. Spectrosc.*, 1975, **29,** 129
R. S. Tipson, 'Infrared spectroscopy of carbohydrates', *U.S. Dept. Commerce Nat. Bur. Stand. Monogr.*, **110,** 1968
R. G. Zhbankov, 'Infrared spectra of polysaccharides', *Faserforsch. Textiltech.*, 1975, **26,** 112
H. Spedding, 'Infrared spectra of carbohydrates', *Adv. Carbohyd. Chem.*, 1964, **19,** 23

R. T. O'Connor, 'Infrared spectra of modified cellulose', *Instrum. Anal. Cotton Cellul. Modif. Cotton Cellul.*, **1972,** 401–462

R. T. O'Connor, 'Infrared spectra of modified cotton', *High Polym.*, 1971, **5,** 51

M. Tsuboi, 'Infrared and Raman spectroscopy of nucleic acids in base residues'. In *Synthetic Procedures in Nucleic Acid Chemistry*, Vol. 2 (W. W. Zorbach and R. S. Timpson, eds). Academic Press, New York, 1973, p. 91

W. W. Zorbach and R. S. Timpson, 'Infrared spectra of nucleic acids', In *Synthetic Procedures in Nucleic Acid Chemistry*, Vol. 2 (W. W. Zorbach and R. S. Timpson, eds). Academic Press, New York, 1973, p. 215

C. Lenzen and L. D. Delcombe, 'Infrared spectroscopy of antibiotics', *Inf. Bull. Int. Cent. Inf. Antibiot.*, 1976, **13,** 178

J. Holubek, *Spectral Data and Physical Constantsof Alkaloids*. Heyden, London, 1972, Vol. 7, cards 801–900

G. P. Zhizhina and E. F. Oleinik, 'Infrared spectra of nucleic acids', *Russ. Chem. Rev.*, 1972, **41,** 258 and 474

L. Wayland and P. J. Weiss, 'Antibiotic substances'. In *Infrared and Ultraviolet Spectra of Some Compounds of Pharmaceutical Interest*, revised ed. Association of Official Analytical Chemists, Washington DC, 1972, pp. 234–241

A. L. Hayden *et al.*, 'Pharmaceutical compounds'. In *Infrared and Ultraviolet Spectra of Some Compounds of Pharmaceutical Interest*, revised ed. Association of Official Analytical Chemists, Washington DC, 1972, pp. 1–101 and 176–219

F. Fazzari *et al.*, 'Pharmaceutical compounds'. In *Infrared and Ultraviolet Spectra of Some Compounds of Pharmaceutical Interest*, revised ed. Association of Official Analytical Chemists, Washington DC, 1972, pp. 220–233

O. R. Sammul *et al.*, 'Pharmaceutical compounds'. In *Infrared and Ultraviolet Spectra of Some Compounds of Pharmaceutical Interest*, revised ed. Association of Official Analytical Chemists, Washington DC, 1972, pp. 102–175

J. L. Brazier, 'Pharmaceutical compounds', *Lyon Pharm.*, 1974, **25,** 639

H. Fritzache, 'Infrared and NMR spectra of nucleic acids', *Zeit. Chem.*, 1972, **12,** 1

M. Tsuboi, 'Infrared and Raman spectroscopy of nucleic acids'. In *Basic Principles in Nucleic Acid Chemistry*, Vol I (P.O.P. Ts'O, ed.). Academic Press, New York, 1974, p. 399

K. A. Hartman *et al.*, 'Infrared and Raman spectroscopy of nucleic acids and polynucleotides'. In *Physico-chemical Properties of Nucleic Acids*, Vol. 2 (J. Duchense, ed.). Academic Press, New York, 1973, p. 1

A. P. Arzamaster and D. S. Yashkina, *UV and IR Spectra of Drugs*, No. 1, *Steroids*. Meditsina, USSR, 1975

J. Bellanato and A. Hidalgo, *Infrared Analysis of Essential Oils*. Sadtler Research Laboratories, 1971

S. K. Freeman, *Applications of Laser Raman Spectroscopy*. Wiley, New York, 1974

F. R. Dollish, W. G. Fateley, and F. F. Bentley, *Characteristic Raman Frequencies of Organic Compounds*. Wiley, New York, 1974

N. V. Madhusudana, 'Infrared spectra of liquid crystals', *App. Spectrosc. Rev.*, 1972, **6,** 189

A. Saupe, 'Infrared and ultraviolet spectra of liquid crystals', *Mol. Cryst. Liquid Cryst.*, 1972, **16,** 87

Index

Page numbers in *italics* refer to detailed treatment of the item listed.